Linear Time-Varying Systems

Control and Adaptation

Kostas S. Tsakalis
Electrical Engineering
Arizona State University

Petros A. Ioannou
Electrical Engineering–Systems
University of Southern California

PRENTICE HALL, Englewood Cliffs, New Jersey 07632

Library of Congress Cataloging-in-Publication Data
Tsakalis, Kostas S.
  Linear time-varying systems: control and adaptation/Kostas S.
  Tsakalis, Petros A. Ioannou
      p.   cm.
  Includes bibliographical references and index.
  ISBN 0-13-012923-2
  1. Adaptive control systems—Mathematical models. 2. Linear systems—Mathematical models. I. Ioannou, P. A. (Petros A.), 1953-.
II. Title.
TJ217.T67 1993
629.8'32—dc20
                                                                92-20842
                                                                    CIP

Acquisitions editor: Karen Gettman
Editorial/production supervision: Linda B. Pawelchak
Cover design: Wanda Lubelska Design
Prepress buyer: Mary McCartney
Manufacturing buyer: Susan Brunke

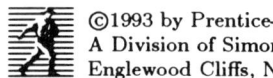
©1993 by Prentice-Hall, Inc.
A Division of Simon & Schuster
Englewood Cliffs, New Jersey, 07632

All rights reserved. No part of this book may be
reproduced, in any form or by any means,
without permission in writing from the publisher.

Printed in the United States of America
10 9 8 7 6 5 4 3 2 1

ISBN 0-13-012923-2

Prentice-Hall International (UK) Limited, *London*
Prentice-Hall of Australia Pty. Limited, *Sydney*
Prentice-Hall Canada Inc., *Toronto*
Prentice-Hall Hispanoamericana, S. A., *Mexico*
Prentice-Hall of India Private Limited, *New Delhi*
Prentice-Hall of Japan, Inc., *Tokyo*
Simon & Schuster Asia Pte. Ltd., *Singapore*
Editora Prentice-Hall do Brasil, Ltda., *Rio de Janeiro*

Στους Σπυρο και Νουλα,
Μικα, Ανδρεα και Κυριακουλα

Κ.Σ.Τ, Π.Α.Ι

Contents

Preface	ix
List of Acronyms	xiii
1 Introduction	**1**
1.1 State Transition Matrix and Stability	2
1.2 Control of LTV Plants	7
1.2.1 Model Reference Control	8
1.2.2 Pole-Placement Control	9
1.3 Adaptive Control of LTV Plants	10
2 Preliminaries	**15**
2.1 Introduction	15
2.2 Time-Varying Differential Operators	15
2.3 Time-Varying I/O Operators	25
2.4 Controllability, Observability and I/O Descriptions of LTV Systems	29
2.5 On the Internal Stability of LTV Feedback Systems	33
2.6 Realizations of TV I/O Operators	35
2.7 Linear Slowly TV Systems	37
2.8 More I/O Properties	41
2.8.1 Elements of L_p Theory	41
2.8.2 Exponentially Weighted L_p Spaces	48
2.8.3 Normalization Signals	53
2.8.4 Swapping and Operator Inversion Lemmas	55
2.8.5 The Bellman-Gronwall Lemma	58
2.8.6 Smooth Approximations of Continuous Functions	59
Appendix II	60

3 The LTV Plant — 72
- 3.1 Introduction 72
- 3.2 Smooth Parameter Variations 73
- 3.3 Non-Smooth Parameter Variations 75
- 3.4 Parametric Models of TV I/O Operators 81
  - 3.4.1 Structured Parameter Variations 86

4 Model Reference Control — 90
- 4.1 Introduction 90
- 4.2 Problem Statement 91
- 4.3 TV MRC Design 93
  - 4.3.1 Design of Over-Parametrized TV MRC's 96
- 4.4 Realization of the TV MRC and Internal Stability of the Closed-Loop Plant 98
- 4.5 Slowly TV Plants 102
- 4.6 Non-Smooth Parameter Variations 104
  - 4.6.1 Slowly TV Plants Revisited 107
- 4.7 Examples 108
- Appendix IV 113

5 TV 'Pole-Placement' Control — 124
- 5.1 Introduction 124
- 5.2 TV PPC Design 125
- 5.3 Realization of the TV PPC and Internal Stability of the Closed-Loop Plant 129
- 5.4 Command Tracking with the TV PPC 131
- 5.5 Non-Smooth Parameter Variations 134
- 5.6 Slowly TV Plants 135
- 5.7 Examples 139
- Appendix V 146

6 On-Line Parametric Identification — 148
- 6.1 Introduction 148
- 6.2 Affine Parametric Models and Estimation of TV Parameters . . 149
- 6.3 Parametric Identification of LTV Plants 157
  - 6.3.1 Smooth Parameter Variations 157
  - 6.3.2 Non-Smooth Parameter Variations 169
- Appendix VI 173

CONTENTS

7 Model Reference Adaptive Control — 181
- 7.1 Introduction — 181
- 7.2 Parameter Estimation in Direct MRAC — 182
  - 7.2.1 Smooth Parameter Variations — 183
  - 7.2.2 Non-Smooth Parameter Variations — 193
- 7.3 MRAC: Design and Stability Analysis — 195
  - 7.3.1 Smooth Parameter Variations — 199
  - 7.3.2 Non-Smooth Parameter Variations — 204
- 7.4 Examples — 206
- Appendix VII — 213

8 Adaptive TV PPC — 231
- 8.1 Introduction — 231
- 8.2 APPC: Design and Stability Analysis — 232
  - 8.2.1 Smooth Parameter Variations — 232
  - 8.2.2 Non-Smooth Parameter Variations — 239
  - 8.2.3 Slowly TV Plants — 242
- 8.3 Command Tracking with the TV APPC — 243
- 8.4 Examples — 247
- Appendix VIII — 256

Bibliography — 262

Index — 269

Preface

The lack of complete knowledge about the plant to be controlled and the unpredictable changes in the environment are among the main reasons for using feedback. The design of feedback control laws to meet the stability and performance requirements for a given plant continues to be a challenging problem in many applications. This is partly due to the complexity and uncertainty associated with real plants and partly due to the scarcity of feedback control laws even when the plant is perfectly known. In practice, plants are nonlinear, time-varying and infinite dimensional and any attempt to represent them precisely by mathematical models is almost futile. Consequently, a common practice adopted by most, if not all, control engineers is to approximate the actual plant by a simplified model allowing the development of feedback control laws which are easily implemented using available technologies. A class of popular simplified models are those represented by a linear, time-invariant, ordinary differential equation. Such models may provide a good approximation of the actual plant around operating points while maintaining a reasonable degree of analytical simplicity. Naturally, a large amount of research effort has been directed over the years to the detailed study of linear time-invariant systems and numerous results have been reported in the literature on the systematic and quantitative analysis and design of control laws for such systems.

On the other hand, in many applications such as aerospace, process control etc., the ever increasing performance demands and more stringent specifications over a wide range of operating conditions diminish the value of linear time-invariant models as good approximations of the actual plant. For such systems, a more accurate description can be provided by considering linear time-varying models which capture the time dependent characteristics of the plant. Time dependence may manifest itself as small in magnitude time-varying perturbations of the nominal parameters of a linear time-invariant model. In such a case, the control problem can still be approached from a linear time-invariant point of view by designing a control law which is robust with respect to the time-varying parameter perturbations. In many applications, however, time dependence is so pronounced that the linear time-invariant approach is no longer adequate for modeling and control. In these situations the plant may be modeled by a linear ordinary differential equation with time-varying coefficients, simply referred to as a linear time-varying plant. Until recently, the lack of technology for implementing complex controllers did not justify an extensive and thorough analysis and design of control laws for linear time-varying plants. However, in view of the current state of the art in electronics and computer technology, the implementation of complex control algorithms is both feasible and inexpensive, motivating a more detailed study of the control problem for linear time-varying plants.

The purpose of this book is to contribute to the advancement of the theory and design techniques of control laws for linear time-varying plants. An input-output framework is adopted, preserving many similarities with the well-understood linear time-invariant case. Polynomial differential operators and time-varying input-output operators serve as generalizations of the polynomials and transfer functions of the time-invariant case. The notion of poles is generalized by an integral operator associated with simple differential equations. In this mathematical framework, algebraic techniques are employed to design and analyze control laws for linear time-varying plants with smooth or piecewise continuous parameters which are known functions of time. The precise knowledge of the plant parameters, however, may be quite restrictive in many applications where the plant parameter variations arise due to either unpredictable or complicated changes in the plant description. The need to counteract such forms of plant uncertainty, also known as parametric uncertainty, has led to the development of the so-called adaptive control laws that require little or no information about the plant parameters. Adaptive control was motivated by the aircraft control problem in the 1950s and early 1960s and continues to be an active area of research, having successfully produced several classes of adaptive control schemes. Despite the intuitive claim, however, that such control schemes should find major applications with linear slowly time-varying plants, it was not until recently that their use in a time-varying environment could be theoretically justified. A considerable part of this book is devoted to the analysis and design of adaptive controllers for a wide class of linear time-varying plants.

A brief outline of the book is as follows: Chapter 1 contains an introduction on the control structures studied in the later chapters. Chapter 2 presents the essential mathematical background for the representation and analysis of linear time-varying plants. The linear time-varying plant and its various

PREFACE

representations used in the subsequent chapters for control and identification purposes are studied in Chapter 3. Chapter 4 deals with the model reference control problem for linear time-varying plants with known parameters. In Chapter 5, the pole-placement control problem is studied after being appropriately generalized to the time-varying case. In Chapter 6, the identification problem of linear time-varying plants is considered using parameter estimation techniques. The results of Chapters 4 and 6 are combined in Chapter 7 to develop model reference adaptive control schemes which relax the requirement of Chapter 4 that the plant parameters are precisely known and are applicable to plants with large parametric uncertainty. An analogous development is presented in Chapter 8 where the results of Chapters 5 and 6 are combined to yield adaptive 'pole-placement' schemes for linear time-varying plants with large parametric uncertainty. In all chapters, the theoretical analysis is complemented by examples and numerical simulations illustrating the discussed controller design techniques and the resulting closed-loop plant behavior.

The book is intended to introduce control researchers and practitioners to certain aspects of theory and design of control systems for linear time-varying plants with full or partial knowledge of parameter variations. The adopted treatment revolves around the use of input-output tools generalizing the corresponding linear time-invariant methods to the linear time-varying case. This approach allows for the development of analysis and design techniques that resemble the more elementary ones used for linear time-invariant plants and makes the first five chapters of the book easy to follow by graduate students and researchers with background on classical control and basic linear system theory. On the other hand, Chapters 6, 7 and 8 assume some familiarity with Lyapunov techniques and stability theory. Such a background may be acquired during a first graduate course on nonlinear systems.

It is a pleasure to acknowledge the contributions of colleagues, researchers and former students who helped and motivated us to write this book. We begin by thanking some of the earlier key researchers in adaptive control Bob Narendra, Karl Astrom, Gerhard Kreissimeier, Yoan Landau, Brian Anderson, Steve Morse, Graham Goodwin, R. Monopoli, Rick Johnson and Bo Egardt whose continuous enthusiasm and work laid the foundations and paved the road towards the solution of several of the problems addressed in this book. We would especially like to express our deepest appreciation to Petar Kokotovic and Laurent Praly. Their constructive criticism, guidance and input held our motivation for research high and helped us refine most of the material in this book.

We are also indebted to many colleagues for stimulating discussions at conferences and other scientific meetings. Their comments were invaluable in enhancing our understanding of the field and improving our work. In particular, we would like to thank Anu Annaswamy, Fouad Giri, Rick Middleton, Bob Narendra, Brian Anderson, Steve Morse, Shankar Sastry, Jim Winkelman, Bob Kosut, Marc Bodson, Eric Ydstie, Ivan Mareels, Robert Bitmead, Rogelio Lozano-Leal, Romeo Ortega and Ken Gousman, to mention a few, for many fruitful discussions.

Closer to home, our co-workers and former students played a vital role in the development of this book, by sharing with us their views and insight on our research. We are appreciative of the comments of Mike Safonov and Gary Rosen who served in the doctoral dissertation committee of the first author on which this book is based. We also thank the former students of the second author Gang Tao, Jing Sun, Aniruddha Datta, John Reed and Farid Ahmed-Zaid for their comments and stimulating discussions. And, on behalf of the first author, Thanasis Sideris, Georgios Giannakis and the late Constantine Economou deserve a special mention.

Finally, we acknowledge the support of several organizations including the National Science Foundation, General Motors Project Trilby, Ford Motor Company, Rockwell International and Lockheed. Special thanks are due to Bob Borcherts, Bill Powers, Roger Fruechte, Neil Schilke, James Rillings and Bob Rooney whose continuous support of our research made this book possible.

Kostas S. Tsakalis

Petros A. Ioannou

List of Acronyms

APPC: Adaptive Pole-Placement Control
BIBO: Bounded-Input, Bounded-Output
BIBS: Bounded-Input, Bounded-State
ES: Exponentially Stable
IMP: Internal Model Principle
I/O: Input-Output
LQ: Linear Quadratic
LQG: Linear Quadratic Gaussian
LQR: Linear Quadratic Regulator
LTI: Linear Time-Invariant
LTR: Loop Transfer Recovery
LTV: Linear Time-Varying
MRAC: Model Reference Adaptive Control
MRC: Model Reference Control
ODE: Ordinary Differential Equation
PE: Persistent Excitation
PDO: Polynomial Differential Operator
PIO: Polynomial Integral Operator
PPC: Pole-Placement Control
PW: Pointwise
RMS: Root-Mean-Square
SISO: Single-Input Single-Output
STM: State Transition Matrix
TV: Time-Varying
UB: Uniformly Bounded
ZIR: Zero-Input Response
ZSR: Zero-State Response

Chapter 1

Introduction

This book is concerned with the analysis, parametrization and control of a class of linear systems represented by a vector differential equation of the form

$$\begin{aligned} \dot{x}(t) &= A(t)x(t) + b(t)u(t) \\ y(t) &= c^T(t)x(t) \end{aligned} \quad (1.1)$$

where $x(t) \in \mathbf{R}^n$ is the state vector, $u(t) \in \mathbf{R}$ is the input, often referred to as the control input, $y(t) \in \mathbf{R}$ is the output and $A(t), b(t), c(t)$ are matrices of appropriate dimensions, referred to as the system parameters, whose elements are piecewise continuous functions of time. We refer to (1.1) as a linear time-varying (LTV) system. When $A(t), b(t), c(t)$ are independent of time, (1.1) is referred to as a linear time-invariant (LTI) system.

Equation (1.1) arises when an approximate model of a physical process is obtained by linearizing a more accurate nonlinear model about a nominal trajectory or generated directly from physical laws. For example, consider a dynamical system described by the ordinary differential equation (ODE)

$$\dot{x}_s(t) = f(x_s(t), u_s(t), t) \; ; \quad t \geq t_0 \quad (1.2)$$

and evolving around a nominal trajectory described by

$$\dot{x}_o(t) = f(x_o(t), u_o(t), t) \; ; \quad t \geq t_0 \quad (1.3)$$

Then, assuming that $f(.,.,.)$ is well-behaved in a neighborhood of (x_o, u_o) and for small deviations $(x, u) = (x_s - x_o, u_s - u_o)$, the behavior of the system (1.2) can be approximately described by

$$\dot{x}(t) = A(t)x(t) + b(t)u(t)$$

where
$$A(t) = \left[\frac{\partial f(x,u,t)}{\partial x}\right]_{x=x_o} ; \quad b(t) = \left[\frac{\partial f(x,u,t)}{\partial u}\right]_{u=u_o}$$

Some examples of such systems are:

- The dynamics of an aircraft where (1.3) may represent a desired flight trajectory with specified nominal velocity and altitude pattern and (1.2) the actual flight trajectory with small velocity and altitude deviations from their prescribed nominal values.

- The motion of a space station acted upon by gravity gradient and aerodynamic torques is described by a set of nonlinear differential equations. Even after linearization, significant variations in the inertia matrix and the system parameters are introduced by large moving payloads.

- In chemical process control, heat transfer rates may be altered by fouling or corrosion. Such phenomena may induce, for example, thermocouple drift or change the characteristics of a heat exchanger resulting in a time-varying process description. Moreover, the dynamics of several chemical reactions are highly nonlinear. Consequently, in processes involving such reactions, varying set points or feed-stream composition and other disturbances in general may cause a significant change in the operating conditions and the parameters of the process model.

1.1 State Transition Matrix and Stability

Consider the vector differential equation

$$\dot{x} = A(t)x ; \quad x(t_0) = x_0, \ t \geq t_0 \tag{1.4}$$

which is of the form of (1.1) with $u(t) \equiv 0$. The point $x_e = 0$ is an equilibrium point of (1.4) at any time t_0 and an isolated equilibrium point of (1.4) if $A(t)$ is nonsingular at some $t \geq t_0$.

The general solution of (1.4) is of the form (e.g., see [Kai.80, Des.69])

$$x(t) = \Phi(t,t_0)x_0$$

where $\Phi(.,.)$ is the unique solution of the matrix ODE

$$\frac{\partial \Phi(t,t_0)}{\partial t} = A(t)\Phi(t,t_0) ; \quad \Phi(t_0,t_0) = I, \ t \geq t_0$$

The matrix $\Phi(.,.)$ is commonly referred to as the state transition matrix (STM) associated with $A(t)$. In general, it may not be possible to derive

1.1. STM, STABILITY

an analytical, closed-form expression of the STM associated with an arbitrary matrix $A(t)$. Such expressions are only available in certain special cases, the most frequently encountered example of which is the case of LTI systems. That is, when $A(t) \equiv A$ is a constant matrix, it is easy to verify that the STM of A is of the form

$$\Phi(t,t_0) = I + A(t-t_0) + \frac{1}{2}A^2(t-t_0)^2 + \cdots \triangleq e^{A(t-t_0)} = \exp\int_{t_0}^{t} A\,d\tau$$

In the time-varying case, the matrix exponential is generalized to the use of a power series expansion for $\Phi(t,t_0)$, called the *Peano-Baker formula*:

$$\begin{aligned}\Phi(t,t_0) &= I + \int_{t_0}^{t} A(\tau_1)\,d\tau_1 + \int_{t_0}^{t}\int_{t_0}^{\tau_1} A(\tau_1)A(\tau_2)\,d\tau_2 d\tau_1 + \cdots \\ &+ \int_{t_0}^{t}\int_{t_0}^{\tau_1}\cdots\int_{t_0}^{\tau_{n-1}} A(\tau_1)A(\tau_2)\ldots A(\tau_n)\,d\tau_n\ldots d\tau_2 d\tau_1 + \cdots\end{aligned}$$

It should be emphasized, however, that the matrix resulting from the Peano-Baker formula is not equal to $\exp\int_{t_0}^{t} A(\tau)\,d\tau$, unless $A(t)$ and $\int_{t_0}^{t} A(\tau)\,d\tau$ commute.

Despite the lack of general closed-form expressions, the state transition matrix is a powerful analytical tool in studying the properties of the solutions of the ODE (1.4), especially those pertaining to the stability of the equilibrium $x_e = 0$. These properties, summarized by the following theorem, are exploited throughout this book in studying the stability properties of several control systems.

1.1 Theorem: The equilibrium point $x_e = 0$ of $\dot{x} = A(t)x$, $t \geq t_0 \geq 0$, is

1. *stable* iff $\sup_{t \geq t_0} \|\Phi(t,t_0)\| \triangleq c(t_0) < \infty$.

2. *asymptotically stable* iff $\lim_{t\to\infty} \|\Phi(t,t_0)\| = 0$.

3. *uniformly stable* iff $\sup_{t_0 \geq 0} c(t_0) = \sup_{t_0 \geq 0} \sup_{t \geq t_0} \|\Phi(t,t_0)\| \triangleq c < \infty$.

4. *uniformly asymptotically stable and exponentially stable* iff there exist constants $k, a > 0$ such that $\|\Phi(t,t_0)\| \leq k\exp[-a(t-t_0)]$, $\forall t \geq t_0$, $\forall t_0 \geq 0$.

for any finite initial conditions $x(t_0)$. ▽▽

Proof: See e.g., [Vid.78, Kai.80].

An important issue raised by the above theorem is that the stability of the zero equilibrium of (1.4) is given in terms of the difficult or even impossible to compute analytically state transition matrix. Based on the LTI case analogue, one may hope that the stability problem can be reduced to examining properties of the matrix $A(t)$ and, in particular, its eigenvalues. In fact, this type of a result has recently been established in [Z.J.89] for a wide class of LTV systems. Without getting into the details, the necessary and sufficient condition for stability of [Z.J.89] is given in terms of the eigenvalues and the so-called co-eigenvalues of a matrix obtained from $A(t)$ via a similarity transformation and closely resembles the condition derived for the scalar case (see Example 1.2 below). Without diminishing the significance of the result, it should be mentioned that the computation of the co-eigenvalues and the similarity transformation, although straightforward, is quite complicated. The complexity of this result justifies the development and study of simpler and more intuitive sufficient conditions for stability relying on assumptions on the structure and/or the speed of the time variations [Ven.77, Vid.78, IOP.87]. The following examples illustrate the difficulty of the stability problem in the time-varying case.

1.2 Example: Let us begin by considering the simplest first order LTV system described by

$$\dot{x} = -a(t)x \ ; \ \ x(0) = x_0 \qquad (1.5)$$

where $a(t)$ is a scalar piecewise continuous function of time. Since $a(t)$ and its integral commute, the corresponding state transition matrix is given by

$$\Phi(t,\tau) = e^{-\int_\tau^t a(s)\,ds}$$

and therefore, the solution of (1.5) is

$$x(t) = e^{-\int_0^t a(s)\,ds} x_0$$

Clearly, if $a(t) = a_0$ is a constant, $x(t)$ converges to zero exponentially fast when $a_0 > 0$ while it diverges to infinity when $a_0 < 0$, for any initial condition x_0. On the other hand, if $a(t)$ varies with time, the necessary and sufficient condition for exponential stability given by Theorem 1.1 is that there exists a positive constant a_0 and a constant c_0 such that for all $t \geq 0$

$$\int_0^t a(\tau)\,d\tau \geq c_0 + a_0 t \qquad (1.6)$$

Note that in contrast to the LTI case, condition (1.6) allows $a(t)$ to assume both positive and negative values as long as its average as $t \to \infty$ is positive, uniformly in t. ▽▽

The situation becomes considerably more complicated in the higher dimensional case, as demonstrated by the following classical examples where the zero

1.1. STM, STABILITY

equilibrium of (1.4) is unstable, even though the matrix $A(t)$ is stable for each fixed time t.

1.3 Example: Consider the second order differential equation

$$\ddot{y} + a_0(t)\dot{y} + a_1(t)y = 0 \tag{1.7}$$

and its state-space realization

$$\dot{x} = \underbrace{\begin{pmatrix} 0 & 1 \\ -a_1(t) & -a_0(t) \end{pmatrix}}_{A(t)} x \; ; \; y = [1, 0]x \tag{1.8}$$

When $a_0(t), a_1(t)$ are constants, a necessary and sufficient condition for the exponential stability of the zero equilibrium of (1.8) is $a_0, a_1 > 0$. On the other hand, consider the case where

$$a_0(t) = a_0 > 0 \; ; \; a_1(t) = \alpha - \beta \cos wt > 0, \; \forall t$$

representing the well known damped Mathieu equation. For $a_0 = 0.1$, $\alpha = 10.1$, $\beta = 10$ and $w = \pi = 3.14159$, both $a_0, a_1(t)$ are positive while the eigenvalues of the state matrix $A(t)$, given by

$$-0.05 \pm j \frac{\sqrt{0.39 + 40(1 - \cos \pi t)}}{2}$$

are in the left half-plane for all t and, in fact, their real part is constant and equal to -0.05.

$$s^2 + 0.1s + 10(1 - \cos \pi t) + 0.1$$

On the other hand, the stability properties of (1.8) are determined by the properties of the associated state transition matrix $\Phi(.,.)$. Since $A(t)$ is periodic with period $T = 2$, one may apply Floquet analysis to study its stability properties [C.L.55]. A numerical evaluation of $\Phi(T, 0)$ yields

$$\Phi(T, 0) = \begin{pmatrix} 1.0988 & 0.1784 \\ 2.0683 & 1.0809 \end{pmatrix}$$

whose eigenvalues have absolute values

$$1.6973, \quad 0.4824$$

Since $x(kT) = \Phi^k(T, 0)x(0)$, $k \in \mathbb{N}$ and $\Phi(T, 0)$ has eigenvalues outside the unit circle, we can always select $x(0)$ such that $\|x(kT)\| \to \infty$ as $k \to \infty$. Hence, the zero equilibrium of (1.8) is unstable. ▽▽

1.4 Example: Consider the differential equation (1.4) with $t_0 = 0$ and

$$A(t) = \begin{pmatrix} \cos t & -\sin t \\ \sin t & \cos t \end{pmatrix} \begin{pmatrix} -1 & -4 \\ 0 & -1 \end{pmatrix} \begin{pmatrix} \cos t & \sin t \\ -\sin t & \cos t \end{pmatrix}$$

which is of the form $e^{\Omega t}Be^{-\Omega t}$ where Ω and B are constant matrices. Note that both the eigenvalues of $A(t)$ are constant and equal to -1. It can be easily verified that the solution of (1.4) is given by

$$x(t) = e^{\Omega t}e^{(-\Omega+B)t}x_0$$

But,

$$A_0 \triangleq -\Omega + B = \begin{pmatrix} -1 & -3 \\ -1 & -1 \end{pmatrix}$$

has eigenvalues $-1 \pm \sqrt{3}$ and, hence, we can find x_0 such that $\|x(t)\| \to \infty$ as $t \to \infty$, implying that the zero equilibrium of $\dot{x} = A(t)x$ is unstable. ▽▽

Conversely, one may construct examples where the matrix $A(t)$ is unstable for each fixed time t but the zero equilibrium of (1.4) is exponentially stable.

1.5 Example: Consider the system of Example 1.4 except that now

$$B = \begin{pmatrix} -1 & 4 \\ 1 & -1 \end{pmatrix}$$

In this case the eigenvalues of $A(t)$ are constant and equal to $1, -3$ and, hence, $A(t)$ is unstable at each fixed time t. However,

$$A_0 = -\Omega + B = \begin{pmatrix} -1 & 5 \\ 0 & -1 \end{pmatrix}$$

has both eigenvalues equal to -1. Since $e^{\Omega t}$ is bounded, we have that for any x_0 $\|x(t)\| \to 0$ exponentially fast as $t \to \infty$, uniformly in t, implying that the zero equilibrium of $\dot{x} = A(t)x$ is exponentially stable. ▽▽

The above examples illustrate that the stability properties of LTV systems are quite complicated and not always compatible with our LTI intuition. Furthermore, the analysis of these properties for a given state matrix $A(t)$ is in general a formidable task. Nevertheless, in several practically interesting special cases, it is possible to develop simple sufficient conditions for stability. For example, in Chapter 2 of this book, we discuss the case of slowly varying systems where a sufficient condition for the exponential stability of the trivial equilibrium of (1.4) is that $A(t)$ is pointwise exponentially stable with a nonzero stability margin, uniformly in time and the variations of $A(t)$ are sufficiently small or small in the mean square. This result is also extended to the case where the dependence of the matrix $A(t)$ has a special form. In particular, when $A(t) = A(t, \theta(t))$ where $\theta(t)$ are parameters which are themselves functions of time, then the zero equilibrium of (1.4) is exponentially stable provided that for each fixed $\hat{\theta}$, $A(t, \hat{\theta})$ is exponentially stable with a nonzero stability margin, uniformly in $\hat{\theta}$ and the variations of $\theta(t)$ are sufficiently small or small in the mean square. In this case, the overall time variations of $A(t)$ are not restricted to be slow. Such stability results are exploited in the rest of this book to analyze and establish the stability properties of various feedback control schemes.

1.2 Control of LTV Plants

Let us consider the LTV plant to be controlled

$$\begin{aligned} \dot{x}(t) &= A(t)x(t) + b(t)u(t) \\ y(t) &= c^\top(t)x(t) \end{aligned} \tag{1.9}$$

where $y(t)$ is the measured output and $u(t)$ is the control input, to be chosen so that (1.9) meets certain prescribed performance objectives.

The design of control laws for LTV plants of the form (1.9) has attracted considerable interest since the early 1970s. Quadratic optimization criteria and geometric arguments yielded compensators whose parameters are computed by solving a Riccati differential equation [Bro.70, Kai.80, B.H.69]. On the other hand, a general algebraic approach motivated by the results of [YJB.76, DLMS.80] was the subject of more recent studies [F.F.84, Man.87, RKMN.88, RNK.90]. These studies established a compensator design methodology, parallel to the popular H_∞ or 'LQG/LTR' robustness/performance approach for LTI plants, where the control objectives are expressed in terms of induced gains of certain closed-loop sensitivity operators. The implementation of these controllers, however, requires great computational effort and becomes less attractive when there are limitations on the time allowed for the control law computations which, for example, is the case in adaptive control.

In this book we overcome this difficulty by considering an input-output (I/O) description of (1.9) which is motivated by the corresponding I/O differential equation. Such description is compatible with the general algebraic formulation of [DLMS.80] and enjoys several algebraic properties, closely resembling the transfer function approach in the LTI case.[1] In this framework we concentrate on two general control objectives, illustrated by simple examples in the following subsections. For these control objectives we derive *explicit* control laws whose realization only requires the solution of linear algebraic equations and, as such, are especially attractive in the design of adaptive control schemes.

---

[1] See, for example, [Sol.66, INS.84, K.H.79, Kam.79] for similar developments.

1.2.1 Model Reference Control

Let us consider the first order LTV plant described by the differential equation

$$\dot{y} = a(t)y + b(t)u \; ; \quad y(0) = y_0 \tag{1.10}$$

where $a(t), b(t)$ are given piecewise continuous, bounded functions of time and $b(t) \neq 0$, $\forall t \geq 0$. The model reference control (MRC) objective is to select the control input $u(t)$ so that all signals in the closed-loop plant are bounded and the output y tracks, asymptotically as $t \to \infty$, the output of the reference model

$$\dot{y}_m = -y_m + r \; ; \quad y_m(0) = 0 \tag{1.11}$$

where r is a piecewise continuous, bounded signal, usually called *the reference input*. Note that the choice of a LTI reference model serves only to simplify the presentation and is not crucial for the rest of our discussion; similar arguments can be made for LTV reference models as well.

The choice of the control input $u(t)$ in (1.10) that meets the model reference control objective is obvious for this simple example, i.e.,

$$u = -\frac{a(t)y + y + r}{b(t)}$$

For this control input, the closed-loop plant is described by

$$\dot{y} = -y + r$$

and the tracking error $e_1 = y - y_m$ satisfies $\dot{e}_1 = -e_1$ i.e., $e_1(t) = e^{-t}e_1(0)$. Therefore, for any bounded reference input, e_1 and y are bounded and $e_1(t) \to 0$ as $t \to \infty$ exponentially fast. Furthermore, if $|b(t)| \geq c > 0$, $\forall t \geq 0$, then u is also bounded.

Next, let us consider the second order LTV plant

$$\ddot{y} + a_1(t)\dot{y} + a_2(t)y = u \; ; \quad y(0) = y_0, \; \dot{y}(0) = \dot{y}_0 \tag{1.12}$$

where $a_1(t), a_2(t)$ are given piecewise continuous, bounded functions of time. The control input $u(t)$ is now to be chosen so that all signals in the closed-loop plant are bounded and the output y tracks the output of the reference model

$$\ddot{y}_m + 3\dot{y}_m + 2y_m = r \tag{1.13}$$

asymptotically as $t \to \infty$, where r is a piecewise continuous, bounded reference input signal.

Assuming that, in addition to y, $\dot{y}$ is available for measurement, the choice of the control input u is again obvious, i.e.,

$$u = (a_1(t) - 3)\dot{y} + (a_2(t) - 2)y + r$$

1.2. CONTROL OF LTV PLANTS

For this control input, the closed-loop plant is

$$\dddot{y} + 3\ddot{y} + 2y = r$$

and the tracking error $e_1 = y - y_m$ satisfies $\ddot{e}_1 + 3\dot{e}_1 + 2e_1 = 0$. Clearly, for any bounded reference input, $e_1, y, \dot{y}$ and u are bounded and $e_1(t) \to 0$ as $t \to \infty$ exponentially fast.

On the other hand, if $\dot{y}$ is not available for measurement, which is often the case in many physical systems, the choice of u that meets the model reference control objective is not as obvious. Chapter 4 is devoted to the analysis and design of time-varying control laws which meet the model reference control objective exactly for a wide class of LTV plants. The last example is then revisited in Section 4.7 where it is employed to illustrate the calculation and properties of various controller designs. As demonstrated in the same section a controller realization that meets the model reference control objective for an LTI plant does not necessarily achieve the same result when the plant is LTV. This observation confirms the intuitive expectation that any plant parameter variations should carefully be accounted for during the design of a controller.

1.2.2 Pole-Placement Control

When the plant is LTI, the pole-placement (or pole-assignment) problem is well defined and extensively studied. It involves the selection of the control input so that the closed-loop poles are placed at some desired locations. Since in the LTI case the location of the poles completely determines the stability properties of the system, pole-placement control (PPC) can be used as an approach to stabilize a plant. Furthermore, a necessary and sufficient condition for such a design to be possible for arbitrary pole locations is that the plant is completely controllable.

In the LTV case, on the other hand, such a simple notion of poles is unavailable. It also goes without saying that in view of the examples of the previous section, pointwise design techniques are only applicable to special classes of LTV plants. Consequently, the extension of the PPC objective to the LTV case requires a suitable formulation of the problem and the development of appropriate analytical tools. One of the goals of this book is to establish such an analytical framework and is achieved in Chapter 2 by considering I/O descriptions of LTV systems in terms of elementary differential and integral operators. The integral operators in particular are defined as solutions of homogenous differential equations and therefore determine the stability of the system. Such descriptions are directly related to canonical forms (controllable or observable) which allow for simple and explicit designs. In this framework, the generalization of the PPC objective can be simply stated in terms of the homogeneous solution of the differential equation describing the closed-loop plant. Furthermore, using the concepts established in [DLMS.80], the design of a PPC scheme is achieved by solving a Diophantine equation involving differential operators. Remarkably, the solution of such an equation can be obtained in closed-form by solving a system of linear algebraic equations and its existence is guaranteed under certain controllability and observability conditions on the LTV plant. In addition, the same approach provides a solution to the problem of exact tracking of a class of reference inputs by incorporating the internal model principle (IMP) in the PPC design. The details of PPC design for LTV plants are discussed in Chapter 5.

It should be mentioned at this point that in both cases of MRC and PPC objectives, the respective designs guarantee that the closed-loop system is stable and meets the control objective exactly independent of the speed of the plant parameter variations. On the other hand, due to its relative importance in applications and the resulting controller simplification, the case of slowly time-varying plants is also discussed separately in Chapters 4 and 5. Indeed, a controller designed in a pointwise sense[2] can be shown to guarantee closed-loop stability and meet its objective approximately, provided that the parameter variations are sufficiently slow, see for example [K.K.79, S.A.91].

1.3 Adaptive Control of LTV Plants

In Section 1.2 we discussed the control problem for a LTV plant of the form (1.9) assuming that the matrices $A(t), b(t), c(t)$ —often referred to as plant parameters— consist of elements that are known functions of time. In many practical problems, however, the plant parameters are not accurately known. The process characteristics may change with time in a predictable or unpredictable fashion due to a variety of reasons including changes in the operating conditions, normal wear and tear etc. Therefore, although the actual process may be modeled as a linear system of the form (1.9), the assumption that the plant parameters are accurately known is unrealistic in many applications such as aircraft control, chemical process control, control of prosthetic devices and robots etc. The often large parametric uncertainty associated with LTV plants was the motivation for developing a class of nonlinear control laws in the late 1950s and early 1960s, known as adaptive controllers [AMS.58, Bel.61, Jac.61]. The intuitive idea behind adaptive control was that the measured input and output of the plant contain information not only about its internal signals but

---

[2]Such design techniques are also referred to as *gain-scheduling*.

1.3. ADAPTIVE CONTROL OF LTV PLANTS

about the plant parameters as well. Therefore, in principle, one can extract information about the plant parameters and their variations by properly processing the input-output signals and use them for control purposes. Starting with the late 1950s, the field of adaptive control evolved into a major area of research. Several different classes of adaptive control schemes have been proposed and analyzed in recent books and manuscripts. [Ega.79, Lan.79, I.K.83, G.S.84, Gaw.87, A++.86, N.A.89, S.B.89, BGW.90, A.W.88].

A widely used methodology for designing adaptive control systems consists of two steps. In the first step, the plant parameters are assumed to be known and a control law is designed to meet a control objective. In the second step, a parameter estimation technique is used to estimate the unknown parameters on-line. The adaptive controller is then implemented by using the parameter estimates as if they were the true parameters; such controllers are commonly referred to as *certainty equivalence adaptive controllers*. In the subsequent chapters we distinguish two classes of algorithms depending on the parameters we choose to estimate. The identification (via parameter estimation) of a suitably parametrized model of the plant, e.g., a canonical form, leads to an *indirect* adaptive controller where the controller parameters are indirectly obtained from the plant parameters using the control law design equations. In some cases, however, it is possible to re-parametrize the plant with respect to the controller parameters which can then be estimated directly; the resulting adaptive controller is commonly referred to as *direct*, a term signifying the direct estimation of the control law parameters. Further, another classification of adaptive control schemes is based on the choice of the control objective and the corresponding control law. In this book we focus our interest on the two most frequently analyzed control strategies, the pole-placement control and the model reference control, primarily due to the simplicity of the associated design equations.

In principle, an adaptive controller has the capability to learn about the changes in the plant parameters and compensate for their effects on the behavior of the closed-loop plant. The component responsible for this property is the on-line parameter estimation (or, in general, system identification) algorithm. However, the interaction of this component with the control law makes the overall closed-loop system nonlinear and difficult to analyze even for LTI plants. For this reason, most hard results in adaptive control theory deal with LTI plants with unknown, constant parameters. Intuitive arguments were often used to claim that an adaptive controller with good performance in the LTI case should be able to cope with slow time-variations of the plant parameters. Initially, however, these arguments were not analytically substantiated for a general class of slow variations due to the lack of sufficient theoretical developments in robust adaptive control. Parameter variations introduce a perturbation in the parameter estimator which acts as an unknown disturbance. Since the early adaptive control schemes could exhibit unbounded response in the presence of modeling errors and disturbances [Ega.79, I.K.83, P.N.82, RVAS.85], there was no significant progress towards the adaptive control of LTV systems until the mid-1980s when some fundamental robustness problems were resolved.

In the early attempts [A.J.83, G.T.83], the persistence of excitation (PE) of certain signals in the adaptive loop was employed to guarantee the exponential stability of the unperturbed error system, which eventually led to the local stability of the closed-loop TV plant. Elsewhere, the restriction of the type of time variations of the plant parameters also led to the conclusion that an adaptive controller could be used in the respective environment. More specifically, in [C.C.84] the parameter variations were assumed to be perturbations of some nominal constant parameters, which are small in the norm and modeled as a martingale process with bounded covariance. A treatment of the parameter variations as small TV perturbations, in an L_2-gain sense, was also considered in [G.C.86] for a restrictive class of LTI-nominal plants. Special models of parameter variations, such as exponential or $1/t$-decaying or finite number of jump-discontinuities, were considered in [GHX.84, Ohk.85, M-S.85].

Some more general results were initiated by the development of adaptive laws that were shown to be robust in the presence of bounded disturbances [Ega.79]. Further modifications and studies in [Pra.83, Pra.84] and later in [K.A.86, I.T.86], established the robustness of these adaptive laws with respect to bounded disturbances as well as unmodeled dynamics and plant uncertainty, without requiring any PE conditions. The use of robust adaptive laws for estimating the TV parameters was shown to guarantee signal boundedness and small in the mean tracking errors in the case of direct model reference adaptive control (MRAC) [T.I.87], and indirect adaptive control [Kre.86, M.G.88] for slowly TV plants, without any PE requirements. Furthermore, an internally generated excitation sequence was developed in [GMDD.87] to show the robustness of a direct adaptive pole-placement controller (APPC) with respect to plant parameter variations. The common key-point of those studies was the intuitive idea that, at each time instant, a linear slowly TV system behaves 'almost' like an LTI one. The effect of the time variations was then expressed as a disturbance, not necessarily bounded but small compared to the useful signals. This treatment led to the proof that the respective robust adaptive controller can operate successfully in such a TV

1.3. ADAPTIVE CONTROL OF LTV PLANTS

environment. We should note, however, that in those studies the effort to design an adaptive controller for linear time-varying plants was concentrated on improving the robustness properties of the estimator while the control law was designed in a pointwise sense, following LTI principles. Such an approach is therefore limited to the practically interesting, nonetheless restrictive class of slowly varying plants.

Further analysis of the available results indicates that in order to devise adaptive controllers for general LTV plants, the design of both the control and parameter estimation algorithms needs to be improved. In particular, the standard control laws, used in the adaptive literature, are derived based on the assumption that the plant parameters are constant. When these control laws are applied to a LTV plant they can only meet the control objective approximately provided that the plant parameters are slowly varying with time. However, for a general LTV plant with not necessarily slowly varying parameters, such control laws may fail to guarantee even closed-loop stability. Furthermore, the standard adaptive laws (parameter estimators) are essentially integrators with a finite gain at all frequencies except zero. That is, only constant parameters can be estimated with asymptotically zero estimation error. Consequently, in order to allow for exact asymptotic identification, at least in cases where the functional form of the parameter variations is known, a suitable parametrization of LTV plants is required.

In this study we deviate from the standard approach in which the design of adaptive controllers begins with the assumption that the plant is LTI and restricts the study of the TV case to a robustness analysis. Instead, we formulate and analyze the control problem by considering that the plant is LTV from the very beginning, an approach consistent with the introduction of the adaptive controllers to be operating in a possibly TV environment. Having established controller design procedures for a wide class of LTV plants, the next step in the development of a certainty equivalence adaptive controller is to estimate the unknown plant or controller parameters. We introduce the notion of structured parameter variations (various forms of which have appeared in [Ohk.85, X.E.84, L.S.88]) whereby any a priori available knowledge of the modes of variation of the plant parameters is explicitly incorporated in the parameter estimates. Such an approach has the advantage that the unknown parameters can be estimated with asymptotically zero estimation error in the case of structured parameter variations e.g., if the frequency or form of variation of the plant parameters is known a priori. Furthermore, the I/O quality of the parameter estimates depends on the speed of variation of the unstructured part only and not on the overall speed of variation of the plant parameters. We discuss the issues pertaining to the estimation of time-varying parameters and the parametric identification of LTV systems in Chapter 6. A by-product of this approach is the development of a plant parametrization which allows for its exact asymptotic identification, under some PE conditions, in the case of structured parameter variations.

In the last step of our study, we employ the certainty equivalence principle to combine the previously designed control laws and estimators to obtain the respective adaptive controller (MRAC, APPC or IMP/APPC). We analyze the stability properties of these adaptive control schemes in Chapters 7 and 8 where we show that the boundedness of all signals in the closed loop depends only on the speed of the unstructured part of the plant parameter variations, while the overall plant may be fast TV. This result is obtained using standard assumptions, analogous to those used for LTI plants and without requiring any PE conditions. In the ideal case of structured parameter variations our adaptive controllers meet the respective control objective asymptotically as $t \to \infty$, independent of the speed of the plant parameter variations. The price paid for this result is the updating of additional parameters. We note that such results cannot be derived by using the standard pointwise designs which, even in the case of completely structured parameter variations, cannot guarantee stability unless the plant is slowly TV. The impact and conceptual difference of this result from the existing ones is that the use of adaptive controllers need not be restricted to slowly TV plants only. The critical quantity is the speed of variation of the unstructured part of the unknown parameters and not the overall speed of their variation.

Chapter 2

Preliminaries

2.1 Introduction

In this chapter we present some definitions and lemmas which comprise the essential mathematical background used in the subsequent chapters. Our aim is to establish a general framework where LTV systems are described by input-output (I/O) operators with fractional representations and take full advantage of the powerful methodology and intuition available in the algebraic design of controllers [DLMS.80]. This development allows us to treat the TV systems using I/O relationships in a similar way to the LTI case where transfer functions are used to describe I/O properties.

2.2 Time-Varying Differential Operators

Let us begin by considering the LTV nth order ordinary differential equation (ODE)

$$y^{(n)} + a_1(t)y^{(n-1)} + \cdots + a_{n-1}(t)y^{(1)} + a_n(t)y = u \qquad (2.1)$$

with initial conditions $y^{(i)}(0)$, $i = 0, 1, \ldots, n-1$, where $y^{(i)} \triangleq \frac{d^i}{dt^i} y(t)$, y, u : $\mathbf{R}_+ \mapsto \mathbf{R}$ and $a_i(t)$, $i = 1, \ldots, n$ are piecewise continuous,[1] uniformly bounded (UB) functions of time. Equation (2.1) may be written in a *left* polynomial form as

$$\{s^n + a_1(t)s^{n-1} + \cdots + a_{n-1}(t)s + a_n(t)\}[y] = u \qquad (2.2)$$

---
[1] In the following, any parameter discontinuity is always of the first kind.

where $s \triangleq \frac{d}{dt}(\cdot)$ is the differential operator [C.L.55], or in the state representation form

$$\dot{X} = \begin{bmatrix} 0 & 1 & 0 & \cdots & 0 \\ 0 & 0 & 1 & \cdots & 0 \\ \vdots & \vdots & \vdots & & 1 \\ -a_n(t) & -a_{n-1}(t) & -a_{n-2}(t) & \cdots & -a_1(t) \end{bmatrix} X + \begin{bmatrix} 0 \\ 0 \\ \vdots \\ 1 \end{bmatrix} u$$

$$y = [1, 0, \ldots, 0]X \qquad (2.3)$$

where $X = [y, y^{(1)}, \ldots, y^{(n-1)}]^T$.

2.1 Definition: An LTV left Polynomial Differential Operator (PDO) of degree n is defined by

$$P(s,t) = a_0(t)s^n + a_1(t)s^{n-1} + \cdots + a_n(t) \qquad (2.4)$$

where $s \triangleq \frac{d}{dt}(\cdot)$; $a_i(t), i = 0, 1, \ldots, n$ are piecewise continuous functions of time, $a_0(t) \neq 0$ for some $t \in \mathbf{R}_+$. When $a_0(t) \equiv 1 \ \forall t \in \mathbf{R}_+$, $P(s,t)$ is referred to as a monic PDO. ▽▽

In view of Definition 2.1 we may now simplify the notation and write (2.1) or (2.3) in the form

$$P(s,t)[y] = u$$

where $P(s,t)$ is the PDO $s^n + a_1(t)s^{n-1} + \cdots + a_{n-1}(t)s + a_n(t)$.

Similarly, an LTV right PDO is defined as:

2.2 Definition: An LTV right PDO of degree n is defined by

$$\hat{P}(s,t) = s^n \hat{a}_0(t) + s^{n-1}\hat{a}_1(t) + \cdots + \hat{a}_n(t) \qquad (2.5)$$

where $s \triangleq \frac{d}{dt}(\cdot)$; $\hat{a}_i(t), i = 0, 1, \ldots, n$ are smooth functions of time, $\hat{a}_0(t) \neq 0$ for some $t \in \mathbf{R}_+$. When $\hat{a}_0(t) \equiv 1 \ \forall t \in \mathbf{R}_+$, $\hat{P}(s,t)$ is referred to as a monic PDO. ▽▽

Thus, the I/O map of a linear ordinary differential equation with state space representation

$$\dot{X} = \begin{bmatrix} -\hat{a}_1(t) & 1 & 0 & \cdots & 0 \\ -\hat{a}_2(t) & 0 & 1 & \cdots & 0 \\ \vdots & \vdots & \vdots & & 1 \\ -\hat{a}_n(t) & 0 & 0 & \cdots & 0 \end{bmatrix} X + \begin{bmatrix} 0 \\ 0 \\ \vdots \\ 1 \end{bmatrix} u$$

$$y = [1, 0, \ldots, 0]X \qquad (2.6)$$

2.2. TIME-VARYING DIFFERENTIAL OPERATORS

can be written as
$$\hat{P}(s,t)[y] = u$$
where now, $\hat{P}(s,t)$ is a right PDO. Notice that, in this framework, PDO's are interpreted as components of a particular state-space realization of an I/O map. In this sense we may extend the definition of the right PDO to admit piecewise continuous parameters, provided of course that the I/O map definition and treatment does not require the differentiation of these parameters.[2]

From the properties of differentiation it follows that if the coefficients $a_i(t)$ or $\hat{a}_i(t)$ are smooth functions of time, a PDO can be written either in the left or the right form, as illustrated by the following example.

2.3 Example: Consider the left PDO $P(s,t) = a_0(t)s^2 + a_1(t)s + a_2(t)$, where $a_i(t)$ are smooth functions of time. Then $P(s,t)$ can be written as a right PDO, i.e., $P(s,t) = \hat{P}(s,t) = s^2 \hat{a}_0(t) + s\hat{a}_1(t) + \hat{a}_2(t)$ where

$$\begin{aligned}
\hat{a}_0(t) &= a_0(t) \\
\hat{a}_1(t) &= a_1(t) - 2\dot{a}_0(t) \\
\hat{a}_2(t) &= a_2(t) - \dot{a}_1(t) + \ddot{a}_0(t)
\end{aligned}$$

Alternatively, if we consider the right PDO $\hat{P}(s,t) = s^2 \hat{a}_0(t) + s\hat{a}_1(t) + \hat{a}_2(t)$, then $P(s,t)$ can be expressed as a left PDO, i.e., $\hat{P}(s,t) = P(s,t) = a_0(t)s^2 + a_1(t)s + a_2(t)$ where

$$\begin{aligned}
a_0(t) &= \hat{a}_0(t) \\
a_1(t) &= \hat{a}_1(t) + 2\dot{\hat{a}}_0(t) \\
a_2(t) &= \hat{a}_2(t) + \dot{\hat{a}}_1(t) + \ddot{\hat{a}}_0(t)
\end{aligned}$$

In both cases the relationships between the coefficients of the two forms follow from the operator identity $as = sa - \dot{a}$. ▽▽

The properties of PDO's follow from the rules of differentiation and can be summarized as follows:[3]

P1. If $P(s,t)$, $Q(s,t)$ are left (right) PDO's of degree n, m and with coefficients $a_i(t)$, $b_j(t)$ respectively, $P(s,t) = Q(s,t)$ iff $n = m$ and $a_i(t) = b_i(t)$, $\forall t$, $i = 0, 1, \ldots, n$.

P2. $P(s,t) + Q(s,t) = Q(s,t) + P(s,t)$.

P3. $P(s,t)c = cP(s,t)$; c is a constant.

---

[2] That is, a right PDO with non-differentiable parameters should be treated as such and should not be converted to a left PDO.

[3] The equality of two operators $P : u \mapsto y$ and $Q : u \mapsto \bar{y}$ with the same domain $\mathcal{D}$ is generally defined as $P = Q$ if $y = \bar{y}$, $\forall u \in \mathcal{D}$.

P4. $P(s,t)x(t) = x(t)P(s,t) + \bar{P}(s,t)$; $x(t)$ is a smooth function of t and $\bar{P}(s,t)$ is a PDO with $\deg[\bar{P}(s,t)] \leq \deg[P(s,t)] - 1$ whose coefficients are $O[\dot{x}(t), \ddot{x}(t), \ldots]$.[4]

The proof of P1, P2 and P3 is immediate from the properties of differentiation. To show P4 we use the operator identity $sx(t) = x(t)s + \dot{x}(t)$ which, repeatedly applied on $P(s,t)x(t)$, yields P4, where[5]

$$\bar{P}(s,t) = \sum_{i=1}^{n}\sum_{k=1}^{i} a_{i-k}(t) \binom{n+k-i}{k} x^{(k)}(t) s^{n-i}$$

and $P(s,t)$ is an nth degree left PDO. In a similar way, P4 can be established for right PDO's. □□

The rest of this section is devoted to the solution and properties of the PDO equation

$$Q(s,t)D(s,t) + P(s,t)N(s,t) = A(s,t).$$

Equations of this form are commonly referred to as *Diophantine* equations (or Bezout equations when $A(s,t) = 1$) and they are frequently encountered in the algebraic controller design ([DLMS.80], [Frn.87]). In order to assess the solvability of the above equation where $D(s,t)$, $N(s,t)$, $A(s,t)$ are given and $Q(s,t)$, $P(s,t)$ are unknown PDO's we need to introduce the notion of *coprimeness* of PDO's. It is exactly at this point where the non-commutativity with respect to multiplication of TV PDO's (see property P4) necessitates a somewhat different treatment of the TV case than the TI one. It should be pointed out that this treatment is expected to have many similarities with the multivariable LTI case where the polynomial matrices also lack commutativity with respect to multiplication [Frn.87]. In our case, however, an additional difficulty is due to the appearance of the differential operator which makes a Diophantine equation to be, at the first glance, a differential equation; consequently, it is of special interest to obtain the solution of a Diophantine equation, whenever possible, as a solution of an algebraic equation. The coprimeness of two TV PDO's is defined as follows.

2.4 Definition: Let $D(s,t)$, $N(s,t)$ be PDO's with smooth coefficients and such that $D(s,t)$ is monic. We say that $D(s,t)$ and $N(s,t)$ are *right coprime* in (t_1, t_2),[6] if there exist PDO's $P_0(s,t)$, $Q_0(s,t)$, $Q_1(s,t)$ and $P_1(s,t)$ with

---
[4] We say that a function $f(x)$ is $O(x)$ if there exists a constant $k \in (0, \infty)$ such that $\|f(x)\| \leq k\|x\|$, $\forall x$.

[5] $\binom{n}{m} \triangleq \frac{n!}{m!(n-m)!}$

[6] The same definition can be used for not necessarily open intervals as long as $t_2 > t_1$.

2.2. TIME-VARYING DIFFERENTIAL OPERATORS

smooth coefficients, such that $P_0(s,t)$ is monic, $\deg[P_0(s,t)] = \deg[D(s,t)]$ and

$$Q_0(s,t)D(s,t) + P_0(s,t)N(s,t) = 0$$

$$Q_1(s,t)D(s,t) + P_1(s,t)N(s,t) = 1$$

for all $t \in (t_1, t_2)$. Dually, we say that $D(s,t)$ and $N(s,t)$ are left coprime in (t_1, t_2), if there exist PDO's $P_0(s,t)$, $Q_0(s,t)$, $Q_1(s,t)$ and $P_1(s,t)$ with smooth coefficients, such that $P_0(s,t)$ is monic, $\deg[P_0(s,t)] = \deg[D(s,t)]$ and

$$D(s,t)Q_0(s,t) + N(s,t)P_0(s,t) = 0$$

$$D(s,t)Q_1(s,t) + N(s,t)P_1(s,t) = 1$$

for all $t \in (t_1, t_2)$. ▽▽

2.5 Example: Let $D(s,t) = [s^2 + a_1(t)s + a_2(t)]$ and $N(s,t) = [s + b(t)]$ where $a_1(t), a_2(t), b(t)$ are smooth functions of t. According to the previous definition, $D(s,t)$ and $N(s,t)$ are right coprime in a time interval J, if we can find $q_{0i}(t), p_{0i}(t), p_{1i}(t)$, $i = 1, 2$, and $q_{11}(t)$ satisfying the Diophantine equations

$$[q_{01}(t)s + q_{02}(t)]D(s,t) + [s^2 + p_{01}(t)s + p_{02}(t)]N(s,t) = 0$$

$$q_{11}(t)D(s,t) + [p_{11}(t)s + p_{12}(t)]N(s,t) = 1.$$

for all $t \in J$. After some straightforward calculations, the above equations can be written as

$$q_{01}(t) = -1$$

$$\underbrace{\begin{bmatrix} 1 & 1 & 0 \\ a_1(t) & b(t) & 1 \\ a_2(t) & \dot{b}(t) & b(t) \end{bmatrix}}_{S_R(t)} \begin{bmatrix} q_{02}(t) & q_{11}(t) \\ p_{01}(t) & p_{11}(t) \\ p_{02}(t) & p_{12}(t) \end{bmatrix} = \begin{bmatrix} a_1(t) - b(t) & 0 \\ a_2(t) + \dot{a}_1(t) - 2\dot{b}(t) & 0 \\ \dot{a}_2(t) - \ddot{b}(t) & 1 \end{bmatrix}$$

which imply that $D(s,t)$ and $N(s,t)$ are right coprime in J if $det[S_R(t)] \neq 0$, $\forall t \in J$. ▽▽

The above example indicates that the coprimeness of two TV PDO's can be checked in a similar way as in the TI case by first extending the definition of the Sylvester matrix to the TV case and then examining its properties.

2.6 Definition: *The right TV Sylvester matrix $S_R(t)$ of the PDO's $D(s,t)$, $N(s,t)$, of degree n, m respectively is defined as*

$$S_R(t) = [C_1, \ldots, C_m, B_1, \ldots, B_n]$$

where C_i, B_j are column vectors of the coefficients of $s^{m-i}D(s,t)$, $i = 1, 2, \ldots, m$ and $s^{n-j}N(s,t)$, $j = 1, 2, \ldots, n$ respectively, expressed as $(n+m-1)$-degree left PDO's, by setting the missing coefficients equal to zero.[7] The left TV Sylvester matrix $S_L(t)$ of $D(s,t)$, $N(s,t)$ is defined in a dual manner as

$$S_L(t) = [\hat{C}_1, \ldots, \hat{C}_m, \hat{B}_1, \ldots, \hat{B}_n]$$

where $\hat{C}_i$, $\hat{B}_j$ are vectors of the coefficients of $D(s,t)s^{m-i}$, $N(s,t)s^{n-j}$ respectively, expressed as $(n+m-1)$-degree right PDO's. ▽▽

The motivation of Definition 2.6 becomes clear if, given the PDO's $D(s,t)$, $N(s,t)$ of degree n, m respectively, we consider the Diophantine equation

$$Q(s,t)D(s,t) + P(s,t)N(s,t) = A(s,t) \tag{2.7}$$

with $\deg[A(s,t)] \leq n + m - 1$ and express both sides as left PDO's. We can then write (2.7) as a system of linear algebraic equations $S_R(t)x(t) = a(t)$ where $x(t)$, $a(t)$ are vectors of the TV coefficients of $Q(s,t)$, $P(s,t)$ and $A(s,t)$ respectively.

Similarly, if we express both sides of the Diophantine equation

$$D(s,t)Q(s,t) + N(s,t)P(s,t) = A(s,t) \tag{2.8}$$

as right PDO's, (2.8) can be written as $S_L(t)x(t) = a(t)$.

2.7 Example: Let $D(s,t) = s^2 + a_1(t)s + a_2(t)$, $N(s,t) = b_0(t)s^2 + b_1(t)s + b_2(t)$ and $A(s,t) = 1$ and consider the equation (2.7) where $Q(s,t) = q_0(t)s + q_1(t)$, $P(s,t) = p_0(t)s + p_1(t)$ are unknown PDO's. Expressing both sides of (2.7) as left PDO's we obtain the following system of algebraic equations for $q_i(t)$, $p_i(t)$:

$$\underbrace{\begin{bmatrix} 1 & 0 & b_0(t) & 0 \\ a_1(t) & 1 & b_1(t) + \dot{b}_0(t) & b_0(t) \\ a_2(t) + \dot{a}_1(t) & a_1(t) & b_2(t) + \dot{b}_1(t) & b_1(t) \\ \dot{a}_2(t) & a_2(t) & \dot{b}_2(t) & b_2(t) \end{bmatrix}}_{S_R(t)} \begin{bmatrix} q_0(t) \\ q_1(t) \\ p_0(t) \\ p_1(t) \end{bmatrix} = \begin{bmatrix} 0 \\ 0 \\ 0 \\ 1 \end{bmatrix}$$

Similarly, if $D(s,t) = s^2 + sa_1(t) + a_2(t)$, $N(s,t) = s^2 b_0(t) + sb_1(t) + b_2(t)$, $A(s,t) = 1$ then equation (2.8) can be written as

$$\underbrace{\begin{bmatrix} 1 & 0 & b_0(t) & 0 \\ a_1(t) & 1 & b_1(t) - \dot{b}_0(t) & b_0(t) \\ a_2(t) - \dot{a}_1(t) & a_1(t) & b_2(t) - \dot{b}_1(t) & b_1(t) \\ -\dot{a}_2(t) & a_2(t) & -\dot{b}_2(t) & b_2(t) \end{bmatrix}}_{S_L(t)} \begin{bmatrix} q_0(t) \\ q_1(t) \\ p_0(t) \\ p_1(t) \end{bmatrix} = \begin{bmatrix} 0 \\ 0 \\ 0 \\ 1 \end{bmatrix}$$

---

[7] A zero degree indicates absence of the corresponding columns.

2.2. TIME-VARYING DIFFERENTIAL OPERATORS

where q_0, q_1, p_0, p_1 are the coefficients of the right PDO's $Q(s,t), P(s,t)$, i.e.,

$$Q(s,t) = sq_0(t) + q_1(t) \;\; ; \;\; P(s,t) = sp_0(t) + p_1(t)$$

This example illustrates that a Diophantine equation of the form (2.7) or (2.8) can be written as a system of linear algebraic equations and therefore, its solution can be obtained without necessarily solving a differential equation. We should emphasize that this result is made possible by expressing the PDO's in (2.7) or (2.8) in the appropriate left or right form. ▽▽

2.8 Lemma: *Two PDO's with smooth coefficients, one of which is monic, are right (left) coprime in (t_1, t_2) iff their right (left) TV Sylvester matrix is nonsingular for all $t \in (t_1, t_2)$.* ▽▽

Proof: In Appendix II.

2.9 Example: Let $D(s,t)$ be a monic PDO of degree 1 and $N(s,t)$ be a PDO of degree zero, i.e., $D(s,t) = s + a(t)$ and $N(s,t) = b(t)$ where $a(t), b(t)$ are smooth functions of t. Following Definition 2.4, $D(s,t), N(s,t)$ are coprime in an interval J if there exist PDO's $q_0(t)$, $[s + p_0(t)]$ and $p_1(t)$ such that

$$q_0(t)[s + a(t)] + [s + p_0(t)]b(t) = 0 \;\; ; \;\; p_1(t)b(t) = 1$$

for all $t \in J$. From the above equation,

$$q_0(t) = -b(t) \;\; ; \;\; p_0(t)b(t) = -\dot{b}(t) - q_0(t)a(t) \;\; ; \;\; p_1(t) = 1/b(t)$$

which implies that $D(s,t)$ and $N(s,t)$ are coprime in J iff $b(t) \neq 0, \forall t \in J$. Further, the right TV Sylvester matrix of $D(s,t)$ and $N(s,t)$ is $S_R(t) = b(t)$ which is nonsingular iff $b(t)$ is non-zero in J. ▽▽

A straightforward extension of Lemma 2.8 in the case of non-monic PDO's can be given as follows:

2.10 Corollary: *Let $k_1(t), k_2(t)$ be smooth functions of time such that $k_1(t), k_2(t) \neq 0, \forall t \in (t_1, t_2)$ and $D(s,t), N(s,t)$ be two PDO's with $D(s,t)$ monic. Then, the right TV Sylvester matrix of $k_1(t)D(s,t)$ and $k_2(t)N(s,t)$ is nonsingular in (t_1, t_2) iff $D(s,t), N(s,t)$ are right coprime PDO's in (t_1, t_2).*

Dually, the left TV Sylvester matrix of $D(s,t)k_1(t)$ and $N(s,t)k_2(t)$ is nonsingular in (t_1, t_2) iff $D(s,t), N(s,t)$ are left coprime PDO's in (t_1, t_2). ▽▽

The coprimeness (right or left) of two PDO's is needed to guarantee, in general, the existence of solutions of the respective Diophantine equation.

Further, since such solutions are subsequently used in the design of control laws for TV systems, we are also interested in deriving conditions that ensure the boundedness of the coefficients of the PDO's, satisfying a Diophantine equation. Such conditions are primarily needed to describe the coprimeness properties of two PDO's in the limit as $t \to \infty$, as demonstrated by the following example.

2.11 Example: Consider the PDO's $D(s,t) = s^2 + s\frac{1}{t+1} - 1$, $N(s,t) = s + 1$. By Definition 2.6, the left TV Sylvester matrix of $D(s,t)$, $N(s,t)$ is given as

$$S_L(t) = \begin{pmatrix} 1 & 1 & 0 \\ \frac{1}{t+1} & 1 & 1 \\ -1 & 0 & 1 \end{pmatrix}$$

whose determinant is $\det[S_L(t)] = -\frac{1}{t+1}$. Hence, $D(s,t)$, $N(s,t)$ are left coprime in any interval $(0,T)$, $T \in \mathbf{R}_+$. However, in the limit as $t \to \infty$, $\lim_{t \to \infty} S_L(t)$ is a singular matrix and therefore $D(s,t)$, $N(s,t)$ are not left coprime in $(0,\infty]$. The 'non-strong' or 'non-uniform' coprimeness of $D(s,t)$, $N(s,t)$ manifests itself if we attempt to solve a Diophantine equation of the form

$$[s^2 + s\frac{1}{t+1} - 1][q_0(t)] + [s + 1][sp_0(t) + p_1(t)] = 1$$

whose solution is $q_0(t) = -(t+1)$, $p_0(t) = t+1$, $p_1(t) = -t$. We observe that, although the coefficients of $D(s,t)$, $N(s,t)$ are bounded, the solution $[q_0(t), p_0(t), p_1(t)]$ is not, when we consider a time interval open at ∞. ▽▽

2.12 Definition: Let $D(s,t)$, $N(s,t)$ be PDO's with smooth,[8] UB coefficients and $D(s,t)$ monic. We say that $D(s,t)$ and $N(s,t)$ are *strongly right (left) coprime* in (t_0,∞), if they are right (left) coprime in (t_0,T), $\forall T > t_0$ and the coefficients of the PDO's $D_0(s,t)$, $N_0(s,t)$, $Q(s,t)$ and $P(s,t)$ of Definition 2.4, are UB $\forall t \in (t_0,\infty)$. ▽▽

With the above definition, the determinant of the TV Sylvester matrix of strongly coprime PDO's is strongly nonsingular, as shown by the following lemma.

2.13 Lemma: Two PDO's with smooth, UB coefficients, one of which is monic, are strongly right (left) coprime in (t_0,∞) iff there exists a constant $c > 0$ such that their right (left) TV Sylvester matrix $S_R(t)$ $(S_L(t))$ satisfies $|\det[S_R(t)]| \geq c$, $(|\det[S_L(t)]| \geq c)$, $\forall t \in (t_0,\infty)$. ▽▽

Proof: In Appendix II.

---

[8] In the sequel, we use the term *smoothness* to signify that the derivatives of the coefficients (or parameters) exist and are bounded functions of time.

2.2. TIME-VARYING DIFFERENTIAL OPERATORS

The above lemma is easily extended to cover cases where both PDO's are non-monic and at least one of them has a leading coefficient bounded away from zero:

2.14 Corollary: Let $k_1(t)$, $k_2(t)$ be smooth, UB functions of time such that $|k_1(t)|, |k_2(t)| \geq c > 0$, $\forall t \in (t_0, \infty)$ and $D(s,t), N(s,t)$ be two PDO's with smooth, UB coefficients and $D(s,t)$ monic. Then, there exists a constant $c' > 0$ such that the right (left) TV Sylvester matrix of $k_1(t)D(s,t)$, $k_2(t)N(s,t)$ $(D(s,t)k_1(t), N(s,t)k_2(t))$ $S_R(t)$ $(S_L(t))$ satisfies

$$|\det[S_R(t)]| \geq c' \quad (|\det[S_L(t)]| \geq c') ; \quad \forall t \in (t_0, \infty)$$

iff $D(s,t)$, $N(s,t)$ are strongly right (left) coprime PDO's in (t_0, ∞). ▽▽

Proof: Straightforward extension of the proof of Lemma 2.13. □□

2.15 Example: Considering the same PDO's as in Example 2.11 we note that the conditions of both Definition 2.12 and Lemma 2.13 are violated since the coefficients of the PDO's that satisfy the Bezout equation are unbounded functions of time over $\mathbf{R}_+$ and $\det[S_L(t)] = -\frac{1}{t+1}$ is not bounded away from zero when $t \in \mathbf{R}_+$. Hence, the PDO's $[s^2 + s\frac{1}{t+1} - 1]$ and $[s+1]$, although left coprime in every finite interval $[0,T)$, are not strongly left coprime in $\mathbf{R}_+$.
▽▽

The previously presented definitions and lemmas can be used to give a general, closed-form characterization of the solutions of Diophantine equations. In the subsequent chapters it is of particular interest to solve the following type of problems:

Given PDO's $D(s,t)$, $N(s,t)$ and $A_*(s,t)$ with smooth, UB coefficients and a smooth, UB and bounded away from zero function $k_1(t)$ where $D(s,t)$, and $A_*(s,t)$ are monic PDO's, $\deg[N(s,t)] \leq \deg[D(s,t)] - 1$, $\deg[A_*(s,t)] \geq \deg[D(s,t)] + \deg[N(s,t)]$, find PDO's $P(s,t), Q(s,t)$ such that

$$Q(s,t)k_1(t)D(s,t) + P(s,t)N(s,t) = k_1(t)A_*(s,t) \tag{2.9}$$

or

$$D(s,t)Q(s,t) + N(s,t)P(s,t) = A_*(s,t) \tag{2.10}$$

For the PDO's $P(s,t), Q(s,t)$, we seek a solution such that $\deg[P(s,t)] \leq \deg[D(s,t)] - 1$ which implies that $Q(s,t)$ must be a monic PDO with

$$\deg[Q(s,t)] = \deg[A_*(s,t)] - \deg[D(s,t)]$$

Under these conditions the following corollary can be used.

2.16 Corollary: Suppose that the PDO's $D(s,t)$, $N(s,t)$ are strongly right coprime in $\mathbf{R}_+$ and $|k_1(t)| \geq c > 0 \; \forall t \in \mathbf{R}_+$. Then the Diophantine equation (2.9) has a unique solution for $Q(s,t)$, $P(s,t)$ with UB coefficients $\forall t \in \mathbf{R}_+$ and such that $\deg[P(s,t)] \leq \deg[D(s,t)] - 1$. The coefficients of $Q(s,t), P(s,t)$ can be calculated by solving a system of linear algebraic equations $\hat{S}(t)x(t) = a(t)$ where $x(t)$ is the vector of coefficients of $Q(s,t)$, $P(s,t)$; $\hat{S}(t)$ is an invertible matrix and $a(t)$ is a vector whose entries depend on $k_1(t)$ and the coefficients of $D(s,t)$, $N(s,t)$, $A_*(s,t)$ and their derivatives. This statement also holds in the case of equation (2.10) provided that $D(s,t)$, $N(s,t)$ are strongly left coprime PDO's in $\mathbf{R}_+$. ▽▽

Proof: In Appendix II.

2.17 Example: Consider the Diophantine equation

$$Q(s,t)D(s,t) + P(s,t)N(s,t) = A_*(s,t) \qquad (2.11)$$

where $D(s,t) = [s^2 + a_1(t)s + a_2(t)]$ and $N(s,t) = [s + b(t)]$ are strongly right coprime PDO's in $\mathbf{R}_+$ and $a_1(t), a_2(t), b(t)$ are smooth, UB functions of t. Further, suppose that we seek a solution of (2.11) for $A_*(s,t) = s^3 + k_1 s^2 + k_2 s + k_3$ and such that $\deg[P(s,t)] \leq 1$.

Observe first that for the degrees of the left and right hand-side of (2.11) to be equal $\deg[Q(s,t)] = 1$. Hence, (2.11) can be written as

$$[s + q(t)]D(s,t) + [p_1(t)s + p_2(t)]N(s,t) = A_*(s,t)$$

$$q(t)D(s,t) + [p_1(t)s + p_2(t)]N(s,t) = A_*(s,t) - sD(s,t) \qquad (2.12)$$

Expressing (2.12) in terms of the right TV Sylvester matrix of $D(s,t)$ and $N(s,t)$ we obtain

$$\underbrace{\begin{bmatrix} 1 & 1 & 0 \\ a_1(t) & b(t) & 1 \\ a_2(t) & b(t) & b(t) \end{bmatrix}}_{S_R(t)} \begin{bmatrix} q(t) \\ p_1(t) \\ p_2(t) \end{bmatrix} = \begin{bmatrix} k_1 - a_1(t) \\ k_2 - a_2(t) - \dot{a}_1(t) \\ k_3 - \dot{a}_2(t) \end{bmatrix}$$

Since $D(s,t)$ and $N(s,t)$ are strongly right coprime in $\mathbf{R}_+$, it follows that $\det[S_R(t)] \geq c > 0$ and therefore (2.11) has a unique solution for $Q(s,t)$ and $P(s,t)$ with UB coefficients $\forall t \in \mathbf{R}_+$ and such that $Q(s,t)$ is monic and $\deg[P(s,t)] \leq 1$. ▽▽

2.18 Remark: The uniqueness of solutions of a Diophantine equation, established in Corollary 2.16, is valid under the constraint that $\deg[P(s,t)] \leq \deg[D(s,t)] - 1$. If this condition is removed, the results of the above corollary

2.3. TIME-VARYING I/O OPERATORS

are still valid except for the part regarding the uniqueness of solutions. It is actually straightforward to see that if the pair $[Q(s,t), P(s,t)]$ is a solution of the Diophantine equation

$$Q(s,t)D(s,t) + P(s,t)N(s,t) = A_*(s,t)$$

so is the pair $[Q(s,t)+W(s,t)Q_0(s,t), P(s,t)+W(s,t)P_0(s,t)]$ where $P_0(s,t)$, $Q_0(s,t)$ are as given in Definition 2.4 and $W(s,t)$ is an arbitrary PDO (compare with [DLMS.80]). This observation, together with its dual counterpart for a Diophantine equation $D(s,t)Q(s,t) + N(s,t)P(s,t) = A_*(s,t)$, plays an important role in the design of controllers which are required to meet certain sensitivity-related objectives, in addition to closed-loop stability. ▽▽

2.3 Time-Varying I/O Operators

The next step in our effort to describe LTV systems by fractional representations is to develop the notion of the 'inverse' operator corresponding to a PDO. Considering the differential equation (2.1), the inverse operator would be an integral operator that maps $u \mapsto y$. The solution of (2.1), customarily defined through the state transition matrix of the corresponding differential equation, is given as (see [D.V.75, Des.69])

$$y(t) = \mathbf{c}^\mathsf{T} \underbrace{\int_{t_0}^{t} \Phi(t,\tau)\mathbf{b}u(\tau)\,d\tau}_{ZSR} + \underbrace{\mathbf{c}^\mathsf{T} \Phi(t,t_0)\mathbf{x}(t_0)}_{ZIR} \qquad (2.13)$$

where $\Phi(t,\tau)$ is the state transition matrix of (2.3) and $\mathbf{c}, \mathbf{b}, \mathbf{x}$ are as in (2.3). The first term of the right-hand side of (2.13) is referred to as the zero-state response (ZSR) while the second term is the zero-input response (ZIR) of (2.1).

2.19 Definition: *An LTV left (right) Polynomial Integral Operator PIO of order n is defined as the operator that maps the input u to the zero-state response of the differential equation $P(s,t)[y] = u$ where $P(s,t)$ is a monic left (right) PDO of degree n.*[9] *We denote a PIO by $P^{-1}(s,t)$ and write*

$$P^{-1}(s,t)[u](t) = \mathbf{c}^\mathsf{T} \int_{t_0}^{t} \Phi(t,\tau)\mathbf{b}u(\tau)\,d\tau$$

▽▽

---
[9] The 'polynomial' term in the PIO is due to the polynomial form of the associated differential operator.

Using Definition 2.19 the solution of (2.1) can be written as

$$y(t) = P^{-1}(s,t)[u](t) + c^T \Phi(t,t_0) x(t_0) \tag{2.14}$$

The motivation to denote a PIO as an inverse PDO can be explained by the following lemma.

2.20 Lemma: Let $P(s,t)$ be a monic nth degree PDO with piecewise continuous, UB coefficients. Suppose that $y = P^{-1}(s,t)[u]$ and x is such that $P(s,t)[x] = u$ where $u : [t_0, \infty) \mapsto \mathbf{R}$ is a piecewise continuous function. Also suppose that $y^{(i)}(t_0) = x^{(i)}(t_0) = 0$ for $i = 0, 1, \ldots, n-1$. Then $y = P^{-1}(s,t)P(s,t)[x] = x$, in $[t_0, \infty)$. Further, if $\bar{y} = P(s,t)P^{-1}(s,t)[u]$ then $\bar{y} = u$, in $[t_0, \infty)$. ▽▽

Proof: (a.) From the PIO definition we have that $P(s,t)[y] = P(s,t)[x]$. Since the coefficients of $P(s,t)$ are UB and piecewise continuous the solution of the differential equation $P(s,t)[y] = u$, on $[t_0, \infty)$, exists and is unique. Furthermore, the solutions of $P(s,t)[y] = u$ and $P(s,t)[x] = u$, on $[t_0, \infty)$, pass through the same point since y, x have the same initial conditions; hence, $y(t) = x(t)$, $\forall t \in [t_0, \infty)$ [C.L.55].

(b.) Let $z = P^{-1}(s,t)[u]$. Then $\bar{y} = P(s,t)[z]$, in $[t_0, \infty)$. Also, from Definition 2.19, $P(s,t)[z] = u$, in $[t_0, \infty)$. Hence, $\bar{y}(t) = x(t)$, $\forall t \in [t_0, \infty)$. □□

Using the notion of PDO's and PIO's we can express the zero-state response of more general LTV differential equations than $P(s,t)y = u$ in a compact form. This is achieved by introducing the notion of an LTV I/O operator which can be written as a combination of PDO's and PIO's and has properties analogous to those of a transfer function in LTI systems. We consider the following general LTV system

$$\dot{z} = A(t)z + b(t)u \ , \ y = c^T(t)z + d(t)u \ ; \ z(t_0) = z_0 \tag{2.15}$$

where $A(t)$, $b(t)$, $c(t)$, $d(t)$ are UB, piecewise continuous functions of time.

2.21 Definition: The operator G that maps the input u to the zero-state response of the differential equation (2.15) is defined as

$$G[u](t) = c^T(t) \int_{t_0}^{t} \Phi(t,\tau) b(\tau) u(\tau) \, d\tau + d(t)u, \quad t \geq t_0 \geq 0.$$

We refer to G as the proper LTV I/O operator of (2.15). When $d(t) = 0\ \forall t \geq 0$ we refer to G as the strictly proper LTV I/O operator of (2.15). ▽▽

2.3. TIME-VARYING I/O OPERATORS

2.22 Example: Let us consider the LTV system described by the linear ordinary differential equation

$$\dot{X} = \begin{bmatrix} 0 & 1 & 0 & \cdots & 0 \\ 0 & 0 & 1 & \cdots & 0 \\ \vdots & \vdots & \vdots & & 1 \\ -a_n(t) & -a_{n-1}(t) & -a_{n-2}(t) & \cdots & -a_1(t) \end{bmatrix} X + \begin{bmatrix} 0 \\ 0 \\ \vdots \\ 1 \end{bmatrix} u$$

$$y = [b_{n-1}(t), b_{n-2}(t), \ldots, b_0(t)] X \qquad (2.16)$$

where $X : \mathbf{R}_+ \mapsto \mathbf{R}^n$ is the state vector and $u, y : \mathbf{R}_+ \mapsto \mathbf{R}$ is the input and the output of the system respectively. It is straightforward to see that by letting

$$D(s,t) = s^n + a_1(t)s^{n-1} + \cdots + a_n(t)$$
$$N(s,t) = b_0(t)s^{n-1} + b_1(t)s^{n-2} + \cdots + b_{n-1}(t)$$

the I/O relationship (2.16) can be described as

$$D(s,t)[x] = u \ ; \ y = N(s,t)[x] \qquad (2.17)$$

where $x = [1, 0, \ldots, 0]X$ and thus, the I/O operator of (2.16) admits a *right factorization* in terms of PDO's and PIO's, that is $G_1(s,t) = N(s,t)D^{-1}(s,t)$.

▽▽

2.23 Example: Let us consider the LTV system described by the linear ordinary differential equation

$$\dot{X} = \begin{bmatrix} -a_1(t) & 1 & 0 & \cdots & 0 \\ -a_2(t) & 0 & 1 & \cdots & 0 \\ \vdots & \vdots & \vdots & & 1 \\ -a_n(t) & 0 & 0 & \cdots & 0 \end{bmatrix} X + \begin{bmatrix} b_0(t) \\ b_1(t) \\ \vdots \\ b_{n-1}(t) \end{bmatrix} u$$

$$y = [1, 0, \ldots, 0] X \qquad (2.18)$$

where $X : \mathbf{R}_+ \mapsto \mathbf{R}^n$ is the state vector and $u, y : \mathbf{R}_+ \mapsto \mathbf{R}$ is the input and the output of the system respectively. By differentiating y in (2.18) and letting

$$\hat{D}(s,t) = s^n + s^{n-1}a_1(t) + \cdots + a_n(t)$$
$$\hat{N}(s,t) = s^{n-1}b_0(t) + s^{n-2}b_1(t) + \cdots + b_{n-1}(t)$$

the I/O relationship (2.18) is described as

$$\hat{D}(s,t)[y] = \hat{N}(s,t)[u] \qquad (2.19)$$

and therefore, the I/O operator of (2.18) admits a *left factorization* in terms of PDO's and PIO's, that is $G_2(s,t) = \hat{D}^{-1}(s,t)\hat{N}(s,t)$. It should be noted that as long as (2.19) is interpreted in the sense of (2.18), the coefficients of the right PDO's $\hat{D}(s,t)$ and $\hat{N}(s,t)$ are not required to be differentiable (e.g., piecewise continuity and boundedness are sufficient). ▽▽

The properties of PIO's follow from the theory of differential equations and the rules of integration and are summarized below:

P1'. If $P(s,t) = Q(s,t)$ then $P^{-1}(s,t) = Q^{-1}(s,t)$.

P2'. $P^{-1}(s,t) + Q^{-1}(s,t) = Q^{-1}(s,t) + P^{-1}(s,t)$.

P3'. $P^{-1}(s,t)c = cP^{-1}(s,t)$; c is a constant.

P4'. $[P(s,t)Q(s,t)]^{-1} = Q^{-1}(s,t)P^{-1}(s,t)$.

where $P(s,t), Q(s,t)$ are monic PDO's and, in P4', have smooth UB coefficients. While the proof of the above properties is straightforward and is omitted, we note that the definition of the PIO as the zero-state response of a differential equation is crucial in concluding P1'. That is, if we consider the differential equation $P(s,t)[y] = u$ with two sets of initial conditions then, not only the corresponding solutions are not necessarily equal but their difference may even be unbounded.[10] If, however, we restrict ourselves to the case of exponentially stable (ES) differential equations, we may conclude that the effect of the initial conditions on the solution of the differential equation (output) vanishes asymptotically with time. This observation is quantified in the following discussion.

2.24 Definition: An LTV PIO, $P^{-1}(s,t)$, is said to be ES (or uniformly asymptotically stable) with rate $-a_1$, $a_1 > 0$, if there exist some positive constants k, a_1 such that the state transition matrix $\Phi(t,\tau)$, associated with the linear differential equation $P(s,t)y = u$, satisfies

$$\|\Phi(t,\tau)\| \leq k\exp[-a_1(t-\tau)] \quad \forall t \geq \tau \geq 0$$

▽▽

2.25 Lemma: Let $P^{-1}(s,t)$ be an ES PIO with rate $-a_1$ and consider the LTV systems with I/O operators $P^{-1}(s,t)P(s,t)$, $P(s,t)P^{-1}(s,t)$ and I/O pairs (y,x), $(\bar{y},x)$ respectively. Then, $y = x + \epsilon_t$ and $\bar{y} = x + \epsilon_t$ where ϵ_t denotes exponentially decaying to zero terms with rate at most $-a_1$. Such terms depend on the initial conditions of the state space realization of the respective I/O operator. ▽▽

---

[10] This problem is due to the nature of the solution of an ODE that is a map from the set of inputs and the set of initial conditions to the set of outputs and it also appears in the LTI case as well.

Proof: Straightforward, by applying Definition 2.24 on the solution of the respective differential equations (see Definition 2.19 and 2.21). □□

Lemma 2.25 shows that If $P^{-1}(s,t)$ is ES, the effect of setting

$$P(s,t)P^{-1}(s,t) = P^{-1}(s,t)P(s,t) = I$$

is the appearance of exponentially decaying to zero terms ϵ_t in the solution of the differential equations associated with these operators. Thus, Lemma 2.25 extends the properties of pole-zero cancellation in the LTI case, to the LTV one.

To assess the stability of proper LTV I/O operators, described by multiple combinations of PDO's and PIO's, we note that from Definitions 2.21, 2.24 it follows that such an I/O operator is ES if all the PIO's in its description are ES. For example, if $P^{-1}(s,t)$, $D^{-1}(s,t)$ are ES PIO's then a proper LTV I/O operator described by $P^{-1}(s,t)N(s,t)D^{-1}(s,t)$ is also ES. The last observation together with Lemma 2.25 also indicates that the equality of two TV I/O operators can be interpreted as equality of the output trajectories of the respective ODE's, modulo terms that depend on the initial conditions and decay to zero exponentially fast provided that the I/O operators at hand are ES.

2.4 Controllability, Observability and I/O Descriptions of LTV Systems

In the following, our attention is specifically focused on I/O operators that are expressed as a combination of PDO's and PIO's. Such a description of an LTV system facilitates the analysis and design of controllers in an I/O framework in both the non-adaptive and adaptive cases. This approach, however, gives rise to two fundamental, nontrivial problems, especially if a general class of LTV systems is to be considered. These problems are associated with the class of LTV systems for which the I/O operator admits a PDO/PIO factorization and their internal stability properties.[11]

In order to address these questions, we first recall some of the basic definitions and properties regarding the controllability and observability of LTV systems. In all cases we denote the system matrices by $[A(t), B(t), C(t)]$ [12] and the corresponding state transition matrix by $\Phi(t,\tau)$. Furthermore, we assume that the system parameters are bounded and differentiable, as many times as

---

[11] The need to consider these problems arises from the fact that in the LTV case, I/O equivalent systems do not necessarily share the same internal stability properties.
[12] $\dot{x} = A(t)x + B(t)u$; $y = C(t)x$.

necessary. Although this assumption can be relaxed, we do not pursue such a generalization at this point.

2.26 Theorem: *[S.A.68] Following from the definition of [Kal.60], a bounded system $[A(t), B(t), C(t)]$ is uniformly completely controllable iff there exist $d_c > 0$ such that for all t*

$$M_c(t - d_c, t) \triangleq \int_{t-d_c}^{t} \Phi(t - d_c, \tau) B(\tau) B^{\mathsf{T}}(\tau) \Phi^{\mathsf{T}}(t - d_c, \tau) \, d\tau \geq a_1(d_c) I > 0$$

*and **uniformly completely observable** iff there exists $d_o > 0$ such that for all t*

$$N_o(t, t + d_o) \triangleq \int_{t}^{t+d_o} \Phi^{\mathsf{T}}(\tau, t) C^{\mathsf{T}}(\tau) C(\tau) \Phi(\tau, t) \, d\tau \geq a_1(d_o) I > 0$$

where $a_1(\sigma)$ is used to denote a constant solely determined by its argument. The matrices $M_c(t, \tau)$ and $N_o(t, \tau)$ are referred to as controllability and observability grammians respectively. ▽▽

2.27 Definition: *[Sil.68] A system representation $[A(t), B(t), C(t)]$ is said to be **uniform** if it is continuous, bounded and uniformly completely controllable and observable. An impulse response $H(t, \tau)$ which can be realized by a uniform system representation, is said to be **uniformly realizable**.* ▽▽

2.28 Definition: *[Sil.68] The representation $[A'(t), B'(t), C'(t)]$ is said to be **algebraically equivalent** to $[A(t), B(t), C(t)]$ if there exists a nonsingular matrix T with continuous derivative such that*

$$A'(t) = (TA(t) + \dot{T})T^{-1}, \quad B'(t) = TB(t), \quad C'(t) = C(t)T^{-1}$$

*If T is a Lyapunov transformation, (i.e., $T, T^{-1}, \dot{T}$ are continuous and UB) then the two representations are said to be **topologically equivalent**.* ▽▽

2.29 Theorem: *[Sil.68] a. If $[A(t), B(t), C(t)], [A'(t), B'(t), C'(t)]$ are uniform realizations of the same impulse response, then they are topologically equivalent.*

b. If $[A(t), B(t), C(t)]$ is a uniform realization of an impulse response and topologically equivalent to $[A'(t), B'(t), C'(t)]$ then $[A'(t), B'(t), C'(t)]$ is a uniform realization of the same impulse response. ▽▽

2.30 Definition: *[S.M.67] The state-space representation $[A(t), B(t), C(t)]$, with $dim[A] = n \times n$ is said to be **uniformly controllable** if the controllability matrix*

$$Q_c = [p_0, p_1, \ldots, p_{n-1}] \quad ; \quad p_{k+1} = -A(t)p_k + \dot{p}_k \quad ; \quad p_0 = B(t)$$

2.4. CONTROLLABILITY, OBSERVABILITY ...

is nonsingular for all t. Dually, $[A(t), B(t), C(t)]$ is said to be **uniformly observable** if the observability matrix

$$Q_o^\mathsf{T} = [q_0, q_1, \ldots, q_{n-1}] \; ; \; q_{k+1} = A^\mathsf{T}(t)q_k + \dot{q}_k \; ; \; q_0 = C^\mathsf{T}(t)$$

is nonsingular for all t. ▽▽

2.31 Theorem: *[Sil.66, S.A.68] For single input systems,*

a. The controllable (phase-variable) canonical form is uniformly controllable.

b. A system representation is algebraically equivalent to the phase variable canonical form iff it is uniformly controllable.

c. A bounded realization for which Q_c is a Lyapunov transformation, is uniformly completely controllable.

d. The phase variable canonical form with bounded coefficients is uniformly completely controllable.

The dual statements are also true for the observable canonical form of single output systems. ▽▽

We therefore note that it is possible to write the I/O operator of a single-input single-output (SISO) LTV system in terms of PDO's and PIO's if $[A(t), B(t), C(t)]$ is algebraically equivalent with the controllable (observable) canonical form, as shown in Example 2.22 (Example 2.23). Under the assumption that the original system is uniformly controllable (uniformly observable), such a transformation can be obtained in terms of the entries of $[A(t), B(t), C(t)]$ and their derivatives after some straightforward recursive computations [Sil.66]. However, we must also require that this transformation is a Lyapunov one —so that the internal stability properties of the system are preserved— and that it leads to a canonical form with UB coefficients, needed for the well-posedness of the control problem studied in the subsequent chapters.

In other words, the system representation whose states are the internal signals of interest must be a uniform realization and, in addition, the corresponding controllable and observable canonical representations must also be uniform realizations of the same impulse response. The difficulty in obtaining a simple characterization of such systems stems from the fact that, in general, uniform controllability/observability (required for the transformation to a canonical form) neither implies, nor is implied by, uniform complete controllability/observability. It is, however, possible to give sufficient conditions which guarantee both the uniform and uniform complete controllability/observability for SISO systems with bounded and smooth parameters.

Although only a finite number of derivatives is required to exist for the analysis, we avoid counting the number of differentiations and assume that all necessary derivatives are UB. A generalization of the results to the case of piecewise continuous parameters is discussed separately in the forthcoming chapters.

2.32 Lemma: *Consider a SISO system representation $[A(t), B(t), C(t)]$ which is bounded and smooth and such that the controllability and observability matrices are strongly nonsingular, i.e., there exists a constant $c > 0$ such that*

$$|\det[Q_c(t)]| \, , \, |\det[Q_o(t)]| \geq c, \quad \forall t \in \mathbf{R}_+$$

Such representations are termed strongly controllable and observable. Then $[A(t), B(t), C(t)]$ is a uniform realization, topologically equivalent to its controllable and observable canonical forms which exist and are uniform realizations with smooth coefficients of the same impulse response. Furthermore, the controllability and observability matrices of the corresponding canonical forms are also strongly nonsingular. ▽▽

Proof: From [Sil.66] the nonsingularity of the controllability matrix guarantees the existence of a transformation that puts $[A(t), B(t), C(t)]$ in the controllable (phase-variable) canonical form. The transformation, given explicitly in terms of the parameters of $[A(t), B(t), C(t)]$, is a Lyapunov one since the controllability matrix is strongly nonsingular and the system parameters are smooth and UB functions of time. Furthermore, the parameters of the canonical form are also smooth and UB. Hence the system realization $[A(t), B(t), C(t)]$ is uniformly completely controllable and topologically equivalent to its canonical form. For the last part of the theorem, observe that if $[A(t), B(t), C(t)]$ and $[\bar{A}(t), \bar{B}(t), \bar{C}(t)]$ are algebraically equivalent state-space realizations with a similarity transformation $x = P(t)\bar{x}$, then the corresponding controllability and observability matrices satisfy

$$\bar{Q}_c(t) = P^{-1}(t) Q_c(t) \; ; \; \bar{Q}_o(t) = Q_o(t) P(t)$$

Since the transformation of $[A(t), B(t), C(t)]$ into the phase-variable form is a Lyapunov one, $P(t), P^{-1}(t)$ are strongly nonsingular and the result follows. (Dually for the transformation into the observable canonical form.) □□

Thus, a strongly controllable and observable SISO LTV system that admits a bounded, smooth realization can be described by an I/O operator with a PDO/PIO factorization that preserves the internal stability properties of the original system. Consequently, the PDO/PIO description can be used for

2.5. INTERNAL STABILITY

a quite general class of LTV systems and does not require overly restrictive assumptions, especially for practical cases where physical considerations assure the boundedness and smoothness of the system parameters at least in a piecewise sense.

It is worth mentioning at this point that the notions of uniform (strong) controllability and observability of linear systems bear a close relationship with the notions of uniform (strong) right and left coprimeness of PDO's. As reported in [INS.84], for analytic systems, uniform left coprimeness is equivalent to uniform controllability in a closed interval (dually for observability and right coprimeness). In our case, however, we are more interested in the relationship between the strong versions of left/right coprimeness and controllability/observability. The reason is that for the systems under consideration[13] these properties ensure that the system realization is uniform and that controllers designed via a Diophantine equation have bounded parameters.

2.33 Lemma: *Let Q_c be the controllability matrix of the system (2.18), which is in the observable canonical form, with smooth parameters and I/O operator $D^{-1}(s,t)N(s,t)$ and let $S_L(t)$ be the left TV Sylvester matrix of $D(s,t)$ and $N(s,t)$. Then, $|\det[Q_c(t)]| = |\det[S_L(t)]|$. (Dually for observability and the right TV Sylvester matrix).* ▽▽

Proof: In Appendix II.

2.34 Corollary: *The LTV system (2.16) with I/O operator $N(s,t)D^{-1}(s,t)$ and smooth UB coefficients, is strongly observable in $\mathbf{R}_+$ iff $D(s,t)$, $N(s,t)$ are strongly right coprime PDO's in $\mathbf{R}_+$. The LTV system (2.18), with I/O operator $\hat{D}^{-1}(s,t)\hat{N}(s,t)$ and smooth UB coefficients, is strongly controllable in $\mathbf{R}_+$ iff $\hat{D}(s,t)$, $\hat{N}(s,t)$ are strongly left coprime PDO's in $\mathbf{R}_+$.* ▽▽

Proof: Straightforward from Lemmas 2.13 and 2.33. □□

2.5 On the Internal Stability of LTV Feedback Systems

Next, we address the issue of internal stability of a closed-loop system. This is necessary in order to establish that a controller designed to guarantee bounded-input bounded-output (BIBO) stability, also guarantees bounded-input bounded-state (BIBS) stability of the closed loop. As in the LTI case, such a design fails to ensure BIBS stability if there are any cancellations of

---
[13] SISO, bounded, smooth realizations.

unstable modes in the closed-loop system. This possibility is avoided by using the techniques of [DLMS.80] to design the controller I/O operator, via fractional representations and an appropriate Diophantine equation. In the LTV case, an additional subtlety of the BIBS stability problem lies in the fact that algebraically equivalent representations do not necessarily share the same internal stability properties unless, for example, they are both uniform realizations of the same impulse response [S.A.68]. However, if the closed-loop system contains uncontrollable/unobservable modes, BIBS stability cannot be simply concluded from BIBO stability as in [S.A.68]. This issue is addressed in the subsequent chapters through the following lemmas.

2.35 Lemma: *Consider an LTV system* $\Sigma_p : u \mapsto y$ *described by a continuous, bounded and uniformly completely observable realization*

$$\Sigma_p : \left\{ \begin{array}{l} \dot{x} = Ax + bu \\ y = c^\mathsf{T} x \end{array} \right\}$$

Also consider the control law

$$\Sigma_c : \left\{ \begin{array}{l} \dot{w} = Fw + \theta[u_1, y]^\mathsf{T} \\ u_1 = p_1^\mathsf{T} w + p_2 y \\ u = p_3^\mathsf{T} w + p_4 [u_1, y]^\mathsf{T} \end{array} \right\} \qquad (2.20)$$

where, the state transition matrix associated with F is ES, Σ_c is a bounded, piecewise continuous realization and the various matrices are assumed to have compatible dimensions. Further, assume that for all bounded initial conditions of Σ_p and Σ_c there exist constants $k, a > 0$, with k depending on the size of the initial conditions, such that

$$\|[u_1, y](t)\| \leq k \exp[-a(t - t_0)]$$

for all $t \geq t_0$ and all initial times $t_0 \geq 0$. Then the feedback system of Σ_p, Σ_c is ES and, therefore, BIBS stable and its state transition matrix satisfies

$$\|\Phi(t, \tau)\| \leq K \exp[-a(t - \tau)]$$

for some positive constant K and for all $t \geq \tau \geq 0$. ▽▽

Proof: In Appendix II.

The value of the above lemma is that, under some mild conditions on the controller realization, it establishes the exponential and therefore internal stability of the closed-loop system from its I/O properties. In other words, it is possible to conclude the internal stability of a closed-loop system despite the possible cancellation of certain modes between the plant and the controller as long as the latter is realized with ES filters in the form of (2.20). A weaker

2.6. REALIZATIONS OF TV I/O OPERATORS

result relating BIBO and BIBS stability of the system considered in Lemma 2.35 can be stated as follows:

2.36 Lemma: *For the closed-loop system described in Lemma 2.35, let x be the state vector of a continuous, bounded and uniformly completely observable realization of Σ_p and assume that u_1 and y are UB. Then x and w are also UB.* ▽▽

Proof: The proof can be obtained along the lines of the proof of Lemma 2.35 and is omitted. □□

In the following chapters we employ a wide class of control laws that can be expressed in the form of (2.20) and use Lemmas 2.35 and 2.36 to assess the internal stability properties of closed-loop systems based on their I/O stability properties.

2.6 Realizations of TV I/O Operators

We now present some examples on the realization of TV I/O operators. This issue deserves some additional attention in the TV case where, due to the non-commutativity of TV I/O operators, some care should be exercised when the realization rules for LTI filters are extended to the TV case.

2.37 Example:

a. *Realization of a left-factorized I/O operator.*

To realize
$$y = D^{-1}(s,t)N(s,t)[u]$$
in state space with $\deg[N(s,t)] < \deg[D(s,t)]$ and $D(s,t)$ monic we first express both $D(s,t)$, $N(s,t)$ as right PDO's and then set
$$\dot{x} = A(t)x + b(t)u \; ; \; y = c^\top x$$
where $A(t)$ contains the coefficients of $D(s,t)$ in the left companion form, $b(t)$ contains the coefficients of $N(s,t)$ and $c^\top = [1,0,\ldots,0]$ (see Example 2.23).

b. *Realization of a right factorized I/O operator.*

To realize
$$y = N(s,t)D^{-1}(s,t)[u]$$
with the same assumptions as in (a.), we first express both $D(s,t)$, $N(s,t)$ as left PDO's and then set
$$\dot{x} = A(t)x + bu \; ; \; y = c^\top(t)x$$

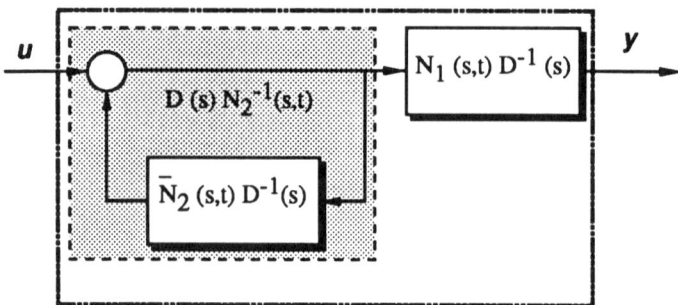

Figure 2.1: Realization of the I/O operator $N_1(s,t)N_2^{-1}(s,t)$ with ES filters.

where $A(t)$ contains the coefficients of $D(s,t)$ in the bottom companion form, $c(t)$ contains the coefficients of $N(s,t)$ and $b^\top = [0,\ldots,0,1]$ (see Example 2.22). ▽▽

Another issue of special interest in our analysis —especially in view of Lemma 2.35— is the realization of an I/O operator by using ES, preferably TI, filters with TV input or/and output vectors and fixed system matrices.[14] The following two examples deal with this problem when the I/O operators are given in a left or right factorization.

2.38 Example: Let us consider the I/O operator $N_1(s,t)N_2^{-1}(s,t)$ with $N_2(s,t)$ monic and $\deg[N_1(s,t)] < \deg[N_2(s,t)] = n$. By Lemma 2.25 we can write $N_1(s,t)N_2^{-1}(s,t) = N_1(s,t)D^{-1}(s)D(s)N_2^{-1}(s,t)$ with $D^{-1}(s)$ being an ES nth order PIO, i.e., the polynomial $D(s)$ being Hurwitz. We may now use the building blocks of Example 2.37 to realize the I/O operators $G_1(s,t) = N_1(s,t)D^{-1}(s)$ and $G_2(s,t) = D(s)N_2^{-1}(s,t)$ in cascade as shown in Fig. 2.1 and described below:

a. $G_1(s,t)$ can be realized in state-space as in Example 2.37.b with A, b being constant and $c(t)$ being TV.

b. $G_2(s,t)$ can be realized with a constant system matrix (A) by considering the feedback system

$$u_1 = u + \bar{N}_2(s,t)D^{-1}(s)[u_1]$$

whose I/O operator $u \mapsto y$ is $D(s)[D(s) - \bar{N}_2(s,t)]^{-1}$. Since $N_2(s,t)$ is a monic nth degree PDO we can select $\bar{N}_2(s,t) = D(s) - N_2(s,t)$ and realize the operator $\bar{N}_2(s,t)D^{-1}(s)$ as in Example 2.37.b with A, b constant and $c(t)$ a TV vector of the coefficients of $\bar{N}_2(s,t)$. ▽▽

---

[14] This problem can also be viewed as a special case of finding a stable, proper factorization of an LTV I/O operator.

2.7. LINEAR SLOWLY TV SYSTEMS

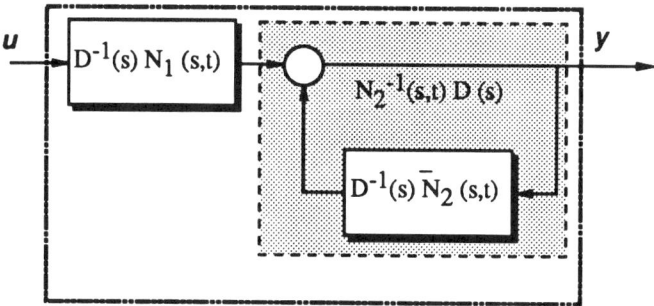

Figure 2.2: Realization of the I/O operator $N_2^{-1}(s,t)N_1(s,t)$ with ES filters.

2.39 Example: Let us now consider the I/O operator $N_2^{-1}(s,t)N_1(s,t)$ and write $N_2^{-1}(s,t)N_1(s,t) = N_2^{-1}(s,t)D(s)D^{-1}(s)N_1(s,t)$ with the operators $D^{-1}(s)$, $N_1(s,t)$ and $N_2(s,t)$, as in Example 2.38. In this case we realize in cascade the I/O operators $G_1(s,t) = N_2^{-1}(s,t)D(s)$ and $G_2(s,t) = D^{-1}(s)N_1(s,t)$ as shown in Fig. 2.2 and described below:

a. $G_2(s,t)$ can be realized as in Example 2.37.a with A, c being constant and $b(t)$ being TV.

b. $G_1(s,t)$ can be realized by considering the feedback system

$$y = u_1 + D^{-1}(s)\bar{N}_2(s,t)[y]$$

whose I/O operator $u_1 \mapsto y$ is $[D(s) - \bar{N}_2(s,t)]^{-1}D(s)$. Since $N_2(s,t)$ is a monic nth degree PDO we can select $\bar{N}_2(s,t) = D(s) - N_2(s,t)$ and realize the operator $\bar{N}_2(s,t)D^{-1}(s)$ as in Example 2.37.a with A, c constant and $b(t)$ a TV vector of the coefficients of $\bar{N}_2(s,t)$. ▽▽

2.7 Linear Slowly TV Systems

In the previous sections the development of the mathematical background was carried out for LTV systems with finite but otherwise arbitrary speed of variation of their parameters. Motivated by the important place the slowly TV systems have in the adaptive control problem, the possibility of a simplification of the previous general results in the case of linear systems with smooth slowly TV parameters deserves special attention. In this case it is intuitively expected that 'robust properties' of LTI systems e.g., exponential stability, strong coprimeness, strong controllability/observability can be extended to slowly TV systems by assuming that the parameter variations are 'sufficiently' slow so that the LTV system behaves at each time instant almost as an LTI one. The benefit of such an approach would be that the relatively simple tests for stability or coprimeness and methods for controller design, available in the LTI case, can be used in a pointwise fashion in the LTV case as well. These arguments and observations are quantified in the rest of this section where, in order to simplify the various statements, we use the notation $\Theta_P(t)$ to denote a vector containing the TV parameters of an I/O operator $P(s,t)$.

Let $D(s,t)$ be a PDO with coefficients $\{a_i(t)\}_0^n$, $t \in \mathbf{R}_+$. Then, at any fixed time instant $\tau \in \mathbf{R}_+$, we can define a polynomial $D_\tau(s)$ with constant coefficients $\{b_i\}_0^n$ such that $b_i = a_i(\tau)$, $i = 0, 1, \ldots, n$. Consequently, the PDO $D(s,t)$ can be associated with a family of polynomials with constant coefficients, denoted by $\{D_t(s)\}_t$, with the definition that the polynomial $D_\tau(s)$ with coefficients $\{b_i\}_0^n$ belongs to $\{D_t(s)\}_t$ if

$$b_i = a_i(\tau), \quad i = 0, 1, \ldots, n$$

for some fixed $\tau \in \mathbf{R}_+$.

2.40 Definition: *We say that the PDO's $D(s,t)$ and $N(s,t)$ are pointwise coprime at time $t = t_0$ if the corresponding members $D_{t_0}(s)$ and $N_{t_0}(s)$ of the families $\{D_t(s)\}_t$ and $\{N_t(s)\}_t$ are coprime polynomials.*

Further, we say that the PDO's $D(s,t)$ and $N(s,t)$ are pointwise strongly coprime in $\mathbf{R}_+$ if there exists a constant $c > 0$ such that the Sylvester matrix $S(\tau)$ of the polynomials $D_\tau(s) \in \{D_t(s)\}_t$ and $N_\tau(s) \in \{N_t(s)\}_t$ satisfies $|\det[S(\tau)]| \geq c, \forall \tau \in \mathbf{R}_+$ ▽▽

The strong right or left coprimeness of two slowly TV PDO's $D(s,t)$, $N(s,t)$ can be deduced from their pointwise strong coprimeness, as described in the following lemma.

2.41 Lemma: *Consider the PDO's $D(s,t)$, $N(s,t)$ of degree n, m respectively, with $\Theta_D(t)$, $\Theta_N(t)$ smooth and UB. Suppose that $D(s,t)$, $N(s,t)$ are pointwise strongly coprime in $\mathbf{R}_+$ and there exists a constant $\mu \geq 0$ such that*

$$\|\frac{d^i}{dt^i}\Theta_D(t)\| \leq \mu \; ; \; \|\frac{d^j}{dt^j}\Theta_N(t)\| \leq \mu \; ; \; \forall t \in \mathbf{R}_+$$

for $i = 1, \ldots, m-1$, $j = 1, \ldots, n-1$. Then there exists a $\mu_o > 0$ such that $\forall \mu \in [0, \mu_o)$, the PDO's $D(s,t)$, $N(s,t)$ are strongly right and left coprime in $\mathbf{R}_+$. ▽▽

Proof: In Appendix II.

2.7. LINEAR SLOWLY TV SYSTEMS

In a similar fashion, sufficient conditions for the stability of slowly TV systems or TV systems that can be considered as perturbations of ES ones can be given in terms of the speed of the parameter variations or the 'size' of the perturbation e.g., see [Vid.78, Pra.85].

2.42 Lemma: *Consider the system $\dot{x} = A(t)x$ with $A(t)$ Lipschitz continuous and UB on $\mathbf{R}_+$ and suppose that there exist constants $\mu, c \geq 0$ such that one of the following conditions is satisfied:*

1. *ess.sup$_{t \geq 0} \|\dot{A}(t)\| \leq \mu$;*

2. *$\int_{t_0}^{t_0+T} \|\dot{A}(t)\| \, dt \leq c + \mu T$, $\forall T \geq 0$, $\forall t_0 \geq 0$;*

3. *$\int_{t_0}^{t_0+T} \|\dot{A}(t)\|^2 \, dt \leq c + \mu T$, $\forall T \geq 0$, $\forall t_0 \geq 0$.*

Furthermore, let $\lambda_i(t)$ denote the ith eigenvalue of $A(t)$ and assume that there exists a constant $a > 0$ such that $Re[\lambda_i(t)] \leq -a$, $\forall i$ and $\forall t \in \mathbf{R}_+$. Then, $\forall \delta \in (0, a)$, there exists $\mu_o(\delta) > 0$ such that $\forall \mu \in [0, \mu_o)$, the system $\dot{x} = A(t)x$ is ES with rate at most $-a + \delta$. ▽▽

Proof: In Appendix II.

A straightforward application of the above lemma can serve to establish stability properties of a PIO with slowly TV coefficients, as follows.

2.43 Corollary: *Consider the PIO $P^{-1}(s,t)$ with $\Theta_P(t)$ Lipschitz continuous and UB on $\mathbf{R}_+$ and suppose that there exist constants $\mu, c \geq 0$ such that one of the following conditions is satisfied:*

1. *ess.sup$_{t \geq 0} \|\dot{\Theta}_P(t)\| \leq \mu$;*

2. *$\int_{t_0}^{t_0+T} \|\dot{\Theta}_P(t)\| \, dt \leq c + \mu T$, $\forall T \geq 0$, $\forall t_0 \geq 0$;*

3. *$\int_{t_0}^{t_0+T} \|\dot{\Theta}_P(t)\|^2 \, dt \leq c + \mu T$, $\forall T \geq 0$, $\forall t_0 \geq 0$.*

Furthermore, let $\lambda_i(\tau)$ denote the ith root of the polynomial $P_\tau(s) \in \{P_t(s)\}_t$ and assume that there exists a constant $a > 0$ such that $Re[\lambda_i(\tau)] \leq -a$, $\forall i$ and $\forall \tau \in \mathbf{R}_+$. Then, $\forall \delta \in (0, a)$, there exists $\mu_o(\delta) > 0$ such that $\forall \mu \in [0, \mu_o)$, $P^{-1}(s,t)$ is ES with rate at most $-a + \delta$. ▽▽

It cannot be overemphasized at this point that a condition on the speed of parameter variations is rather essential, in order for the pointwise asymptotic stability to imply uniform asymptotic stability of a TV linear system. In view of the state-space interpretation of PIO's, fast parameter variations may cause a pointwise stable operator to be unstable, as mentioned in Section 1.1. Consider for example the LTV system

$$P(s,t)[y] = 0$$

where

$$P(s,t) = s^2 + 0.1s + 10(1 - \cos \pi t) + 0.1$$

It can be verified that the PDO $P(s,t)$ is associated with the family of polynomials $\{P_t(s)\}_t$ for which

$$P_t(s) = [s^2 + 0.1s + 10(1 - \cos \pi t) + 0.1] \ ; \ \forall t$$

Clearly, the roots of $P_t(s)$

$$-0.05 \pm j\frac{\sqrt{0.39 + 40(1 - \cos \pi t)}}{2}$$

are in the left half-plane for all t and, in fact, their real part is constant and equal to -0.05. However, as it can be seen from Example 1.3, the PIO $P^{-1}(s,t)$ is unstable since the zero equilibrium of its state-space realization (1.8) is unstable.

Lemma 2.42 can be further generalized to linear systems with TV elements which, in addition, are functions of slowly TV parameters. In this case, if the LTV system is (uniformly) ES for every frozen value of the parameters, then it is ES provided that the speed of variation of the additional parameters is small or small in the mean-square sense. Notice, however, that the overall LTV system is not restricted to be slowly TV. More precisely,

2.44 Lemma: Consider the system $\dot{x} = A(t,\theta)x$ with $A(t)$ Lipschitz continuous in both t, θ and UB on $\mathbf{R}_+$ and θ is a vector of UB, Lipschitz continuous TV parameters (i.e., $\theta(t) \in \mathcal{M}, \forall t \in \mathbf{R}_+$ where $\mathcal{M}$ is a bounded set). Suppose that for every frozen $\bar{\theta} \in \mathcal{M}$ the system $\dot{x} = A(t,\bar{\theta})x$ is ES, uniformly in $\bar{\theta}$, i.e., there exist positive constants k, a, independent of the value of $\bar{\theta}$ such that the state transition matrix $\bar{\Phi}(.,.)$ associated with $A(t,\bar{\theta})$ satisfies

$$\|\bar{\Phi}(t,\tau)\| \leq ke^{-a(t-\tau)}, \ t \geq \tau$$

Further, suppose that there exist constants $\mu, c \geq 0$ such that one of the following conditions is satisfied:

1. ess.sup$_{t \geq 0} \|\dot{\theta}(t)\| \leq \mu$;

2. $\int_{t_0}^{t_0+T} \|\dot{\theta}(t)\| \, dt \leq c + \mu T, \forall T \geq 0, \forall t_0 \geq 0$;

2.8. MORE I/O PROPERTIES

3. $\int_{t_0}^{t_0+T} \|\dot{\theta}(t)\|^2 \, dt \leq c + \mu T, \, \forall T \geq 0, \, \forall t_0 \geq 0.$

Then, $\forall \delta \in (0, a)$, there exists $\mu_o(\delta) > 0$ such that $\forall \mu \in [0, \mu_o)$, the system $\dot{x} = A(t, \theta(t))x$ is ES with rate at most $-a + \delta$. ▽▽

Proof: In Appendix II.

Finally, the stability of TV perturbations of LTV or LTI systems can be established in a similar manner, provided that the perturbation is small or small in the mean.

2.45 Lemma: Consider the system $\dot{x} = A(t)x + \Delta(t)x$ where $A(t), \Delta(t)$ are UB and piecewise continuous, $\dot{x} = A(t)x$ is ES with rate $-a$, and suppose that there exist constants $\mu, c \geq 0$ such that one of the following conditions is satisfied:

1. ess.$\sup_{t \geq 0} \|\Delta(t)\| \leq \mu, \, \forall t \geq 0;$

2. $\int_{t_0}^{t_0+T} \|\Delta(t)\| \, dt \leq c + \mu T, \, \forall T \geq 0, \, \forall t_0 \geq 0;$

3. $\int_{t_0}^{t_0+T} \|\Delta(t)\|^2 \, dt \leq c + \mu T, \, \forall T \geq 0, \, \forall t_0 \geq 0.$

Then $\forall \delta \in (0, a)$ there exists $\mu_o(\delta) > 0$ such that $\forall \mu \in [0, \mu_o)$, the system $\dot{x} = A(t)x + \Delta(t)x$ is ES with rate at most $-a + \delta$. ▽▽

Proof: In Appendix II.

The above results indicate that pointwise techniques can be used in the design of compensators for slowly or slowly in the mean LTV plants. In the subsequent chapters, their use allows us to conclude that such compensators can guarantee closed-loop stability and good performance provided that the plant is slowly TV most of the time.

2.8 More I/O Properties

In this section we focus our interest in certain I/O properties of linear time-varying systems which relate the magnitude of the output with that of the input. We begin our discussion with some standard results on L_p theory, exponentially weighted L_p spaces and operator gains.

2.8.1 Elements of L_p Theory

Let us consider a function $f : \mathbf{R}_+ \mapsto \mathbf{R}$ which is locally integrable. Following standard definitions (e.g., [D.V.75]), for any fixed $p \in [1, \infty)$, we say that

$f \in L_p$ iff $\int_0^\infty |f(t)|^p\, dt < \infty$. It follows that L_p, the space of such functions, with norm

$$\|f\|_p = \left[\int_0^\infty |f(t)|^p\, dt\right]^{1/p}$$

is a Banach space. Further, we say that $f \in L_\infty$ iff ess.$\sup_{t \geq 0} |f(t)| < \infty$ or, in other words, $\inf\{a : |f(t)| \leq a$ almost everywhere$\} < \infty$. Again, L_∞ with norm[15]

$$\|f\|_\infty = \operatorname*{ess.\,sup}_{t \geq 0} |f(t)|$$

is a Banach space. Similar definitions and properties are also applicable for vector valued functions, i.e., $f : \mathbf{R}_+ \mapsto \mathbf{R}^n$, with $|\cdot|$ being replaced (or denoting) a vector norm in $\mathbf{R}^n$. Unless otherwise stated, a vector norm is the standard Euclidean norm.

Next, let us consider the normed linear spaces $(E, \|\cdot\|_E)$ and $(F, \|\cdot\|_F)$ and let A be a linear map from E to F. With appropriately defined addition and scalar multiplication, the space of such maps is a linear space [D.V.75]. Then the function $\|\cdot\|_i$ (or simply $\|\cdot\|$ when its meaning is clear from the context):

$$\|A\|_i = \sup_{x \neq 0} \frac{\|Ax\|_F}{\|x\|_E}$$

is called the induced norm (or gain) of A, induced by $\|\cdot\|_E$, $\|\cdot\|_F$.

The concept of the gain of an operator can be generalized as follows. Let $f : \mathbf{R}_+ \mapsto \mathbf{R}$, $T \in \mathbf{R}_+$ and let P_T denote the truncation operator:

$$P_T f(t) \triangleq f_T(t) = \left\{ \begin{array}{ll} f(t), & t \leq T \\ 0, & t > T \end{array} \right\}$$

Then, for any $p \in [1, \infty]$ we say that f belongs to the extended L_p space, denoted as L_p^e, iff $\forall T \in \mathbf{R}_+$, $f_T \in L_p$. Note that the p-norms, defined above, satisfy the properties

1. $\forall f \in L_p^e$, the map $T \mapsto \|f_T\|_p$ is a nondecreasing function of T;

2. $\forall f \in L_p$, $\|f_T\|_p \to \|f\|_p$, as $T \to \infty$.

The truncation operator is used to define the *causality* of an operator as follows: A map $H : L_p^e \mapsto L_p^e$ is said to be causal iff $P_T H P_T = P_T H$, $\forall T \in \mathbf{R}_+$.

For a causal operator H, the only kind of operators considered here, we define the L_p-gain of H by

$$\gamma_p(H) = \inf\{\gamma \in \mathbf{R}_+ : \exists \beta \in \mathbf{R}_+ \text{ s.t. } \|Hx\|_p \leq \gamma \|x\|_p + \beta,\ \forall x \in L_p\}$$

---

[15] In this context we simply write sup instead of ess.sup.

2.8. MORE I/O PROPERTIES

Such operators are said to be L_p stable if they have finite L_p-gain. In the following we use the notation $\gamma_p(H)$ to denote the L_p gain of H, that is the induced norm $\|H\|_i$ from L_p to L_p.

Similarly, we define the gain of an operator induced by $\|\cdot\|_p, \|\cdot\|_q$. In particular we are interested in gains of operators which map an L_p space into L_∞. In this case we use the notation $g_p(H)$ to denote the gain of a causal operator $H : L_p \mapsto L_\infty$, that is

$$g_p(H) = \inf\{\gamma \in \mathbf{R}_+ : \exists \beta \in \mathbf{R}_+ \text{ s.t. } \|Hx\|_\infty \leq \gamma\|x\|_p + \beta, \, \forall x \in L_p\}$$

An immediate consequence of causality of an L_p stable operator $H : u \mapsto y$ is that the p-norm of the truncated output is directly related with the p-norm of the truncated input through the L_p gain of the operator. That is, $\forall t \geq 0$

$$\|y_t\|_p = \|P_t H u\|_p = \|P_t H P_t u\|_p \leq \gamma_p(H)\|u_t\|_p + \beta$$

where we used the facts that $u_t \in L_p$ and that the restriction of P_t on L_p has induced norm less than or equal to one (similarly for operators $H : L_p \mapsto L_\infty$).

Some frequently used expressions for the gains of LTV integral operators are given in the following lemma.

2.46 Lemma: *[Vid.78] Assume that H is a causal operator defined by*

$$H : u \mapsto y \; ; \; y(t) = \int_0^t h(t,\tau) u(\tau) \, d\tau \qquad (2.21)$$

where $u : \mathbf{R}_+ \mapsto \mathbf{R}$, $h : \mathbf{R}_+^2 \mapsto \mathbf{R}$ is measurable and locally integrable. U.t.c.:
.a. *Suppose that*

$$t \mapsto \int_0^t |h(t,\tau)| \, d\tau \in L_\infty^e \quad (\tau \mapsto h(t,\tau) \in L_1, \; \forall t \geq 0)$$

Then $H : L_\infty^e \mapsto L_\infty^e$. Further, H is L_∞ stable iff

$$\gamma_\infty(H) \triangleq \sup_{t \geq 0}\left\{\int_0^t |h(t,\tau)| \, d\tau\right\} < \infty$$

In such a case, $\|y\|_\infty \leq \gamma_\infty(H)\|u\|_\infty, \, \forall u \in L_\infty$.
.b. *Suppose that*

$$\tau \mapsto \int_\tau^\infty |h(t,\tau)| \, dt \in L_\infty^e \quad (t \mapsto h(t,\tau) \in L_1, \; \forall \tau \geq 0)$$

then $H : L_1^e \mapsto L_1^e$. Further, H is L_1 stable iff

$$\gamma_1(H) \triangleq \sup_{\tau \geq 0}\left\{\int_\tau^\infty |h(t,\tau)| \, dt\right\} < \infty$$

In such a case, $\|y\|_1 \leq \gamma_1(H)\|u\|_1$, $\forall u \in L_1$.

.c. Suppose that for the operator H, both $\gamma_\infty(H)$ and $\gamma_1(H)$ are finite. Then H is L_p stable for all $p \in [1,\infty]$ and

$$\|y\|_p \leq \gamma_1^{1/p}(H)\gamma_\infty^{1/q}(H)\|u\|_p, \quad \forall u \in L_p$$

where $q = p/(p-1)$ is the conjugate exponent of p. ▽▽

These results extend, in a quite straightforward manner, to the case of operators with direct throughput, i.e., $y(t) = (H'u)(t) \triangleq (Hu)(t) + h_0(t)u(t)$ where $h_0(t) \in L_\infty^e$; in this case the corresponding 1 and ∞ gains of the operator are given by

$$\gamma_\infty(H') = \sup_{t \geq 0}\left\{|h_0(t)| + \int_0^t |h(t,\tau)|\,d\tau\right\}$$

$$\gamma_1(H') = \sup_{\tau \geq 0}\left\{|h_0(\tau)| + \int_\tau^\infty |h(t,\tau)|\,dt\right\}$$

(For details and proofs see [Vid.78, D.V.75].)

In the case of LTI operators, simpler expressions are obtained, as indicated by the next lemma.

2.47 Lemma: [D.V.75] Let $p \in [1,\infty]$ and consider the linear map H defined on L_p in terms of an integrable function $h : \mathbf{R}_+ \mapsto \mathbf{R}$ as

$$H : u \mapsto y\,;\ y(t) = \int_0^t h(t-\tau)u(\tau)\,d\tau, \quad \forall t \in \mathbf{R}_+ \qquad (2.22)$$

and assume that $\|h\|_1 < \infty$. Then, $H : L_p \mapsto L_p$ and $\forall u \in L_p$

$$\|y\|_p \leq \|h\|_1 \|u\|_p$$

▽▽

Proof: The proof is quite straightforward, using either Tonelli's theorem ($p=1$), or Holder's inequality ($1 < p < \infty$), or, for $p = \infty$, by just pulling $\|u\|_\infty$ out of the convolution integral (see [D.V.75] for details). □□

It should be noted that the inequality $\|y\|_p \leq \|h\|_1\|u\|_p$ is sharp only for $p = 1,\infty$, i.e., $\gamma_1(H) = \gamma_\infty(H) = \|h\|_1$. For $p = 2$ and using Parseval's theorem we have a sharp bound in terms of the Fourier transform of h, denoted by $\hat{h}$, that is $\gamma_2(H) = \sup_w\{|\hat{h}(jw)|\}$. In other words, if the operator H is defined via a transfer function $\hat{h}(s)$, analytic and bounded in the open right half-plane, then the induced L_2-norm of H is equal to the H_∞ norm of $\hat{h}(s)$.

$$\|H\|_{i2} = \gamma_2(H) = \|\hat{h}(s)\|_\infty \triangleq \sup_{Re\ s > 0}\{|\hat{h}(s)|\}$$

2.8. MORE I/O PROPERTIES

The above results can be generalized to the vector case, i.e., when u and y are vector valued [D.V.75]. For example, consider the causal LTI operator $H : u \mapsto y \triangleq h * u$ where $u, y : \mathbf{R}_+ \mapsto \mathbf{R}^n$ and h is a matrix impulse response with elements $h_{ij} \in L_1$. Slightly abusing the notation, when $x : \mathbf{R}_+ \mapsto \mathbf{R}^n$ we denote by $|x|_p$ the function $t \mapsto |x(t)|_p$ where $|\cdot|_p$ is the usual p-norm in $\mathbf{R}^n$. Then, the following induced norms of H can be obtained:

1. For $H : L_\infty^n \mapsto L_\infty^n$ with the L_∞^n norm $\|(|x|_\infty)\|_\infty$ for a vector valued function x,

$$\gamma_\infty(H) = \max_i \int_0^\infty \sum_{j=1}^n |h_{ij}(\tau)|\, d\tau = \max_i \|(|h_i|_1)\|_1$$

where h_i is the vector valued function denoting the ith row of h.

2. For $H : L_2^n \mapsto L_2^n$ with the L_2^n norm $\|(|x|_2)\|_2$ for a vector valued function x,

$$\gamma_2(H) = \left[\max_w \max_i \lambda_i[\hat{h}(jw)^* \hat{h}(jw)]\right]^{1/2} \triangleq \|\hat{h}(s)\|_\infty$$

where $\lambda_i(M)$ is the necessarily real ith eigenvalue of the Hermitian matrix M.

3. For $H : L_1^n \mapsto L_1^n$ with the L_1^n norm $\|(|x|_1)\|_1$ for a vector valued function x,

$$\gamma_1(H) = \max_j \int_0^\infty \sum_{i=1}^n |h_{ij}(\tau)|\, d\tau = \max_j \|(|h_j|_1)\|_1$$

where h_j is the vector valued function denoting the ith column of h.

Let us now suppose that we would like to describe the properties of the operator H, considered in Lemma 2.46, when viewed as a map $L_p \mapsto L_\infty$, $p \in [1, \infty)$. Such a description is particularly useful in adaptive control where, due to the nonlinear nature of the adaptive closed-loop system, it is often necessary to relate the magnitude of the output signals with the energy of the input. Expressions for the corresponding operator gains are given by the next lemma.

2.48 Lemma: Consider the causal operator H given by (2.21) and
.a. Suppose that $t \mapsto \int_0^t |h(t,\tau)|^p\, d\tau \in L_\infty^e$, $p \in (1, \infty)$. Then $H : L_p^e \mapsto L_\infty^e$ and

$$|y(t)| \leq \|h(t,\cdot)\|_q \|u_t\|_p, \quad \forall t \geq 0$$

where $(\cdot)$ is used to indicate the integration argument and q is the conjugate exponent of p. Furthermore, if the quantity

$$g_p(H) \triangleq \sup_{t\geq 0} \left[\int_0^t |h(t,\tau)|^q \, d\tau\right]^{1/q}$$

is finite, we have that $\forall\, u \in L_p$,

$$\|y\|_\infty \leq g_p(H)\|u\|_p$$

and $g_p(H)$ is exactly the induced gain of $H : L_p \mapsto L_\infty$.

.b. Suppose that $t \mapsto \sup_{\tau\geq 0} |h(t,\tau)| \in L_\infty^e$. Then $H : L_1^e \mapsto L_\infty^e$ and[16]

$$|y(t)| \leq \sup_{\tau\geq 0} |h(t,\tau)| \|u_t\|_1, \quad \forall\, t \geq 0$$

Furthermore, if the quantity

$$g_1(H) \triangleq \sup_{t,\tau\geq 0} |h(t,\tau)|$$

is finite, we have that $\forall\, u \in L_1$,

$$\|y\|_\infty \leq g_1(H)\|u\|_1$$

and $g_1(H)$ is exactly the induced gain of $H : L_1 \mapsto L_\infty$. ▽▽

Proof: From the Holder's inequality we have that for any fixed t

$$|y(t)| \leq \left[\int_0^t |h(t,\tau)|^q \, d\tau\right]^{1/q} \left[\int_0^t |u(\tau)|^p \, d\tau\right]^{1/p}$$

where the equality holds for $u(\tau)$ s.t. $h(t,\tau)u(\tau) = |h(t,\tau)u(\tau)|$ and $|u(\tau)|^p = \text{const.}|h(t,\tau)|^q$. Similarly, for part (.b), by pulling the $\sup_{\tau\geq 0} |h(t,\tau)|$ out of the integral we obtain that $|y(t)| \leq \sup_\tau |h(t,\tau)| \|u_t\|_1$. Taking the supremum with respect to t, we obtain the inequalities of the lemma. Note that in both cases, it is quite straightforward to construct $u \in L_p$ such that $\|y\|_\infty$ is arbitrarily close to $g_p(H)\|u\|_p$ and therefore, $g_p(H)$ is actually the $L_p \mapsto L_\infty$ induced gain of H. □□

Although in our analysis we are primarily interested in the properties of operators mapping L_2 or L_1 to L_∞ it is worth mentioning that other maps and the corresponding gains may have interesting applications as well (e.g., infinite dimensional systems). Further, in the special case of LTI systems $h(t,\tau) = h(t-\tau)$ and $g_2(H) = \|h\|_2$ can be calculated in a straightforward

---

[16] Note that the causality of H implies $h(t,\tau) = 0$ almost everywhere when $\tau > t$.

2.8. MORE I/O PROPERTIES

manner from the controllability (or observability) grammian which is found as the solution of a Lyapunov equation. Using Parseval's theorem, we may also obtain g_2 as a quantity related to the frequency response of the operator, i.e., if $\|h\|_2 < \infty$ then it is equal to the H_2 norm of $\hat{h}(s)$ that is,

$$g_2(H) = \|h\|_2 = \|\hat{h}(s)\|_2 \triangleq \left[\frac{1}{2\pi}\int_{-\infty}^{\infty} |\hat{h}(jw)|^2 \, dw\right]^{1/2}.$$

As in the previous case of $L_p \mapsto L_p$-gains, similar statements can be made for vector valued inputs and outputs, using appropriate induced norms for the matrix valued impulse response of the operator H. For example, the following induced norms can be obtained for the causal LTI operator $H : u \mapsto y \triangleq h * u$ where $u, y : \mathbf{R}_+ \mapsto \mathbf{R}^n$ and h is a matrix impulse response with elements $h_{ij} \in L_2$ (see [Wil.88]).

1. For $H : L_2^n \mapsto L_\infty^n$ with the L_2^n norm $\|(|x|_2)\|_2$ and the L_∞^n norm $\|(|x|_\infty)\|_\infty$ for a vector valued function x,

$$g_2(H) = d_{max}^{1/2}\left(\int_0^\infty h(t)h^\top(t)\,dt\right)$$

where $d_{max}(M)$ is the maximum diagonal entry of the nonnegative matrix M.

2. For $H : L_2^n \mapsto L_\infty^n$ with the L_2^n norm $\|(|x|_2)\|_2$ and the L_∞^n norm $\|(|x|_2)\|_\infty$ for a vector valued function x,

$$g_2(H) = \lambda_{max}^{1/2}\left(\int_0^\infty h(t)h^\top(t)\,dt\right)$$

where $\lambda_{max}(M)$ is the maximum eigenvalue of the nonnegative matrix M.

It should be noted that although the g_p-gains are not submultiplicative, they do, however, possess some interesting properties in relation to the corresponding γ_p (L_p) gains.

2.49 Lemma: Let $H, G : L_\infty \mapsto L_\infty$, $p \in [1, \infty)$.
.a. If $G : L_p \mapsto L_p$ and $H : L_p \mapsto L_\infty$ then

$$g_p(HG) \leq g_p(H)\gamma_p(G) \leq g_p(H)[\gamma_1(G)]^{1/p}[\gamma_\infty(G)]^{1/q}$$

where q is the conjugate exponent of p.
.b. If $G : L_p \mapsto L_\infty$ then

$$g_p(HG) \leq \gamma_\infty(H)g_p(G)$$

whenever the various right-hand side gains exist. ▽▽

Proof: Observe first that $(HG)u = H(Gu)$. Thus, for the first case, $u \in L_p \Rightarrow Gu \in L_p \Rightarrow y \in L_\infty$ and therefore $\|y\|_\infty \leq g_p(H)\|Gu\|_p \leq g_p(H)\gamma_p(G)\|u\|_p$. Hence, the inequalities follow from the fact that $g_p(HG)$ is the induced norm of HG from L_p to L_∞. Further, from Lemma 2.46, the quantity $\sqrt{\gamma_1(G)\gamma_\infty(G)}$ is simply an upper bound of $\gamma_p(G)$ for which an exact value is given for $p = 2$ and LTI plants only, in terms of the H_∞ norm of the associated transfer function. Next, for the second case, $u \in L_p \Rightarrow Gu \in L_\infty \Rightarrow y \in L_\infty$ and therefore $\|y\|_\infty \leq \gamma_\infty(H)\|Gu\|_\infty \leq \gamma_\infty(H)g_p(G)\|u\|_p$ and the inequalities follow from the fact that $g_p(HG)$ is the $(L_p \to L_\infty)$-gain of HG. □□

2.8.2 Exponentially Weighted L_p Spaces

In the spirit of the above, well-known results, several variations of the L_p theory can be found —or formulated. In our development we are primarily interested in the so-called exponentially weighted L_p spaces and their relation to L_∞ and **R**. Several of the results mentioned below can be found in the literature e.g., [D.V.75, Vid.81, Zam.65] to name a few. However, due to the important role these methods play in our analysis and in order to establish the notation, some of the key results and their proofs are repeated here.

Consider a function $f : \mathbf{R}_+ \mapsto \mathbf{R}$ and let $\mathcal{E}_\delta$, $\delta \in \mathbf{R}$, denote the multiplication operator defined by

$$(\mathcal{E}_\delta f)(t) = e^{\delta t} f(t), \quad t \in \mathbf{R}_+$$

With this definition, we say that a locally integrable function $f : \mathbf{R}_+ \mapsto \mathbf{R}$ belongs to $L_p(\delta)$ (or $L_p^e(\delta)$) iff $\mathcal{E}_\delta f \in L_p$ (L_p^e) where $p \in [1, \infty]$.[17] It follows that $(L_p(\delta), \|\cdot\|_{p,\delta})$ with

$$\|f\|_{p,\delta} \triangleq \|\mathcal{E}_\delta f\|_p$$

is a normed linear space. We now show that it is actually a Banach space. Let $f_n \in L_p(\delta)$, $n = 1, 2, \ldots$ be a Cauchy sequence with respect to $\|\cdot\|_{p,\delta}$; then $\mathcal{E}_\delta f_n \in L_p$ is Cauchy with respect to $\|\cdot\|_p$. Hence, $\mathcal{E}_\delta f_n$ converges in L_p to, say, f_*. It follows that $\mathcal{E}_{-\delta} f_* \in L_p(\delta)$ and $\lim_{n \to \infty} \|f_n - \mathcal{E}_{-\delta} f_*\|_{p,\delta} = 0$, i.e., f_n converges in $L_p(\delta)$ with respect to $\|\cdot\|_{p,\delta}$ and thus $(L_p(\delta), \|\cdot\|_{p,\delta})$ is a Banach space.

Following the previous development, it is easy to verify that, with respect to the truncation operator, the (p, δ)-norms inherit the properties of the p-norms and that if $H : L_p^e \mapsto L_p^e$ is a causal operator then $\mathcal{E}_{-\delta} H \mathcal{E}_\delta : L_p^e(\delta) \mapsto$

---
[17] Note that since $e^{\delta t}$ is continuous, if $x \in L_p^e$ then $x \in L_p^e(\delta)$ and vice-versa.

2.8. MORE I/O PROPERTIES

$L_p^e(\delta)$ is also causal. Furthermore, if $H : L_p \mapsto L_q$ then $\mathcal{E}_{-\delta} H \mathcal{E}_\delta : L_p(\delta) \mapsto L_q(\delta)$.

Next, for $p \in [1, \infty]$, the $L_p(\delta)$-gain of an operator $H : L_p(\delta) \mapsto L_p(\delta)$, is similarly defined by

$$\gamma_{p,\delta}(H) = \inf\{\gamma \in \mathbf{R}_+ : \exists \beta \text{ s.t. } \|Hx\|_{p,\delta} \leq \gamma \|x\|_{p,\delta} + \beta, \, \forall x \in L_p(\delta)\}$$

and the gain of $H : L_p(\delta) \mapsto L_\infty(\delta)$ by

$$g_{p,\delta}(H) = \inf\{\gamma \in \mathbf{R}_+ : \exists \beta \text{ s.t. } \|Hx\|_{\infty,\delta} \leq \gamma \|x\|_{p,\delta} + \beta, \, \forall x \in L_p(\delta)\}$$

Some useful properties of operators in exponentially weighted L_p spaces and the relationship between L_p and $L_p(\delta)$-gains are discussed in the following lemmas.

2.50 Lemma:

.a. Let $p \in [1, \infty]$, $H : L_p^e \mapsto L_p^e$ and assume that $\mathcal{E}_\delta H \mathcal{E}_{-\delta} : L_p \mapsto L_p$ and $\gamma_p(\mathcal{E}_\delta H \mathcal{E}_{-\delta}) < \infty$. Then $H : L_p(\delta) \mapsto L_p(\delta)$ and

$$\gamma_{p,\delta}(H) = \gamma_p(\mathcal{E}_\delta H \mathcal{E}_{-\delta})$$

.b. Let $p \in [1, \infty)$, $H : L_p^e \mapsto L_\infty^e$ and assume that $\mathcal{E}_\delta H \mathcal{E}_{-\delta} : L_p \mapsto L_\infty$ and $g_p(\mathcal{E}_\delta H \mathcal{E}_{-\delta}) < \infty$. Then $H : L_p(\delta) \mapsto L_\infty(\delta)$ and

$$g_{p,\delta}(H) = g_p(\mathcal{E}_\delta H \mathcal{E}_{-\delta})$$

.c. The converse (.a) and (.b) statements are also true, i.e., if $H : L_p^e \mapsto L_q^e$ and $H : L_p(\delta) \mapsto L_p(\delta)$ ($L_\infty(\delta)$) has finite gain, then $\mathcal{E}_\delta H \mathcal{E}_{-\delta} : L_p \mapsto L_q$ (L_∞) and the L_p-gain of $\mathcal{E}_\delta H \mathcal{E}_{-\delta}$ is equal to the $L_p(\delta)$-gain of H. ▽▽

Proof: Consider $u \in L_p(\delta)$. Then $y = Hu \in L_p^e$ and $\mathcal{E}_\delta y = \mathcal{E}_\delta H \mathcal{E}_{-\delta} \mathcal{E}_\delta u$; since $u \in L_p(\delta)$ we have that $\mathcal{E}_\delta u \in L_p$ which implies that $\mathcal{E}_\delta y \in L_p$ and therefore $y \in L_p(\delta)$. Furthermore, $\|\mathcal{E}_\delta y\|_p \leq \gamma_p(\mathcal{E}_\delta H \mathcal{E}_{-\delta}) \|\mathcal{E}_\delta u\|_p + \beta$ which can be rewritten as $\|y\|_{p,\delta} \leq \gamma_p(\mathcal{E}_\delta H \mathcal{E}_{-\delta}) \|u\|_{p,\delta} + \beta$. Since $u \in L_p(\delta)$ is arbitrary, $\gamma_{p,\delta}(H) \leq \gamma_p(\mathcal{E}_\delta H \mathcal{E}_{-\delta})$. Next, for $u \in L_p$, let $y = \mathcal{E}_\delta H \mathcal{E}_{-\delta} u (\in L_p)$. It follows that $\mathcal{E}_{-\delta} y = H \mathcal{E}_{-\delta} u$ and $\|\mathcal{E}_{-\delta} y\|_{p,\delta} \leq \gamma_{p,\delta}(H) \|\mathcal{E}_{-\delta} u\|_{p,\delta} + \beta'$. Hence $\|y\|_p \leq \gamma_{p,\delta}(H) \|u\|_p + \beta'$ and $\gamma_p(\mathcal{E}_\delta H \mathcal{E}_{-\delta}) \leq \gamma_{p,\delta}(H)$ which, together with the previously obtained converse inequality, completes the proof of part (.a); the proof of parts (.b) and (.c) follow along the same lines and are omitted. □□

2.51 Lemma: Let $\delta \geq 0$ and suppose $H : L_p(\delta) \mapsto L_p(\delta)$ is a causal, $L_p(\delta)$-stable operator and such that $\|Hu\|_{p,\delta} \leq \gamma_{p,\delta}(H) \|u\|_{p,\delta}$, for any $u \in L_{p,\delta}$. Then $H : L_p \mapsto L_p$ and $\gamma_p(H) \leq \gamma_{p,\delta}(H)$.

Similarly, if $H : L_p(\delta) \mapsto L_\infty(\delta)$, $\delta \geq 0$ is a causal operator and such that $g_{p,\delta}(H) < \infty$ then $H : L_p \mapsto L_\infty$ and $g_p(H) \leq g_{p,\delta}(H)$. ▽▽

Proof: Let $u \in L_p$, $p < \infty$. Then $u \in L_p^e(\delta)$ and therefore, $y = Hu \in L_p^e(\delta)$ and for any $t \geq 0$, $\|y_t\|_{p,\delta} \leq \gamma_{p,\delta}(H)\|u_t\|_{p,\delta}$. Multiplying both sides by $e^{-\delta t}$, raising to the p-power and integrating in an interval $[0, T]$ we obtain after some straightforward calculations,

$$\|y_T\|_p^p - e^{-p\delta T}\|y_T\|_{p,\delta}^p \leq \gamma_{p,\delta}^p(H)\left(\|u_T\|_p^p - e^{-p\delta T}\|u_T\|_{p,\delta}^p\right)$$

Since $\|y_T\|_{p,\delta}^p \leq \gamma_{p,\delta}^p(H)\|u_T\|_{p,\delta}^p$, we obtain that $\|y_T\|_p \leq \gamma_{p,\delta}(H)\|u_T\|_p$. Thus, $u \in L_p$ implies $y \in L_p$ and by the definition of the gain $\gamma_p(H) \leq \gamma_{p,\delta}(H)$.

Further, for $p \in [1, \infty]$ and from $\|y_t\|_{\infty,\delta} \leq g_{p,\delta}(H)\|u_t\|_{p,\delta} + \beta$ we obtain

$$|y(t)| \leq g_{p,\delta}(H)e^{-\delta t}\|u_t\|_{p,\delta} + e^{-\delta t}\beta \leq g_{p,\delta}(H)\|u_t\|_p + \beta$$

Since $u \in L_p$ we have that $H : L_p \mapsto L_\infty$ and $g_p(H) \leq g_{p,\delta}(H)$ (note that $g_\infty(H)$ is, by definition, identical to $\gamma_p(H)$). □□

For an application of Lemma 2.50, let us consider the SISO LTV system

$$\dot{x} = A(t)x + b(t)u \ ; \ y = c(t)x \qquad (2.23)$$

where $A(t), b(t), c(t)$ have entries UB, piecewise continuous functions of time. Furthermore, suppose that (2.23) is ES, i.e., there exist constants $K, a > 0$ such that the state transition matrix associated with $A(t)$ satisfies

$$\|\Phi(t, \tau)\| \leq K \exp[-a(t - \tau)], \ \forall t \geq \tau \geq 0$$

It follows that the I/O operator $H : u \mapsto y$ is given by

$$y(t) = (Hu)(t) = \int_0^t h(t, \tau)u(\tau)\,d\tau$$

where $h(t, \tau) = c(t)\Phi(t, \tau)b(\tau)$ is the impulse response of the operator H.

2.52 Corollary: Under these conditions, for any $\delta \in [0, a)$, $p \in (1, \infty)$ and $\frac{1}{p} + \frac{1}{q} = 1$, the $L_p(\delta)$-gains of H are well defined and given by:

$$\gamma_{\infty,\delta}(H) = \sup_{t \geq 0} \int_0^t |h(t, \tau)|e^{\delta(t-\tau)}\,d\tau$$

$$\gamma_{1,\delta}(H) = \sup_{\tau \geq 0} \int_0^t |h(t, \tau)|e^{\delta(t-\tau)}\,d\tau$$

$$g_{p,\delta}(H) = \sup_{t \geq 0} \left\{\int_0^t \left[|h(t, \tau)|e^{\delta(t-\tau)}\right]^q d\tau\right\}^{1/q}$$

$$g_{1,\delta}(H) = \sup_{t \geq \tau} \left\{|h(t, \tau)|e^{\delta(t-\tau)}\right\}$$

2.8. MORE I/O PROPERTIES

If, in addition, we have that $A(t), b(t), c(t)$ are constant, the impulse response of H is of the form $h(t - \tau)$ and the corresponding transfer function $\hat{h}(s)$ is rational and analytic in the half-plane $Re(s) > -a$. In this case the following simpler expressions are obtained:

$$\gamma_{\infty,\delta}(H) = \gamma_{1,\delta}(H) = \int_0^\infty |h(t)|e^{\delta t}\, dt = \|\mathcal{E}_\delta h\|_1$$

$$\gamma_{2,\delta}(H) = \|\hat{h}(s-\delta)\|_\infty$$

$$g_{p,\delta}(H) = \left\{\int_0^\infty [|h(t)|e^{\delta t}]^q\, d\tau\right\}^{1/q} = \|\mathcal{E}_\delta h\|_q$$

$$g_{2,\delta}(H) = \|\mathcal{E}_\delta h\|_2 = \|\hat{h}(s-\delta)\|_2$$

$$g_{1,\delta}(H) = \sup_{t \geq 0}\left\{|h(t)|e^{\delta t}\right\} = \|\mathcal{E}_\delta h\|_\infty$$

▽▽

Proof: Straightforward application of Lemma 2.50 and the lemmas of the previous subsection. □□

The properties of operators in $L_p(\delta)$ spaces are subsequently used to obtain L_∞^e bounds of an output signal using $L_p^e(\delta)$ bounds of the input signal. Such an approach has the advantage that past values of the input and the output are exponentially de-weighted and consequently the results are more appropriate to describe the 'steady-state' behavior of the signals. An additional advantage is that, in general, it is easier to solve design and optimization problems in $L_2(\delta)$, which is a Hilbert space, rather than L_∞. It should be pointed out, though, that the transformation of an L_∞ stability problem to an $L_2(\delta)$ one introduces a degree of conservatism in the final solution, a trade-off that should be addressed in a quantitative design.

2.53 Lemma: Consider a causal operator $H : L_p^e \mapsto L_\infty^e$, $p \in [1, \infty)$, for which $g_{p,\delta}(H) < \infty$ and let $y = Hu$. Then, for any $u \in L_p^e$ (or $u \in L_\infty^e$),

$$|y(t)| \leq g_{p,\delta}(H)\,(\mathcal{E}_{-\delta}\|u_t\|_{p,\delta})\,(t) + \beta\mathcal{E}_{-\delta}(t), \quad \forall t \geq 0 \tag{2.24}$$

$$(\mathcal{E}_{-\delta}\|y_t\|_{p,\delta})\,(t) \leq \gamma_{p,\delta}(H)\,(\mathcal{E}_{-\delta}\|u_t\|_{p,\delta})\,(t) + \beta\mathcal{E}_{-\delta}(t), \quad \forall t \geq 0 \tag{2.25}$$

where the subscript t denotes truncation. ▽▽

Proof: The proof follows directly from

$$|(\mathcal{E}_\delta y)(t)| \leq \|P_t y\|_{\infty,\delta} \leq g_{p,\delta}(H)\|P_t u\|_{p,\delta} + \beta$$

and the definition and properties of the (p, δ)-norms. □□

The expressions (2.24) and (2.25) can also be written in an easier to visualize form

$$|y(t)| \leq g_{p,\delta}(H) \left\{ \int_0^t \left[e^{-\delta(t-\tau)} |u(\tau)| \right]^p d\tau \right\}^{1/p} + \beta e^{-\delta t}, \quad \forall t \geq 0 \quad (2.26)$$

$$\left\{ \int_0^t \left[e^{-\delta(t-\tau)} |y(\tau)| \right]^p d\tau \right\}^{1/p} \leq \gamma_{p,\delta}(H) \left\{ \int_0^t \left[e^{-\delta(t-\tau)} |u(\tau)| \right]^p d\tau \right\}^{1/p}$$
$$+ \beta e^{-\delta t}, \quad \forall t \geq 0 \quad (2.27)$$

2.54 Corollary: *Let $H : L_p^e \mapsto L_\infty^e$, $p \in [1, \infty)$, be a causal operator and suppose that $g_{p,\delta}(H) < \infty$ for some $\delta > 0$. Then $H : L_\infty \mapsto L_\infty$ and*

$$\gamma_\infty(H) \leq \frac{1}{(p\delta)^{1/p}} g_{p,\delta}(H)$$

i.e., H is L_∞-stable. ▽▽

Proof: Immediate from (2.26) by taking the supremum of both sides with respect to t and pulling $\|u\|_\infty$ out of the integral. □□

Note that the inequalities (2.24) and (2.25) are valid when u, y exist locally, i.e., $u, y \in L_{\infty[0,T]}$ provided that t is restricted to the interval $[0, T]$. This observation can be useful in establishing existence and uniqueness of solutions of differential equations. Furthermore, Corollary 2.54 offers an upper bound of the γ_∞ gain of an operator in terms of the —easier to calculate— $g_{2,\delta}$ gain.

Finally, it is sometimes necessary to establish the L_∞ boundedness of a signal based on information related to its (p, δ)-norm and the (p, δ)-norm of its derivative. There are several ways to approach this problem, one of which is described by the following lemma.

2.55 Lemma: *Let $x : \mathbf{R}_+ \mapsto \mathbf{R}$ be an absolutely continuous function on $[0, T]$, $T > 0$ and let $p \in [1, \infty)$. Then, $\forall t \in [0, T)$,*

$$|x(t)|^p \leq [|x(0)|\mathcal{E}_{-\delta}]^p(t) + p\delta[\mathcal{E}_{-\delta}\|x_t\|_{p,\delta}]^p(t)$$
$$+ p[\mathcal{E}_{-\delta}\|x_t\|_{p,\delta}]^{p-1}(t)[\mathcal{E}_{-\delta}\|\dot{x}_t\|_{p,\delta}](t)$$

where the subscript t denotes, as usual, truncation at t. ▽▽

Proof: In Appendix II.

It should be noted that some applications of special interest of the above lemma include the cases of x being locally Lipschitz on $[0, T]$ and $x, \dot{x} \in L_\infty^e$ where T may be taken as ∞.

2.8.3 Normalization Signals

An immediate application of Lemma 2.53 is the construction of the so-called normalization signals. The use of such signals has been an instrumental part of most of the recent studies of the robustness properties of adaptive controllers. Qualitatively speaking, normalization signals should have the following properties:

- they should provide an upper bound of the output of an operator;

- the normalized output should be small in some sense when the operator gain is small in some sense;

- they should have exponentially fading memory of past inputs.

To illustrate the reasoning behind the last property, let us consider $\|u_t\|_\infty$ as a normalization signal. This signal would indeed normalize the output of an operator $H : u \mapsto y = Hu$, i.e., $|y(t)|/\|u_t\|_\infty \leq \gamma_\infty(H) + \beta/\|u_t\|_\infty$. Suppose also that, in a control systems setup, it is desired that $y \to \infty$ and that, due to initial conditions, y and u may attain some 'large' values during the transient period. In such a case, it is quite apparent that $\|u_t\|_\infty$ is a poor normalizing signal for $y(t)$ as $t \to \infty$, although it does provide a sharp bound for $\|y\|_\infty$.

Normalization signals of the form described below have been used quite frequently in adaptive control to guarantee boundedness of the parameter estimates at least as early as [Ega.79], while their robustness properties with respect to unmodeled dynamics and, in general, state dependent disturbances were first observed by Praly [Pra.83]. The typical properties and design guidelines of such signals are given as follows.

2.56 Lemma: *Consider a causal operator $H : L_p^e \mapsto L_\infty^e$, $p \in [1, \infty)$, and let $y = Hu$ and $m_p(t)$ be the signal defined by*

$$\dot{m}_p = -p\delta m_p + |u|^p + q_e; \ t \geq 0; \ m(0) > 0$$

where δ, q_e are positive constants and assume that $u \in L_p^e$ on an interval $J \subseteq \mathbf{R}_+$, $(0 \in J)$.

a. *If $g_{p,\delta}(H)$ is finite then there exists a constant $\beta_0 > 0$ such that*

$$\frac{|y(t)|}{[m_p(t)]^{1/p}} \leq g_{p,\delta}(H) + \beta_0 e^{-\delta t}, \ \forall t \in J$$

b. *If $\gamma_{p,\delta}(H)$ is finite then there exists a constant $\beta_0 > 0$ such that*

$$\left\{ \frac{\int_0^t \left[e^{-\delta(t-\tau)} |y(\tau)| \right]^p d\tau}{m_p(t)} \right\}^{1/p} \leq \gamma_{p,\delta}(H) + \beta_0 e^{-\delta t}, \ \forall t \in J \qquad \triangledown\triangledown$$

Proof: The proof follows directly from (2.26) and (2.27) with the observation that the positive constant q_e and the initial condition $m(0) > 0$ ensure that $1/m_p(t)$ is well defined for all $t \in J$. □□

It goes without saying that the results of the lemma are also applicable in the vector case with $|\cdot|$ denoting vector norms[18] and the $L_p(\delta)$ gains of the operator H calculated using the appropriate induced norm of the impulse response h (or its Laplace transform).

The normalization signal m_p has the previously described properties with the bound of the normalized output being a $L_p(\delta)$-gain of the operator H. The requirement that this gain should be finite imposes some restrictions on the class of operators for which the above lemma can be applied; a linear system whose state transition matrix has exponential rate of decay greater than $-\delta$ and an operator containing arbitrarily large time delays are typical examples violating the assumptions of the lemma. These restrictions are the price paid for normalization with fading-memory signals. In the same vein, however, it should be pointed out that variations of the previous lemma can be employed to somewhat relax the assumptions on the operator H as indicated by the following corollary.

2.57 Corollary: *Let $\hat{H}$ be a causal operator $\hat{H} : L_p^e \mapsto L_\infty^e$ and suppose that there exists a causal operator $F : L_p^e \mapsto L_\infty^e$ such that $(1 + F)^{-1} : L_\infty^e \mapsto L_\infty^e$ exists $(p \in [1, \infty))$. Further, let $y = \hat{H}u$. Then Lemma 2.56 can be applied with $H = [(1+F)^{-1}\hat{H}, (1+F)^{-1}F]$, $u = [\hat{u}, y]^\mathsf{T}$ and $g_{p,\delta}(\hat{H}), \gamma_{p,\delta}(\hat{H})$ evaluated according to the vector norm $|u|$ used in the definition of m_p.* ▽▽

Proof: Immediate, from $y = \hat{H}u + Fy - Fy$ and the observation that $x \in L_\infty^e$ implies $x \in L_p^e$. □□

2.58 Example: To demonstrate the results obtained by the application of the last corollary, consider the LTI $y = \hat{H}u$ defined by

$$\dot{x} = Ax + bu \ ; \ y = cx$$

and assume that (c, A) is completely observable. Hence, there exists a vector k such that the eigenvalues of $L = A + kc$ have real parts less than $-\delta$. It follows that $y = \hat{H}u$ can be expressed as

$$\dot{x} = (A + kc)x + bu - ky \ ; \ y = cx$$

Thus, we can apply Corollary 2.57 with $\hat{H}$, F, $(1 + F)^{-1}H$ and $(1 + F)^{-1}F$ being given in terms of their transfer functions as $c(sI - A)^{-1}b$, $-c(sI -$

---

[18] Such norms can be weighted by a symmetric, positive definite matrix Q e.g., $|u| \triangleq (u^\mathsf{T} Q u)^{1/2}$.

2.8. MORE I/O PROPERTIES

$A)^{-1}k$, $c(sI-L)^{-1}b$ and $-c(sI-L)^{-1}k$ respectively. This example illustrates how the assumptions of Lemma 2.56 can be relaxed, using some observability conditions on the state representation of the operator $\hat{H}$. ▽▽

2.8.4 Swapping and Operator Inversion Lemmas

In this subsection we present two technical lemmas which are used to simplify the derivations of the subsequent analysis. The first one involves compositions of an LTV dynamical operator and a TV multiplier. Consider for example the case of an LTV operator $H : u \mapsto y = H(u)$ and the multiplier $\Theta : w \mapsto u = w\theta$. If θ is constant, it follows that $H\Theta - \Theta H = 0$. This is not necessarily true, however, if θ is a function of time. The precise nature of the difference between the operators $H\Theta$ and ΘH —caused by the 'swapping' between the dynamical operator H and the multiplier Θ— is assessed by the following lemma.

2.59 Lemma: *('Swapping') Consider the LTV system with piecewise continuous, bounded representation*

$$\dot{x}(t) = A(t)x(t) + B(t)u(t) \; ; \; y(t) = C(t)x(t) \; ; \; x(t_0) = 0$$

and suppose that $u(t) = w(t)\theta(t)$ where θ is absolutely continuous, w is (absolutely) integrable on an interval $J = [t_0, t_0 + T]$, $T > 0$ and A, B, C, u, w, θ are TV matrices of compatible dimensions. Further, define the (causal) operators

$$H : v \mapsto z \; : \; z(t) = \int_{t_0}^{t} C(t)\Phi(t,\tau)B(\tau)v(\tau)\,d\tau$$

$$H_1 : v_1 \mapsto z_1 \; : \; z_1(t) = \int_{t_0}^{t} C(t)\Phi(t,\tau)v_1(\tau)\,d\tau$$

$$H_2 : v_2 \mapsto z_2 \; : \; z_2(t) = \int_{t_0}^{t} \Phi(t,\tau)B(\tau)v_2(\tau)\,d\tau$$

where $\Phi(\cdot, \cdot)$ is the state transition matrix associated with $A(\cdot)$ and $t \in J$. Then,

$$y(t) = (H[w\theta])(t) = (H[w])(t)\theta(t) - \left(H_1\left\{H_2[w]\dot{\theta}\right\}\right)(t)$$

for all $t \in J$. ▽▽

Proof: Let $N(\cdot)$ denote a fundamental solution of $\dot{X} = A(t)X$; then, N is continuous and nonsingular on J and $\Phi(t,\tau) = N(t)N^{-1}(\tau)$. Further,

$$y = C(t)N(t)\int_{t_0}^{t} N^{-1}(\tau)B(\tau)w(\tau)\theta(\tau)\,d\tau$$

Since w is integrable on J, so is $f = N^{-1}Bw$; then, $F(t) = \int_{t_0}^{t} f(\tau)\, d\tau$ is absolutely continuous on J and $F(t_0) = 0$. Next, since F, θ are absolutely continuous on a compact interval $F\theta$ is also absolutely continuous and $\frac{d}{dt}(F\theta)(t) = \dot{F}(t)\theta(t) + F(t)\dot{\theta}(t)$ almost everywhere on J. Further, $\dot{F}(t) = f(t)$ almost everywhere on J. Hence,

$$\int_{t_0}^{t} \frac{d}{d\tau}(F\theta)(\tau)\, d\tau = \int_{t_0}^{t} f(\tau)\theta(\tau)\, d\tau + \int_{t_0}^{t} F(\tau)\dot{\theta}(\tau)\, d\tau$$

for all $t \in J$. Also, $F\theta$ being absolutely continuous on J, the left hand-side of the above equation is equal to $(F\theta)(t) - (F\theta)(t_0) = F(t)\theta(t)$. Hence, y can be written as

$$\begin{aligned} y(t) &= C(t)N(t)\left[\int_{t_0}^{t} N^{-1}(\tau)B(\tau)w(\tau)\, d\tau\right]\theta(t) \\ &\quad - C(t)N(t)\int_{t_0}^{t} N^{-1}(\tau)N(\tau)\left[\int_{t_0}^{\tau} N^{-1}(s)B(s)w(s)\, ds\right]\dot{\theta}(\tau)\, d\tau \end{aligned}$$

from which the result follows. (Notice that $H[w\theta]$ and $H[w]$ need not be of the same dimension.) □□

In other words, the Swapping Lemma shows that the difference $H\Theta - \Theta H$ can be expressed as an operator $H_1 \dot{\Theta} H_2$ where $\dot{\Theta}$ denotes a multiplier $\dot{\Theta} : w \mapsto w\dot{\theta}$ and H_1, H_2 are operators whose stability properties are directly related to those of H. Moreover, invoking the results of the previous subsections, one may easily verify that if the original LTV system is ES with rate $-a$ then for $\delta < a$, the $\gamma_{p,\delta}$ and $g_{p,\delta}$-gains of the swapping operator $H_1 \dot{\Theta} H_2$ are $O(\dot{\theta})$.

The Swapping Lemma can be further generalized to cases where θ contains an additional *jump-function* component. The result follows quite easily as an extension of the previous lemma and is stated as a corollary below.

2.60 Corollary: *Suppose that in Lemma 2.59, $\theta = \theta^s + \theta^J$ where θ^s is absolutely continuous on J and θ^J is a jump function, i.e.,*

$$\theta^J(t) = \sum_{t_j \leq t} \theta^J_j$$

where t_j is a strictly increasing sequence such that $\lim_{j \to \infty} t_j = \infty$ and θ^J_j are constant matrices of appropriate dimension. Then

$$\begin{aligned} (H[w\theta])(t) &= (H[w])(t)\theta(t) - \left(H_1\left\{H_2[w]\dot{\theta}^s\right\}\right)(t) \\ &\quad - \sum_{t_j \leq t} C(t)\Phi(t, t_j)(H_2[w])(t_j)\theta^J_j \end{aligned}$$

for all $t \in J$. ▽▽

2.8. MORE I/O PROPERTIES

Proof: Straightforward, following the same steps as in Lemma 2.59 and using the linearity of H. □□

The second lemma of this subsection deals with the derivation of properties and bounds for the input of a linear filter based on information given on its output. This inversion problem can be handled in a straightforward manner (e.g., see [Kai.80]) when the filter has a non-zero throughput term. In our case, however, we are interested in the inversion of strictly proper filters which, strictly speaking, involves differentiation of the output. Since differentiators do not possess finite L_p gains, we follow a different approach whereby an anticipated bound on the derivative of the input is used to essentially limit the frequency range of the filter inversion and yield finite gains for the restriction of the inverse filter. This procedure proves sufficient for our purposes and is summarized in the following lemma (for simplicity we only consider the LTI version of this result).

2.61 Lemma: *('Operator Inversion') Suppose $W(s)$ is a stable, minimum phase transfer function of relative degree n^* and let $\rho \in [0,1]$ be an arbitrary constant and*

- $\Lambda(s)$ *denote an arbitrary transfer function such that $\Lambda(s)$ is stable, with relative degree $\geq n^*$ and unity D.C. gain $(\Lambda(0) = 1)$;*

- $\Lambda_1(s)$ *denote the transfer function such that $\Lambda_1(s)s = 1 - \Lambda(s)$;*[19]

Further, let (u,y) be the I/O pair of $W(s)$ and suppose that in an interval $[0,T]$, u is absolutely continuous. Then, in the same interval,

$$u = \Lambda_1(s)[\dot{u}] + (1-\rho)\Lambda(s)W^{-1}(s)\{W(s)[u]\} + \rho\Lambda(s)[u]$$

▽▽

Proof: Straightforward from the operator identity

$$[1 - \Lambda(s)] + (1-\rho)\Lambda(s)W^{-1}(s)W(s) + \rho\Lambda(s) = 1$$

□□

From the above lemma it immediately follows that for any $t \in [0,T]$

$$\|u_t\|_{p,\delta} \leq \gamma_{p,\delta}(\Lambda_1)\|\dot{u}_t\|_{p,\delta} + (1-\rho)\gamma_{p,\delta}(\Lambda W^{-1})\|y_t\|_{p,\delta} + \rho\gamma_{p,\delta}(\Lambda)\|u_t\|_{p,\delta}$$

where, for simplicity, we use Λ, Λ_1, W to denote the respective operators and assume that δ is chosen such that the corresponding gains are finite. The

---
[19] Notice that from the assumptions on $\Lambda(s)$ it follows that $\Lambda_1(s)$ and $\Lambda(s)$ have the same finite poles.

last inequality shows that the (p, δ)-norm of the input can be written as a weighted sum of the (p, δ)-norms of $\dot{u}$ and y. The two weights can be made arbitrarily small, but not simultaneously, as it can be readily seen with the simple choice

$$\Lambda(s) = [a/(s+a)]^{n^*}$$

Then $\gamma_{p,\delta}(\Lambda_1) = O(1/a)$ $\gamma_{p,\delta}(\Lambda W^{-1}) = O(a^{n^*})$. The freedom to choose the parameter a is used in the subsequent chapters to reduce the conservatism in estimating an upper bound of $\|u_t\|_{p,\delta}$. Intuitively speaking, a represents the frequency range over which $\dot{u}$ is large and the operator W needs to be inverted.

Finally, the free parameter ρ serves the same purpose (reduction of conservatism) and approaches one when additional considerations allow us to conclude that u itself is small. Further details and one application of this lemma are given in Chapter 7.

2.8.5 The Bellman-Gronwall Lemma

The following lemma, due to Bellman [Bel.53], is extensively used throughout the following chapters.

2.62 Lemma: *Suppose $c \geq 0$, $r(\cdot)$ and $k(\cdot)$ are nonnegative valued continuous functions and*

$$r(t) \leq c + \int_0^t k(\tau) r(\tau) \, d\tau, \quad \forall t \in [0, T]$$

Then

$$r(t) \leq c \exp\left[\int_0^t k(\tau) \, d\tau\right], \quad \forall t \in [0, T]$$

▽▽

Proof: [Vid.78] Let $s(t) = c + \int_0^t k(\tau) r(\tau) \, d\tau$. Then $r(t) \leq s(t)$, $\forall t \in [0, T]$. Further,

$$\dot{s}(t) = k(t) r(t) \leq k(t) s(t), \quad \forall t \in [0, T]$$

Hence, $\dot{s}(t) - k(t)s(t) \leq 0$, $\forall t \in [0, T]$ and therefore,

$$[\dot{s}(t) - k(t)s(t)] \exp\left[\int_0^t -k(\tau) \, d\tau\right] \leq 0, \quad \forall t \in [0, T]$$

$$\frac{d}{dt}\left\{s(t) \exp\left[\int_0^t -k(\tau) \, d\tau\right]\right\} \leq 0, \quad \forall t \in [0, T]$$

2.8. MORE I/O PROPERTIES

$$s(t) \exp\left[\int_0^t -k(\tau)\,d\tau\right] \leq s(0) = c$$

$$s(t) \leq c \exp\left[\int_0^t k(\tau)\,d\tau\right]$$

The proof now follows immediately from the last inequality and the definition of $s(t)$, whereby $r(t) \leq s(t)$. □□

Note that, using analogous arguments, the lemma is also applicable for initial times other than 0 (replacing, of course, 0 by t_0) and for piecewise continuous, UB functions $k(\cdot)$.[20] A more general version of this lemma, given in [D.V.75], is stated as follows.

2.63 Lemma: Let

1. f, g, k: $\mathbf{R}_+ \mapsto \mathbf{R}$ and locally integrable;

2. $g \geq 0$, $k \geq 0$;

3. $g \in L_\infty^e$;

4. gk is locally integrable on $\mathbf{R}_+$.

Under these conditions, if $u : \mathbf{R}_+ \mapsto \mathbf{R}$ satisfies

$$u(t) \leq f(t) + g(t) \int_0^t k(\tau) u(\tau)\,d\tau, \quad \forall t \in \mathbf{R}_+$$

then

$$u(t) \leq f(t) + g(t) \int_0^t k(\tau) f(\tau) \left[\exp \int_\tau^t k(\tau_1) g(\tau_1)\,d\tau_1\right] d\tau, \quad \forall t \in \mathbf{R}_+$$

▽▽

Proof: In [D.V.75].

2.8.6 Smooth Approximations of Continuous Functions

We conclude this section of mathematical preliminaries with a lemma on the smooth approximation of continuous functions. Smooth approximations allow us to extend the results given in this chapter for smooth functions, to a class of functions that are not everywhere differentiable encompassing most of the practically interesting cases.

---

[20] In this case the previous arguments are applied inside each interval where $k(\cdot)$ is continuous and the final expression is obtained by using the continuity of $\int k(\cdot)$.

2.64 Lemma: Let $u(t) : \mathbf{R}_+ \mapsto \mathbf{R}$ be an absolutely continuous function on an interval $[0,T]$ for some $T > 0$. Then for any $a \in \mathbf{R}_+$, there exist functions $u_n(t)$, $n = 1, 2, \ldots$, such that $u_n(t)$ is n-times continuously differentiable in $[0,T)$ and for any $t \in [0,T)$, $p \in [1,\infty]$

$$\|(u_n - u)_t\|_p \le \left(\frac{n}{a}\right)\|(\dot{u})_t\|_p$$

$$\|(u_n^{(i)})_t\|_p \le (2a)^{i-1}\|(\dot{u})_t\|_p \; ; \quad i = 1, 2, \ldots, n$$

where, as usual, the subscript t denotes truncation and the superscript (i) denotes the ith derivative of the function. ▽▽

Proof: In Appendix II.

We note that smooth approximations of piecewise continuous functions with discontinuities of the first kind (in this case, delta-distributions appear in $\dot{u}$) or continuous functions with *cusp* points (where $\dot{u}$ is square integrable but not UB) can be derived in a similar manner. The resulting approximations, however, are not uniform —in the sense that $|u_n(t) - u(t)|$ may not be arbitrarily small everywhere; for this reason it is more convenient in our analysis to treat any such cases in a different way.

APPENDIX II

Proof of Lemma 2.8:

Let $D(s,t)$, $N(s,t)$ be the PDO's of degree n, m respectively with $D(s,t)$ monic and $S_R(t)$ be the corresponding right TV Sylvester matrix.

(if) Consider the equation

$$Q_1(s,t)D(s,t) + P_1(s,t)N(s,t) = 1$$

in the interval (t_1, t_2), which by Definition 2.6 can be written as $S_R(t)x(t) = a$ where $a = [0,\ldots,0,1]^T$ and x is a vector of the coefficients of $Q_1(s,t)$, $P_1(s,t)$. Assuming that $S_R(t)$ is nonsingular in (t_1, t_2), the last equation can be solved uniquely for the coefficients of $Q_1(s,t)$, $P_1(s,t)$ which are smooth functions of time.

Further, since $S_R(t)$ is nonsingular in (t_1, t_2), we can similarly solve the equation

$$\hat{Q}_0(s,t)D(s,t) + \hat{P}_0(s,t)N(s,t) = -s^n N(s,t) - n_0(t)s^m D(s,t)$$

with $\deg[\hat{Q}_0(s,t)] = m-1$, $\deg[\hat{P}_0(s,t)] = n-1$ and $n_0(t)$ being the leading coefficient of $N(s,t)$ respectively. Then the PDO's $Q_0(s,t) = \hat{Q}_0(s,t) + n_0(t)s^m$,

APPENDIX II.

$P_0(s,t) = \hat{P}_0(s,t) + s^n$ have smooth coefficients and satisfy

$$Q_0(s,t)D(s,t) + P_0(s,t)N(s,t) = 0$$

in (t_1, t_2). Hence, $D(s,t)$ and $N(s,t)$ are right coprime in the same interval.

(only if) Assume that there exist PDO's $P_0(s,t)$ —monic of degree n— $Q_0(s,t)$, $P_1(s,t)$ and $Q_1(s,t)$ with smooth coefficients such that

$$Q_1(s,t)D(s,t) + P_1(s,t)N(s,t) = 1 \qquad (2.28)$$

$$Q_0(s,t)D(s,t) + P_0(s,t)N(s,t) = 0 \qquad (2.29)$$

for $t \in (t_1, t_2)$. From (2.28), (2.29) it follows that $\deg[P_1(s,t)] - \deg[Q_1(s,t)] = n - m$, $\deg[Q_0(s,t)] = m$ and

$$n_0(t) = -q_{00}(t) \;\; ; \;\; p_{10}(t)n_0(t) = -q_{10}(t)$$

where $n_0(t)$, $q_{00}(t)$, $q_{10}(t)$, $p_{10}(t)$ denote the leading coefficients of the PDO's $N(s,t)$, $Q_0(s,t)$, $Q_1(s,t)$, $P_1(s,t)$ respectively. Hence, we can assume without loss of generality that $\deg[P_1(s,t)] < n$, $\deg[Q_1(s,t)] < m$. Furthermore, we can construct the PDO's

$$Q_{k+1}(s,t) = sQ_k(s,t) + \lambda_k(t)Q_0(s,t) \;\; ; \;\; P_{k+1}(s,t) = sP_k(s,t) + \lambda_k(t)P_0(s,t) \qquad (2.30)$$

where $k = 1, \ldots, n+m$ and $\lambda_k(t)$ is such that $\deg[Q_{k+1}(s,t)] < m$ and $\deg[P_{k+1}(s,t)] < n$ respectively. Operating on (2.28) with s^k from the left, we obtain

$$s^k Q_1(s,t)D(s,t) + s^k P_1(s,t)N(s,t) = s^k$$

inside (t_1, t_2). Using (2.30) it follows that the PDO's $Q_k(s,t)$, $P_k(s,t)$ satisfy

$$Q_k(s,t)D(s,t) + P_k(s,t)N(s,t) = s^{k-1} \;\; ; \;\; k = 1, \ldots, n+m \qquad (2.31)$$

inside (t_1, t_2). Comparing (2.31) with $S_R(t)x(t) = a$ and using Definition 2.6 we can write (2.31) as $S_R(t)\underline{X}(t) = I$ where $\underline{X}(t)$ is a matrix of the coefficients of $Q_k(s,t)$, $P_k(s,t)$ and I is the $(n+m) \times (n+m)$ identity matrix. Hence, in the interval (t_1, t_2), $rank[S_R(t)] = n+m$ and therefore $\det[S_R(t)] \neq 0$.

The proof for the case of left coprime PDO's follows similar steps and is omitted. □□

Proof of Lemma 2.13:

Using the same notation as in the proof of Lemma 2.8 we have:

(if) Since $|\det[(S_R(t)]| \geq c > 0$, $S_R(t)$ is nonsingular and therefore $D(s,t)$, $N(s,t)$ are right coprime. Furthermore, the PDO's $P_0(s,t)$, $Q_0(s,t)$ and

$P_1(s,t)$, $Q_1(s,t)$ can be found as in the proof of Lemma 2.8, by solving equations of the form $S_R(t)x_i(t) = a_i(t)$, $i = 0, 1$; hence, the uniform boundedness of their coefficients is guaranteed by the smoothness and uniform boundedness of the coefficients of $D(s,t)$, $N(s,t)$ and the strong nonsingularity of $S_R(t)$.

(only if) Following the second part of the proof of Lemma 2.8, the smoothness and uniform boundedness of the coefficients of all the considered PDO's guarantees that the PDO's $Q_k(s,t)$, $P_k(s,t)$ have also smooth, UB coefficients. Hence, from (2.31), $|\det([S_R(t)]^{-1})| = |\det[\underline{X}(t)]| \leq c_1$ for some $c_1 > 0$, which implies that $|\det[S_R(t)]| \geq 1/c_1 \triangleq c > 0$. The proof for the case of strongly left coprime PDO's follows similarly. □□

Proof of Corollary 2.16:

(a.) Let us first consider (2.9) with $m > 0$ and all the PDO's in the left form. Expressing the left-hand side as a single PDO in the left form and equating the coefficients of equal powers of s (2.9) can be written as a system of linear algebraic equations $\hat{S}(t)X(t) = A(t)$ where $X(t)$ is the vector of unknown coefficients and $A(t)$ is the vector of coefficients of $k_1(t)A_*(s,t) - s^{n_1}k_1(t)D(s,t)$, where $n_1 = \deg[Q(s,t)] = \deg[A_*(s,t)] - \deg[D(s,t)]$. The matrix $\hat{S}(t)$ is of the form

$$\hat{S}(t) = [\Lambda_1, \ldots, \Lambda_{n_1}, \Pi_1, \ldots, \Pi_n]$$

where Λ_i, Π_j are vectors of the coefficients of $s^{n_1-i}k_1(t)D(s,t)$, $s^{n-j}N(s,t)$ respectively, expressed as $(n + n_1 - 1)$-degree left PDO's. Comparing $\hat{S}(t)$ with $S_R(t)$, given in Definition 2.6, it follows that

$$\hat{S}(t) = \begin{bmatrix} L(t) & 0 \\ \star & S_R(t) \end{bmatrix}$$

where $L(t)$ is a lower triangular matrix with diagonal entries $k_1(t)$. Since $D(s,t)$ and $N(s,t)$ are strongly right coprime in $\mathbf{R}_+$ and $|k_1(t)| \geq c > 0, \forall t \in \mathbf{R}_+$, it follows by Corollary 2.14 that $S_R(t)$ is strongly nonsingular, i.e., there exists a constant $c_1 > 0$ such that $|\det[S_R(t)]| \geq c_1, \forall t \in \mathbf{R}_+$. Furthermore, $L(t)$ is strongly nonsingular since $|k_1(t)| \geq c > 0, \forall t \in \mathbf{R}_+$. Hence, $\hat{S}(t)$ is strongly nonsingular and (2.9) has a unique solution for $Q(s,t)$, $P(s,t)$ with smooth, UB coefficients and $\deg[P(s,t)] \leq \deg[D(s,t)] - 1$.

Notice that in the case $m = 0$ the coefficients of $Q(s,t)$ can be simply found by forward substitution so that $k_1(t)A_*(s,t) - Q(s,t)k_1(t)D(s,t)$ is a $n-1$-degree PDO. $P(s,t)$ is then chosen to satisfy $P(s,t) = k_1(t)A_*(s,t) - Q(s,t)k_1(t)D(s,t)$. It follows by inspection that the uniform boundedness

APPENDIX II.

of $k_1(t)$ and the coefficients of $D(s,t)$, $N(s,t)$, $A_*(s,t)$ together with the assumption $|k_1(t)| \geq c > 0$, $\forall t \in \mathbf{R}_+$, guarantee the uniform boundedness of the coefficients of $Q(s,t)$ and consequently of $P(s,t)$.

(b.) The proof of Corollary 2.16 for the case of equation (2.10) follows by using similar arguments as in (a.) where the PDO's are now expressed in the right form instead of the left. $\square\square$

Proof of Lemma 2.33:

The idea of the proof is to use column operations to write the left TV Sylvester matrix as[21]

$$S_L U = \begin{bmatrix} L & 0 \\ \star & Q_c \end{bmatrix} \quad (2.32)$$

where U is a unimodular matrix with $|\det[U]| = 1$, L is an $m \times m$ lower triangular matrix with ones in the diagonal and $\star$ denotes irrelevant terms. The appearance of L is quite obvious from the definition of the Sylvester matrix (2.6) since $D(s,t)$ is a monic PDO. For the rest, we need to take advantage of the recursiveness in the Sylvester and Controllability matrices. First, the controllability matrix is written as

$$Q_c = [p_0, p_1, \ldots, p_{n-1}] \ ; \quad p_{k+1} = -Ap_k + \dot{p}_k \ ; \quad p_0 = B$$

and A is in the left companion form

$$A = \begin{bmatrix} -a & I \\ & 0 \end{bmatrix}$$

and $[1, a^\mathsf{T}]$, B^T are the vectors of coefficients of $D(s,t)$ and $N(s,t)$ respectively. On the other hand, the left TV Sylvester was defined as

$$S_L = [C_{m-1}, \ldots, C_0, B_{n-1}, \ldots, B_0]$$

where C_i, B_j are vectors of the coefficients of $D(s,t)s^i$, $N(s,t)s^j$ respectively, expressed as $(n + m - 1)$-degree right PDO's (the order of the subscripts is reversed for convenience). Since $x(t)s = sx(t) - \dot{x}(t)$, we may write the following recursions for the non-identically zero parts of the columns of S_L

$$B_{k+1} = \begin{bmatrix} B_k \\ 0 \end{bmatrix} + \begin{bmatrix} 0 \\ \dot{B}_k \end{bmatrix} \ ; \quad B_0 = B \quad (2.33)$$

$$C_{k+1} = \begin{bmatrix} C_k \\ 0 \end{bmatrix} + \begin{bmatrix} 0 \\ \dot{C}_k \end{bmatrix} \ ; \quad C_0 = \begin{bmatrix} 1 \\ a \end{bmatrix} \quad (2.34)$$

---
[21] The time dependence arguments are suppressed for simplicity.

The proof of the lemma now follows by induction.
1. We have that $p_0 = B_0$
2. Assume that[22]

$$B_k = \sum_{0}^{k-1} \lambda_i \begin{bmatrix} 0_i \\ C_{k-1-i} \end{bmatrix} + \begin{bmatrix} 0_k \\ p_k \end{bmatrix} \qquad (2.35)$$

where 0_i denotes a column of i-zeros and λ_i are some functions of time. This form of B_k indicates that it can be reduced to p_k by adding to this column a linear combination of the C_i columns, the latter being represented in a matrix notation as a left multiplication with a unimodular matrix with determinant ± 1.

3. Show that

$$B_{k+1} = \sum_{0}^{k} \lambda_i \begin{bmatrix} 0_i \\ C_{k-i} \end{bmatrix} + \begin{bmatrix} 0_{k+1} \\ p_{k+1} \end{bmatrix} \qquad (2.36)$$

It is now straightforward to obtain equation (2.36) by substituting (2.35) in the recursion of B_k (2.34) and using the recursions of A_k and p_k to regroup the various terms. Note that a useful identity in this procedure is

$$\begin{bmatrix} p_k \\ 0 \end{bmatrix} + \begin{bmatrix} 0 \\ \dot{p}_k \end{bmatrix} - \begin{bmatrix} p_{k0}C_0 \end{bmatrix} = -\begin{bmatrix} 0 \\ Ap_k \end{bmatrix} + \begin{bmatrix} 0 \\ \dot{p}_k \end{bmatrix} = \begin{bmatrix} 0 \\ p_{k+1} \end{bmatrix}$$

where p_{k0} denotes the top element of p_k. Thus, equation (2.32) holds and consequently $|\det[Q_c]| = |\det[S_L]|$. The proof of the dual statement for the observability and right TV Sylvester matrices follows similarly and is omitted.
□□

Proof of Lemma 2.35:

Let x, $A(t), B(t), C(t)$ be the state vector and matrices of a uniform realization of Σ_p with a corresponding STM $\Phi(t,\tau)$. It then suffices to show that there exist constants $K, a > 0$ such that for all bounded initial conditions (e.g., $\|x(t_0)\|, \|w(t_0)\| \le K_0$ for some $K_0 > 0$) and all $t_0 \ge 0$, $\|x(t)\|, \|w(t)\| \le K_1 \exp[-a(t-t_0)]$.

We have that for all $t \ge t_0$ and $\delta > 0$

$$k^2 e^{-2a(t-t_0)} \delta \ge \int_t^{t+\delta} \|y(\tau)\|^2 \, d\tau = \int_t^{t+\delta} x^T(\tau) C^T(\tau) C(\tau) x(\tau) \, d\tau \qquad (2.37)$$

$$x(\tau) = \Phi(\tau,t)x(t) + \int_t^\tau \Phi(\tau,s)B(s)u(s)\,ds \;,\quad \tau \in [t, t+\delta]. \qquad (2.38)$$

---
[22] The form of the recursion is easily deduced by performing a couple of iterations.

Furthermore, since $A(t)$ is UB, we have that[23]

$$\|\Phi(t,\tau)\| \leq f_1(\delta), \quad \tau \in [t, t+\delta]$$

and since $B(t)$ and p_3, p_4 are UB,

$$\left\| \int_t^\tau \Phi(\tau, s) B(s) u(s)\, ds \right\| \leq f_2(\delta) k e^{-a(t-t_0)} \delta$$

for all $\tau \in [t, t+\delta]$. Thus, using the previous inequalities in (2.38), we obtain

$$\int_t^{t+\delta} x^\top(\tau) C^\top(\tau) C(\tau) x(\tau)\, d\tau \geq x^\top(t) N_o(t, t+\delta) x(t) \qquad (2.39)$$
$$- 2\delta^2 \sup_t \left[\|C(t)\|^2 \right] f_1(\delta) f_2(\delta) k e^{-a(t-t_0)} \|x(t)\|$$

where $N_o(t, \tau)$ is the observability grammian of Σ_p. Since Σ_p is assumed to be uniformly completely observable, there exists a constant $d_o > 0$ such that for $\delta = d_o$ its observability grammian satisfies $N(t, t+\delta) \geq f_0(\delta) I > 0$, for all t. Completing the squares in the right-hand side of (2.39) and substituting the result in (2.37) we obtain that

$$\|x(t)\| \leq K_1 e^{-a(t-t_0)}$$

for some constant K_1, independent of t_0. Note that the size of the initial conditions enters K_1 implicitly, through the bounds on u and y. That is, K can be expressed as $K'k$ where K' is a constant depending on the system parameters and k is the proportionality constant in the exponential bound of u_1 and y, which reflects the size of the initial conditions.

The proof of the lemma now follows by observing that the same inequality also holds for the states $w(t)$ of Σ_c since the STM of F is ES, Σ_c is a bounded realization and its inputs u_1, y are bounded by decaying exponential. □□

Proof of Lemma 2.41:

Let $S_0(t)$ denote the Sylvester matrix of the polynomials $D_t(s) \in \{D_t(s)\}_t$ and $N_t(s) \in \{N_t(s)\}_t$ at time t. Thus, the entries of $S_0(t)$ depend on $\Theta_D(t), \Theta_N(t)$ but not on their derivatives. Also let $S_R(t)$ denote the right TV Sylvester matrix of the PDO's $D(s,t), N(s,t)$ and let $S_1(t) = S_R(t) - S_0(t)$ (similar arguments can be used for the left TV Sylvester matrix). Then the matrix $S_1(t)$ has entries depending only on the derivatives, at time t, of $\Theta_D(t), \Theta_N(t)$ up to $m-1, n-1$ order respectively. Hence, we can write $S_1(t) = \mu \hat{S}_1(t)$ where $\hat{S}_1(t)$ is UB. Assuming that $D(s,t), N(s,t)$ are pointwise

---
[23] We use the notation $f_i(x)$ to denote a nonnegative constant, solely determined by x.

strongly coprime in $\mathbf{R}_+$, we have that, for some constant $c_1 > 0$, $|\det[S_0(t)]| \geq c_1$, $\forall t \in \mathbf{R}_+$ and therefore,

$$|\det[S_R(t)]| = |\det[S_0(t)]| |\det[I + \mu \hat{S}_1(t) S_0^{-1}(t)]| \qquad (2.40)$$

By taking $\mu \in [0, \mu_o)$ with $0 < \mu_o < 1/\sup_t \|\hat{S}_1(t) S_0^{-1}(t)\|$ we have that, for some constant $c_2 > 0$, $|\det[I + \mu \hat{S}_1(t) S_0^{-1}(t)]| \geq c_2$, $\forall t \in \mathbf{R}_+$. Hence, using Lemma 2.13, the proof follows. □□

Proof of Lemma 2.42:

Let $\hat{A}(t) = A(t) + (a - \delta)I$ for some fixed $\delta \in (0, a)$. Then, the eigenvalues $\hat{\lambda}_i(t)$ of $\hat{A}(t)$ satisfy $Re(\hat{\lambda}_i(t)) \leq -\delta$ $\forall t \geq 0$. Furthermore, if $\Phi(t, \tau)$, $\hat{\Phi}(t, \tau)$ are the STM's corresponding to $A(t)$, $\hat{A}(t)$ respectively then, due to the commutativity of $\hat{A}(t)$ and $(a - \delta)I$, we have

$$\Phi(t, \tau) = e^{(-a+\delta)(t-\tau)} I \hat{\Phi}(t, \tau) \qquad (2.41)$$

Let us now consider the system

$$\dot{z} = \hat{A}(t) z \qquad (2.42)$$

From Lemma 2.64 and since $\hat{A}(t)$ is Lipschitz continuous and UB, for any $\epsilon > 0$ there exists a UB, continuously differentiable approximation, say $\bar{A}(t)$, such that $\|\bar{A}(t) - \hat{A}(t)\| \leq \epsilon$, $\forall t \in \mathbf{R}_+$. It follows that for sufficiently small ϵ, the eigenvalues of $\bar{A}(t)$ —being ϵ-close to those of $\hat{A}(t)$— are in the left half of the complex plane. Hence, there exists a UB matrix $R(t) = R(t)^\mathsf{T} > 0$ which satisfies the Lyapunov equation

$$\bar{A}^\mathsf{T}(t) R(t) + R(t) \bar{A}(t) = -I \qquad (2.43)$$

for all $t \geq 0$. The matrix $R(t)$ can be evaluated as [Vid.78]

$$R(t) = \int_0^\infty \exp[\bar{A}^\mathsf{T}(t)s] I \exp[\bar{A}(t)s] \, ds$$

Further, by differentiating both sides of (2.43), we have that $\dot{R}(t)$ satisfies a similar Lyapunov equation, namely

$$\bar{A}^\mathsf{T}(t) \dot{R}(t) + \dot{R}(t) \bar{A}(t) = -\dot{\bar{A}}^\mathsf{T}(t) R(t) - R(t) \dot{\bar{A}}(t)$$

which, again, has a unique solution given by

$$\dot{R}(t) = -\int_0^\infty \exp[\bar{A}^\mathsf{T}(t)s] \left[\dot{\bar{A}}^\mathsf{T}(t) R(t) + R(t) \dot{\bar{A}}(t) \right] \exp[\bar{A}(t)s] \, ds$$

APPENDIX II.

Hence, $\|\dot{R}(t)\|$ is UB and $O(\|\dot{\bar{A}}(t)\|)$.

Consider next the positive definite function $V = z^T R(t) z$. Taking the derivative of V along the trajectories of (2.42) with respect to t we get

$$\dot{V} = -z^T z + z^T \dot{R}(t) z + z^T E z \tag{2.44}$$

where $E = (\hat{A} - \bar{A})^T R + R(\hat{A} - \bar{A})$ is a matrix that can be made arbitrarily small since $\|E\| = O(\epsilon)$.

If condition 1 of the lemma holds, it follows from the pointwise properties of $\dot{\bar{A}}(t)$ (see Lemma 2.64) that there exist constants $\lambda, \beta_1 > 0$, depending on the choice of δ and ϵ, such that

$$\dot{V} \leq -(\lambda - \beta_1 \mu) z^T z \tag{2.45}$$

Hence, V is UB and exponentially decaying (0 is a uniformly asymptotically stable equilibrium of (2.42)) $\forall \mu \in [0, \mu_o)$ with $\mu_o = \lambda/\beta_1$.

Further, from (2.44), there exist constants $\lambda, \beta_2 > 0$, depending on the choice of δ and ϵ, such that

$$V \leq V_0 e^{-\lambda(t-t_0)} + \beta_2 \int_{t_0}^{t} e^{-\lambda(t-\tau)} V \|\dot{R}(\tau)\| \, d\tau \tag{2.46}$$

where $V_0 = V(t_0)$ is bounded for bounded initial conditions.

If condition 2 of the lemma holds, (2.46) and the Bellman-Gronwall Lemma yield

$$V \leq V_0 e^{-\lambda(t-t_0)} \exp\left[\beta_2 \int_{t_0}^{t} \|\dot{R}(\tau)\| \, d\tau\right] \tag{2.47}$$

which, using the mean-absolute properties of $\dot{\bar{A}}(t)$, becomes

$$V \leq V_0 e^{\beta \beta_2 c} e^{-(\lambda - \mu \beta \beta_2)(t-t_0)}$$

for some constant $\beta > 0$. Thus, V is UB and exponentially decaying $\forall \mu \in [0, \mu_o)$ with $\mu_o = \lambda/\beta\beta_2$.

If condition 3 of the lemma holds, then by using the Cauchy inequality to square both sides of (2.46) and applying the Schwarz inequality we obtain

$$\begin{aligned} V^2 &\leq p_1(\epsilon') V_0^2 e^{-2\lambda(t-t_0)} \\ &\quad + p_2(\epsilon') \beta_2^2 \left(\int_{t_0}^{t} e^{-\lambda(t-\tau)} \, d\tau \int_{t_0}^{t} e^{-\lambda(t-\tau)} V^2 \|\dot{R}(\tau)\|^2 \, d\tau \right) \\ &\leq p_1(\epsilon') V_0^2 e^{-2\lambda(t-t_0)} + \frac{p_2(\epsilon') \beta_2^2}{\lambda} \int_{t_0}^{t} e^{-\lambda(t-\tau)} V^2 \|\dot{R}(\tau)\|^2 \, d\tau \end{aligned} \tag{2.48}$$

where ϵ' is an arbitrary positive constant and $p_1(\cdot), p_2(\cdot)$ are the Cauchy constants

$$p_1(\epsilon') = \left(1 + \frac{1}{\epsilon'}\right) \quad ; \quad p_2(\epsilon') = 1 + \epsilon' \qquad (2.49)$$

Again, applying the Bellman-Gronwall Lemma on (2.48) and using the mean-square properties of $\dot{\hat{A}}(t)$, we obtain that V^2 and therefore V is UB and exponentially decaying provided that $\mu < \lambda^2/\beta'\beta_2^2(1+\epsilon')$, where β' is a positive constant; since ϵ' is arbitrary, the latter holds $\forall \mu \in [0, \mu_o)$ with $\mu_o = \lambda^2/\beta'\beta_2^2$.

Thus, in all three cases, $\hat{\Phi}(t, \tau)$ is UB for μ sufficiently small and the proof of Lemma 2.42 follows immediately from (2.41). □□

Proof of Lemma 2.44:

The proof of the lemma is obtained along the lines of the proof of Lemma 2.42 with the following differences.

As in Lemma 2.42, we define $\hat{A}(t, \theta) = A(t, \theta) + (a - \delta)I$ for some fixed $\delta \in (0, a)$, and consider the frozen-θ system

$$\dot{z} = \hat{A}(t, \bar{\theta})z \qquad (2.50)$$

where $\bar{\theta} \in \mathcal{M}$ is constant. From the assumptions of the lemma we have that (2.50) is ES, so there exists a UB matrix $R(t, \bar{\theta}) = R^\top(t, \bar{\theta}) > 0$ satisfying the TV Lyapunov equation

$$\frac{\partial R}{\partial t}(t, \bar{\theta}) + \hat{A}^\top(t, \bar{\theta})R(t, \bar{\theta}) + R(t, \bar{\theta})\hat{A}(t, \bar{\theta}) = -I$$

and can be evaluated as [Vid.78]

$$R(t, \bar{\theta}) = \int_t^\infty \hat{\Phi}^\top(\tau, t) I \hat{\Phi}(\tau, t) \, d\tau$$

where $\hat{\Phi}(.,.)$ is the STM corresponding to $\hat{A}(t, \bar{\theta})$. Moreover, $\frac{\partial R}{\partial \theta}(t, \bar{\theta})$ satisfies a similar TV Lyapunov equation, with the solution being UB and given as

$$\frac{\partial R}{\partial \theta}(t, \bar{\theta}) = -\int_t^\infty \hat{\Phi}^\top(\tau, t) \left[\frac{\partial \hat{A}^\top}{\partial \theta} R + R \frac{\partial \hat{A}}{\partial \theta}\right](\tau, \bar{\theta}) \hat{\Phi}(\tau, t) \, d\tau$$

Hence, considering the positive definite function $V = z^\top R(t, \theta)z$, its derivative along the trajectories of (2.50) is

$$\dot{V} = -z^\top z + z^\top \frac{\partial R}{\partial \theta}(t, \theta)\dot{\theta}(t)z$$

The rest of the arguments are identical to the corresponding ones in the proof of Lemma 2.42 and are omitted. □□

APPENDIX II.

Proof of Lemma 2.45:

The proof of Lemma 2.45 can also be obtained as an application of the Bellman-Gronwall Lemma, similar to the proof of Lemma 2.42. We start by considering the ODE

$$\dot{x} = A(t)x + \Delta(t)x \tag{2.51}$$

Since $\dot{x} = A(t)x$ is ES with rate $-a$ we have that there exists a constant $\beta > 0$ such that the STM $\Phi_A(t,\tau)$, corresponding to $A(t)$ satisfies

$$\|\Phi_A(t,\tau)\| \leq \beta e^{-a(t-\tau)}, \quad \forall t \geq \tau \geq 0$$

Hence,[24]

$$\|x(t)\| \leq \beta \|x(t_0)\| e^{-a(t-t_0)} + \beta \int_{t_0}^{t} e^{-a(t-\tau)} \|x(\tau)\| \|\Delta(\tau)\| \, d\tau \tag{2.52}$$

Furthermore, applying the Cauchy and Schwarz inequalities on (2.52) we obtain

$$\|x(t)\|^2 \leq p_1(\epsilon)\beta^2 \|x(t_0)\|^2 e^{-2a(t-t_0)} + \frac{p_2(\epsilon)\beta^2}{a} \int_{t_0}^{t} e^{-a(t-\tau)} \|x(\tau)\|^2 \|\Delta(\tau)\|^2 \, d\tau \tag{2.53}$$

where $p_1(\cdot), p_2(\cdot)$ are as defined in (2.49) and ϵ is an arbitrary positive constant. Using the Bellman-Gronwall Lemma and condition 1 or 2 with (2.52) or condition 3 with (2.53) we get that $\|x(t)\| \leq k\|x(t_0)\| \exp[(-a + \mu\beta_1)(t-t_0)]$ where k is a positive constant and $\beta_1 = 1/\beta$ if 2 holds or $\beta_1 = a/\beta^2 + \epsilon'$ if 3 holds, with $\epsilon' > 0$ being arbitrarily small. The proof of the lemma now follows directly from the last observation and the STM property $x(t) = \Phi(t,t_0)x(t_0)$.
□□

Proof of Lemma 2.55:

Let $z = |x|^p$ for $p > 1$ and $z = x$ for $p = 1$. Notice that z is also absolutely continuous on $[0,T]$ and therefore $\dot{z}$ is bounded almost everywhere on $[0,T]$ [K.F.70] (for $p > 1$, $\dot{z} = p\,sign(x)|x|^{p-1}\dot{x}$ a.e.). Further, z being bounded on $[0,T]$, $e^{p\delta t}z(t)$ is absolutely continuous on $[0,T]$. Thus, for any $t \in [0,T)$,

$$e^{p\delta t}z(t) = z(0) + \int_0^t p\delta e^{p\delta\tau} z(\tau)\, d\tau + \int_0^t e^{p\delta\tau} \dot{z}(\tau)\, d\tau$$

---

[24] Since $A(t)$ is piecewise continuous, x and $\Phi_A(\cdot,\cdot)$ are only piecewise differentiable; consequently, for a formal derivation of (2.52), we should consider the solution inside the intervals where $A(t)$ is continuous and combine the pieces using the continuity of x as a boundary condition.

Hence,

$$|x(t)|^p \leq [|x(0)|\mathcal{E}_{-\delta}]^p(t) + p\delta[\mathcal{E}_{-\delta}\|x_t\|_{p,\delta}]^p(t)$$
$$+\mathcal{E}_{-\delta}\int_0^t pe^{p\delta\tau}|x(\tau)|^{p-1}|\dot{x}(\tau)|\,d\tau$$

Using Holder's inequality, the last term of the above expression becomes

$$p\mathcal{E}_{-\delta}\left[\int_0^t e^{(p-1)q\delta\tau}|x(\tau)|^{(p-1)q}\,d\tau\right]^{1/q}\left[\int_0^t e^{p\delta\tau}|\dot{x}(\tau)|^p\,d\tau\right]^{1/p}$$

where q is the conjugate index of p and therefore $(p-1)q = p$ from which the inequality of the lemma follows. □□

Proof of Lemma 2.64:

Observe first that since u is absolutely continuous on $[0, T]$, $\dot{u}$ exists and is bounded almost everywhere in $[0, T]$ [K.F.70]. It follows that the right-hand sides of the inequalities of the lemma are well defined.

Next, define the functions $u_n : \mathbf{R}_+ \mapsto \mathbf{R}$ by the recursion

$$\dot{u}_n = -au_n + au_{n-1} \;;\; u_n(0) = u_{n-1}(0),\; t \in [0, T] \qquad (2.54)$$

where $n = 1, 2, \ldots$ and $u_0 = u$. It follows that

$$u_n(t) = e^{-at}u_n(0) + a\int_0^t e^{-a(t-\tau)}u_{n-1}(\tau)\,d\tau \;;\; t \in [0, T]$$

and since u_{n-1} is continuous and therefore bounded in $[0, T]$, the integrand in the previous equation is absolutely continuous on the same interval. Hence, [K.F.70],

$$u_n(t) = e^{-at}u_n(0) + u_{n-1}(t) - e^{-at}u_{n-1}(0) - \int_0^t e^{-a(t-\tau)}\dot{u}_{n-1}(\tau)\,d\tau \;;\; t \in [0, T]$$

and, since $u_n(0) = u_{n-1}(0)$,

$$u_n(t) - u_{n-1}(t) = -\int_0^t e^{-a(t-\tau)}\dot{u}_{n-1}(\tau)\,d\tau \;;\; t \in [0, T]$$

Next, invoking Lemma 2.47 we obtain

$$\|(u_n - u_{n-1})_t\|_p \leq \frac{1}{a}\|(\dot{u}_{n-1})_t\|_p$$

while, from the recursive definition of u_n (2.54), we have

$$\|(\dot{u}_n)_t\|_p \leq \|(\dot{u}_{n-1})_t\|_p$$

APPENDIX II.

Thus, using the triangle inequality for $\|\cdot\|_p$ the first inequality of the lemma follows.

Further, since both sides of (2.54) can be differentiated at least $n-1$ times with continuous derivatives, u_n is n-times continuously differentiable. Hence, from (2.54),

$$\begin{aligned}\|(u_n^{(i)})_t\|_p &\leq a\|(u_{n-1}^{(i-1)})_t\|_p + a^2\|(u_{n-1}^{(i-2)})_t\|_p + \cdots + a^{i-1}\|(u_{n-1}^{(1)})_t\|_p \\ &\quad + a^{i-1}\|(u_n^{(1)})_t\|_p\end{aligned}$$

The second inequality of the lemma now follows with a simple induction argument. □□

Chapter 3

The LTV Plant

3.1 Introduction

In this chapter we define the class of LTV plants for which we subsequently formulate and study certain control problems in the progressively harder cases of complete and incomplete a priori knowledge of parameter variations.

The choice of the control objective, for which the control problem has a meaningful solution, depends of course on the underlying structure of the plant and, conversely, for the control problem to be well-posed the plant should satisfy certain assumptions depending on the selected control objective. An issue of particular interest in our study is the control of plants whose TV parameters are only partially known. In such a case, one may attempt to meet the control objective by combining a parameter estimator or adaptive law with a particular controller structure. The underlying intuitive idea behind this approach is that the parameter estimator uses I/O information to estimate the unknown parameters on-line while the updated parameter estimates are used in the calculation of the control input signal.

Inherent in this approach is the concept of plant parametrization, that is, a description of the plant I/O operator in terms of some parameters. However, not all possible parametrizations of LTV I/O operators are convenient for the identification of the I/O operator via a parameter estimation algorithm. For example, if we consider an arbitrary state-space description of an LTV system $\dot{x} = Ax + bu$, $y = c^\mathsf{T} x$ and its natural parametrization in terms of the triple $[A, b, c]$, it may not be possible to design an estimator which determines the parameters $[A, b, c]$ uniquely from I/O information. In order to achieve such an objective, we need to impose certain conditions on the values and structure of the triple $[A, b, c]$.

3.2. SMOOTH PARAMETER VARIATIONS

In this chapter we provide the conditions ensuring that a general class of LTV plants admits a convenient parametrization for estimation and control purposes. We begin with Section 3.2 where we discuss the analytically simpler case of plant representations with smooth parameter variations. In Section 3.3 we consider LTV plants with non-smooth, possibly discontinuous parameter variations. Finally, convenient for parameter estimation plant parametrizations are studied in Section 3.4, for both smooth and non-smooth parameter variations.

3.2 Smooth Parameter Variations

Consider a SISO LTV plant described by the differential equation

$$\begin{aligned} \dot{x}(t) &= A(t)x(t) + b(t)u_p(t) \; ; \quad x(t_0) = x_0 \\ y_p(t) &= c^\top(t)x(t) \end{aligned} \qquad (3.1)$$

where $t_0 \in \mathbf{R}_+$, $(u_p, y_p)(t) \in \mathbf{R} \times \mathbf{R}$ and $x(t) \in \mathbf{R}^n$, satisfying the following assumptions:

3.1 Assumption: *$A(t), b(t), c(t)$ are smooth, UB functions of time with UB derivatives.* ∎

3.2 Assumption: *The triple $[A(t), b(t), c(t)]$ is strongly controllable and observable in $[t_0, \infty)$.* ∎

3.3 Assumption: *The order of the plant, denoted by n, is constant and finite.* ∎

Under these assumptions, Lemma 2.32 ensures that the plant I/O map is described by a topologically equivalent state-space representation which is in either the controllable canonical form, i.e.,

$$\dot{x}_c = \begin{bmatrix} 0 & 1 & 0 & \cdots & 0 \\ 0 & 0 & 1 & \cdots & 0 \\ \vdots & \vdots & \vdots & & 1 \\ -a_n(t) & -a_{n-1}(t) & -a_{n-2}(t) & \cdots & -a_1(t) \end{bmatrix} x_c + \begin{bmatrix} 0 \\ 0 \\ \vdots \\ 1 \end{bmatrix} u_p$$

$$y_p = [b_{n-1}(t), b_{n-2}(t), \ldots, b_0(t)] x_c \qquad (3.2)$$

or the observable canonical form, i.e.,

$$\dot{x}_o = \begin{bmatrix} -a'_1(t) & 1 & 0 & \cdots & 0 \\ -a'_2(t) & 0 & 1 & \cdots & 0 \\ \vdots & \vdots & \vdots & & 1 \\ -a'_n(t) & 0 & 0 & \cdots & 0 \end{bmatrix} x_o + \begin{bmatrix} b'_0(t) \\ b'_1(t) \\ \vdots \\ b'_{n-1}(t) \end{bmatrix} u_p$$

$$y_p = [1, 0, \ldots, 0]x_o \tag{3.3}$$

We refer to (3.2) as the P_R form and to (3.3) as the P_L form of the plant. As noted in Example 2.22, a plant in the P_R form has I/O operator

$$y_p = G_p^R(s,t)[u_p] \; ; \; G_p^R(s,t) = N_p(s,t)D_p^{-1}(s,t) \tag{3.4}$$

where

$$D_p(s,t) = s^n + a_1(t)s^{n-1} + \cdots + a_n(t)$$
$$N_p(s,t) = b_0(t)s^{n-1} + b_1(t)s^{n-2} + \cdots + b_{n-1}(t)$$

are PDO's in the left form with smooth, UB coefficients. Furthermore, due to Assumption 3.2 and from Corollary (2.34), $D_p(s,t)$, $N_p(s,t)$ are strongly right coprime PDO's in $[t_0, \infty)$.

Similarly, using Example 2.23 we obtain that a plant in the P_L form has I/O operator

$$y_p = G_p^L[u_p] \; ; \; G_p^L(s,t) = D_p^{-1}(s,t)N_p(s,t) \tag{3.5}$$

where

$$D_p(s,t) = s^n + s^{n-1}a_1'(t) + \cdots + a_n'(t)$$
$$N_p(s,t) = s^{n-1}b_0'(t) + s^{n-2}b_1'(t) + \cdots + b_{n-1}'(t)$$

are PDO's in the right form with smooth, UB coefficients. Also, due to Assumption 3.2 and from Corollary (2.34), $D_p(s,t)$, $N_p(s,t)$ are strongly left coprime PDO's in $[t_0, \infty)$.

For simplicity, we use the same notation for the PDO's and PIO's of the plant I/O operator in either the P_R or P_L form; their meaning is clear from the context. Moreover, we denote the coefficients of the plant PDO and PIO ($a_i(t), b_i(t)$ or $a_i'(t), b_i'(t)$) by the vector $\Theta_p(t)$.

Further, the *relative degree* of the plant G_p^R or G_p^L

$$n^* \triangleq \deg[D_p(s,t)] - \deg[N_p(s,t)]$$

is equal to one on any interval where $b_0(t) \neq 0$. If, for some constant integer $m \leq n-1$ we have that

$$b_0(t) = b_1(t) = \cdots = b_{n-m-2}(t) = 0 \; ; \; b_{n-m-1} \neq 0$$

for all $t \in [t_0 \infty)$, then the relative degree of the plant is $n-m$. In this case, the I/O description (3.4) can be written as

$$y_p = G_p^R[u_p] \; ; \; G_p^R(s,t) = k_p(t)N_p(s,t)D_p^{-1}(s,t) \tag{3.6}$$

3.3. NON-SMOOTH PARAMETER VARIATIONS

where, now, $N_p(s,t)$ is a monic PDO of degree m and with coefficients $\bar{b}_i(t) = b_i(t)/b_{n-m-1}(t)$ and $k_p(t) = b_{n-m-1}(t)$ is the so called *high-frequency gain*. Similarly, if the plant has relative degree $n - m$, the I/O description (3.5) can be written as

$$y_p = G_p^L[u_p] \; ; \quad G_p^L(s,t) = D_p^{-1}(s,t) N_p(s,t) k_p(t) \tag{3.7}$$

where $N_p(s,t)$ is a monic PDO of degree m and with coefficients $\bar{b}'_i(t) = b'_i(t)/b'_{n-m-1}(t)$ and $k_p(t) = b'_{n-m-1}(t)$.

The above assumptions capture some of the essential properties that are required for the well-posedness of the control problem. To clarify the meaning and give an interpretation of the assumptions we invoke the results of Chapter 2 to make the following observations.

In the plant representation (3.1) we have tacitly assumed that the state $x(t)$ includes the physical quantities, e.g., voltage, displacement, temperature etc., whose properties, such as continuity, boundedness etc., are of interest. Since in the LTV case algebraically equivalent systems do not necessarily share the same internal stability properties, the properties of the state $x(t)$ may not be deduced, in general, from the properties of the I/O operator of the plant. For this purpose, Assumption 3.2 introduces a notion of minimality and ensures that the plant is uniformly controllable and observable. Consequently, all states can be accessed with bounded inputs and no state can grow unbounded without being observed at the output.

Assumption 3.2 also guarantees that the plant can be brought to any of the canonical forms via a Lyapunov transformation and hence described by an I/O operator, factorized in terms of PDO's with smooth, UB coefficients (eqns. (3.4)–(3.7)). In addition, Corollary 2.34 shows that these PDO's are strongly (right or left) coprime.

3.3 Non-Smooth Parameter Variations

Among the critical assumptions about the plant, discussed in the previous section, were the smoothness of the plant parameters and the strong controllability/observability of the plant. Despite their generality, such assumptions exclude an important class of LTV plants with discontinuous or non-differentiable parameters.[1] In addition to the discontinuities, it may be possible that our assumptions are violated inside 'short' time intervals, e.g., during a transition between two different modes of operation of the plant. In such

---

[1] Although it is straightforward —and tedious— to derive the exact number of required differentiations, such calculations are omitted since they depend critically on the assumed state-space representation of the plant.

cases of non-smooth parameter variations one of the major points of concern, particularly for continuous time systems, is that it may not be possible at all to describe the LTV system by a convenient PDO/PIO factorization and/or perform the necessary PDO operations. The reason is that parameter smoothness is a property associated with the physical variables that affect the system behavior. Such variables are not necessarily related in a simple way with the parameters of a canonical form. For example, consider the PDO's $[s + a(t)]$ and $[s + b(t)]$ where $a(t), b(t)$ are discontinuous functions. Any attempt to express the product $[s + a(t)][s + b(t)]$ as a single PDO of degree two would cause the appearance of delta distributions in its coefficients.

In our formulation we avoid any unnecessary further complications caused by such a description, by considering plants in the general state-sace representation (3.1). In order to handle discontinuous as well as non-differentiable system parameters we decompose, without loss of generality, the representation (3.1) as

$$\begin{aligned} \dot{x} &= A_o(t)x + b_o(t)u_p + \tilde{A}(t)x + \tilde{b}(t)u_p \\ y_p &= c_o^T(t)x + \tilde{c}^T(t)x \end{aligned} \qquad (3.8)$$

where $A_o(t)$, $b_o(t)$, $c_o(t)$ are referred to as the 'nominal' part of the plant and $\tilde{A}(t)$, $\tilde{b}(t)$, $\tilde{c}(t)$ as the 'perturbation' part of the plant. Thus, instead of Assumptions 3.1–3.3, the LTV plant (3.8) is assumed to satisfy:

3.4 Assumption: *The entries of $A_o(t), b_o(t), c_o(t)$ are piecewise smooth, UB functions of time and the entries of $\tilde{A}(t)$, $\tilde{b}(t)$, $\tilde{c}(t)$ are piecewise continuous,[2] UB functions of time, satisfying*

$$\int_{t_0}^{t_0+T} \|\tilde{\cdot}\|^2 \leq C + \mu'T$$

where $C, \mu' \geq 0$ are some constants, for all $t_0 \geq 0$ and $T \geq 0$.[3] Furthermore, let us denote by t_j, $j = 1, \ldots, \infty$, the points of discontinuity of $A_o(t)$, $b_o(t)$, $c_o(t)$ where $\{t_j\}_1^\infty$ is a strictly increasing sequence $\in \mathbf{R}_+$ with $t_j \to \infty$ as $j \to \infty$.[4] That is, $A_o(t), b_o(t), c_o(t)$ are smooth for all t except t_j. ∎

3.5 Assumption: *The triple $[A_o(t), b_o(t), c_o(t)]$ satisfies Assumptions 3.1–3.3 inside each interval (t_j, t_{j+1}), $j \in \mathbf{N}$, uniformly in j.* ∎

---

[2] Without loss of generality, both the nominal and perturbation part are assumed to be continuous from the right.

[3] Similar results can be obtained if $\int_{t_0}^{t_0+T} \|\tilde{\cdot}\| \leq C + \mu'T$ holds.

[4] If the number of discontinuities is finite, we may consider a sequence $\{t_j\}$ padded with points at infinity.

3.3. NON-SMOOTH PARAMETER VARIATIONS

3.6 Assumption: *There exist constants C, ν such that in any interval $(t_0, t_0 + T)$ the number of discontinuities n_I of the nominal plant parameters satisfies*
$$n_I \leq C + \nu T$$
$\forall t_0 \geq 0$, $\forall T \geq 0$. ∎

In other words, we consider LTV plants which are perturbations of 'well-behaved' plants, i.e., plants with smooth parameters satisfying our strong controllability and observability assumption as stated in the previous section. Such perturbations can have the form of a small-in-the-mean non-differentiable part, or infrequent discontinuities (jumps) added to the nominal parameters.

Of course, given a general non-smooth state-space representation of a plant, its decomposition into a nominal and a perturbation part is not unique and can be performed in several different ways. For example, a discontinuity in a parameter (or its derivative) can be included directly in the nominal part of the plant or a smooth approximation can be considered as a nominal part and the difference as a perturbation. Needless to say, although different representations of the same plant affect the conservatism in estimating regions of stability, the final result is qualitatively the same.

We must emphasize at this point that for the class of plants described by (3.8) to be a non-trivial extension of the smooth parameter case, we must allow for discontinuities in the nominal parameters. This is necessary in order to admit plants whose parameters cross a controllability/observability boundary. In such a case, the nominal part of the state-space description should either contain a jump or a loss in strong controllability/observability for a short time interval. As a simple example to illustrate this concept, let us consider the plant
$$\dot{x} = b(t)u \quad ; \quad y = x$$
where $b(t) = 1$, $t \in [2n, 2n+1)$, $n \in \mathbf{N}$ and $b(t) = -1$ otherwise. Clearly, although $b(t)$ is discontinuous, it can be approximated within a small error in the mean-square sense resulting in an alternative description for the plant
$$\dot{x} = b_o(t)u + \tilde{b}(t)u \quad ; \quad y = x$$
where b_o is a smooth function and $\tilde{b} = b - b_o$ is small in the mean-square. For this example, the controllability matrix of the nominal plant is $Q_c(t) = b_o(t)$ and must be equal to zero at some time instants t_j since $b_o(t)$ is continuous. This implies loss of uniform controllability at those time instants and consequently loss of strong controllability in an interval around each t_j.

Thus, in our formulation, we admit a quite general class of plants with piecewise Lipschitz continuous parameters, including cases where smoothness and/or strong controllability/observability are lost during short time periods. Moreover, inside each interval (t_j, t_{j+1}) the nominal part of the plant $[A_o, b_o, c_o]$ satisfies Assumptions 3.1–3.3, uniformly in j. Therefore, inside each (t_j, t_{j+1}), the nominal plant admits an I/O operator description of the form (3.4) or (3.5) in a piecewise sense. This enables us to extend any results obtained for smooth parameters to the case of non-smooth ones, by expressing the effects of the perturbation part $[\tilde{A}, \tilde{b}, \tilde{c}]$ and the discontinuities at t_j as a small in-the-mean-square error.

We note, however, that the stability analysis for systems with discontinuous parameters becomes considerably harder since piecewise stability does not, in general, guarantee closed-loop stability unless some additional conditions are imposed on the average frequency of the discontinuity points.[5] To establish and make this statement precise we use the following notation.

$\{t_j\}_1^\infty$, a strictly increasing sequence in $\mathbf{R}_+$ with $t_j \to \infty$ as $j \to \infty$;

J, a subset of the natural numbers $\mathbf{N}$;

$\mathcal{U}_J(t)$, the characteristic function of the set $\bigcup_{j \in J} [t_j, t_{j+1})$, defined as:

$$\mathcal{U}_J(t) = \left\{ \begin{array}{l} 1 \text{ if } t \in [t_j, t_{j+1}) \text{ for some } j \in J \\ 0 \text{ otherwise} \end{array} \right\}$$

$n_J, \bar{n}_J \in \mathbf{N}$, the number of subintervals (t_j, t_{j+1}) of an interval $[t_0, t_0 + T]$ for which $j \in J$ and the number of transitions from an interval for which $j \in J$ to one for which $j \notin J$, respectively.[6] More precisely, for $T, t_0 \geq 0$, let $m, n \in \mathbf{N}$ such that $t_0 \in [t_m, t_{m+1})$ and $t_0 + T \in (t_n, t_{n+1}]$. Then,

$$n_J = \sum_{j=m}^{n} \mathcal{U}_J(t_j) \; ; \; \bar{n}_J = [1 - \mathcal{U}_J(t_m)] + \sum_{j=m+1}^{n} \max\{[\mathcal{U}_J(t_{j-1}) - \mathcal{U}_J(t_j)], 0\}$$

3.7 Lemma: *Consider the system $\dot{x} = A(t)x$ where $A(t)$ is a matrix with piecewise continuous, UB elements. Further, assume that there exist two positive constants k, a such that $\forall j \in J \subset \mathbf{N}$*

$$\|\Phi(t, \tau)\| \leq k e^{-a(t-\tau)}, \quad \forall t, \tau \in (t_j, t_{j+1}); \; t \geq \tau$$

where $\Phi(.,.)$ is the state transition matrix associated with $A(t)$. Then, the system $\dot{x} = A(t)x$ is ES with rate $-\lambda$, for some constant $\lambda > 0$, if there exists

---

[5] This issue is of particular interest in our study where we intend to design controllers for systems with jump parameters in a piecewise sense, i.e., design the controller as to make the closed-loop ES, inside every interval (t_j, t_{j+1}).

[6] That is, consecutive intervals for which $j \notin J$ count as one.

3.3. NON-SMOOTH PARAMETER VARIATIONS

a constant $C \geq 0$ such that

$$-a \int_{t_0}^{t_0+T} \mathcal{U}_J(t)\,dt + b \int_{t_0}^{t_0+T} [1 - \mathcal{U}_J(t)]\,dt + n_J \ln(k) + \bar{n}_J \ln(k') \leq -\lambda T + C \quad (3.9)$$

$\forall t_0 \geq 0, \forall T \geq 0.$ ▽▽

Proof: It suffices to show that, under the conditions stated above, if x satisfies the ODE $\dot{x} = A(t)x$, then for some constant $K > 0$ and for all $T \geq 0$, $t_0 \geq 0$,

$$\|x(t_0 + T)\| \leq K e^{-\lambda T} \|x(t_0)\| \quad (3.10)$$

Let $t \in (t_j, t_{j+1})$. If $j \in J$ we have that

$$\|x(t)\| \leq k e^{-a(t-t_j)} \|x(t_j)\| \quad (3.11)$$

On the other hand, since $A(t)$ is UB, there exist constants k', b such that

$$\|\Phi(t,\tau)\| \leq k' e^{b(t-\tau)} \;;\; \forall t \geq \tau$$

Hence, if $j \notin J$,

$$\|x(t)\| \leq k' e^{b(t-t_j)} \|x(t_j)\| \quad (3.12)$$

Further, from (3.11), (3.12) and the continuity of $x(t)$ we obtain, grouping together consecutive intervals for which $j \notin J$,

$$\|x(t_0+T)\| \leq k^{n_J} k'^{\bar{n}_J} \exp\left[b\int_{t_0}^{t_0+T}[1-\mathcal{U}_J(t)]\,dt - a\int_{t_0}^{t_0+T}\mathcal{U}_J(t)\,dt\right]\|x(t_0)\|$$

In view of (3.9), the last inequality implies (3.10) with $K = \exp[C]$. □□

Despite its complicated appearance, the condition of Lemma 3.7 is nothing more than an upper bound on the average size (measure) of intervals where $\Phi(.,.)$ may not be exponentially decaying with rate $-a$ and the average number of discontinuities, such that the overall state transition matrix $\Phi(.,.)$ is exponentially decaying with rate $-\lambda$. This situation may arise in a control systems framework when, for example, there is loss of strong controllability/observability of the plant state-space representation inside short time intervals. In such a case, there may not exist a control law which internally stabilizes the plant in those intervals. However, in view of (3.8) and in order to simplify the presentation, we may describe such a situation by an appropriate selection of the modeled and perturbation parts of the plant. That is, without loss of generality, we may select the nominal part of the plant to be piecewise strongly controllable and observable even if the actual plant fails to be so and incorporate the difference in the perturbation part. Consequently, the following simpler version of Lemma 3.7 is adequate for our purposes.

3.8 Corollary: *The result of Lemma 3.7 holds if* $J = \mathbf{N}$ *and for some constant* $\lambda > 0$, *there exists a constant* $C \geq 0$ *such that*

$$n_J \ln(k) \leq (a - \lambda)T + C$$

$\forall t_0 \geq 0, \forall T \geq 0.$ ▽▽

Again, a condition on the average number of discontinuities is essential in order to guarantee that piecewise ES implies ES (compared with the slowly TV case where pointwise stability implies stability). It is actually quite straightforward to construct counter-examples of piecewise LTI systems which are ES inside every interval but overall unstable, if there is no constraint on the number of discontinuities. A typical and quite illustrative example is given below.

3.9 Example: Consider the system

$$\dot{x} = A(t)x \; ; \; A(t) = \begin{cases} A_0 & \text{if } t \in [2k, 2k+1) \\ A_0^T & \text{if } t \in [2k+1, 2k+2) \end{cases}, k = 0, 1, \ldots$$

where

$$A_0 = \begin{bmatrix} -1 & 5 \\ 0 & -1 \end{bmatrix}$$

Applying Floquet analysis on the above system, we have that

$$x(2k) = \left(e^{A_0^T} e^{A_0}\right)^k x(0)$$

and the eigenvalues of the matrix $e^{A_0^T} e^{A_0}$ are easily found to be 0.005 and 3.649. Hence the system is unstable, despite the fact that it is ES inside every interval $(k, k+1)$.

At this point, it is very interesting to perform a simulation of the response of this system. As shown in Fig. 3.1, starting with initial conditions $x(0) = [1, 1]^T$, inside every interval $(k, k+1)$, one of the states decays as e^{-t+k} while the other decays as e^{-t+k} and $(t-k)e^{-t+k}$. Since, for small positive values of $(t-k)$, the latter is an increasing function of $(t-k)$, the corresponding state increases in magnitude at the beginning of every interval. Thus, if the frequency of discontinuities is too high, the decrease in the magnitude of the states inside each interval may be insufficient to counteract the magnitude increase at the beginning of the interval and instability may occur. ▽▽

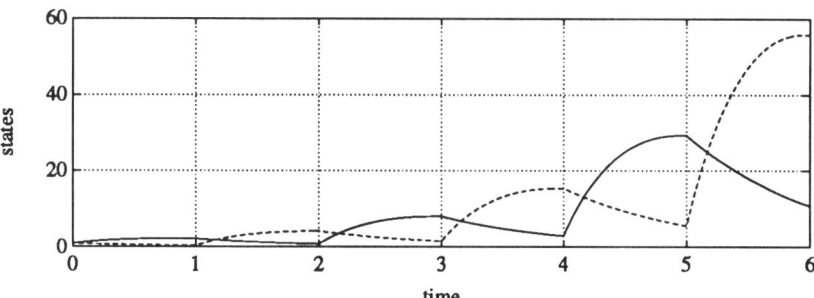

Figure 3.1: Example of instability, occuring when the average number of discontinuities is too large.

3.4 Parametric Models of TV I/O Operators

An issue of particular interest in the case of plants with partially known parameters is the design of parameter estimation algorithms, used to identify an unknown I/O operator on-line, from I/O measurements. Such estimators, discussed in more detail in Chapter 6, rely on the ability to describe the unknown operator in an inner product form between a vector of unknown parameters and a vector of signals, often referred to as the regressor vector, which are available for measurement. In this section, our objective is to establish a basic parametrization of certain types of I/O operators having the inner product form that is convenient for parameter estimation.

The following lemma gives a parametric model of a plant in the P_L-form that allows for the identification of the plant I/O operator via parameter estimation.

3.10 Lemma: *Consider a plant described by a strictly proper I/O operator $D_p^{-1}(s,t)N_p(s,t)$, i.e.,*

$$D_p(s,t)[y_p] = N_p(s,t)[u_p] \qquad (3.13)$$

where (u_p, y_p) is the I/O pair, $D_p(s,t), N_p(s,t)$ are PDO's in the right form with piecewise continuous, UB coefficients and $D_p(s,t)$ is monic of degree n. Then there exists $\theta_* : \mathbf{R}_+ \mapsto \mathbf{R}^{2n}$ such that, with zero initial conditions,

$$y_p = G(s)[u_p \theta_{1*}] + G(s)[y_p \theta_{2*}] \qquad (3.14)$$

where

$$G(s) = q^\top (sI - F)^{-1} \; ; \; \theta_*^\top = [\theta_{1*}^\top, \theta_{2*}^\top] \qquad (3.15)$$

F is an $n \times n$ Hurwitz matrix and $(q^\top, F)$ is a completely observable pair.

▽▽

Proof: Let $D_F(s) = \det(sI - F)$ and operate on (3.13) from the left with $D_F^{-1}(s)$. Then,

$$y_p = D_F^{-1}(s)N_p(s,t)[u_p] + D_F^{-1}(s)\{D_F(s) - D_p(s,t)\}[y_p]$$

Noting that $\deg[D_F(s) - D_p(s,t)] \leq n-1$, a realization of the above equation as in Example 2.37 yields (3.14) and (3.15).

At this point it is worthwhile to perform the state-space analog of this proof. Consider the state-space realization of (3.13) as in Example (2.37):

$$\dot{x} = A(t)x + b(t)u_p \; ; \; y_p = c^\top x \qquad (3.16)$$

where $A(t)$ contains the coefficients of $D_p(s,t)$ in the left-companion form, $b(t)$ contains the coefficients of $N_p(s,t)$ and $c^\top = [1, 0, \ldots, 0]$. Further, suppose that F is in the left companion form. (There is no loss of generality in such an assumption since $(q^\top, F)$ is a completely observable pair and hence we can always find a (constant) similarity transformation to put it in the left companion form.) We may therefore rewrite (3.16) as

$$\begin{aligned}\dot{x} &= A(t)x + b(t)u_p + \kappa(t)y_p - \kappa(t)c^\top x \\ y_p &= c^\top x\end{aligned}$$

where $\kappa(t)$ is arbitrary. Since $(c^\top, A(t))$ is uniformly observable (Theorem 2.31), there exists $\kappa(t)$ such that $A(t) - \kappa(t)c^\top = F$; in fact it is quite straightforward to construct κ component-wise by taking $\kappa_i(t) = A_{i1}(t) - F_{i1}$, with obvious notation. Using again Example 2.37 and noting that I/O operators are invariant under similarity transformations the proof follows. Finally, it is interesting to observe that a by-product of this analysis is that any initial conditions of the original system are transferred to the modified one, indicating that for arbitrary initial conditions, (3.14) is still valid modulo an exponentially decaying term, depending on the initial conditions. □□

We should note that the direct analog of this result for plants in the P_R-form is not convenient for estimation purposes since in that case the parametric model contains an internal signal which, in general, is not available for measurement.

A generalization of Lemma 3.10 to systems with the general state-space description (3.8) is given below.

3.11 Lemma: *Consider an LTV plant satisfying Assumptions 3.4–3.6. Then, for any completely observable n-dimensional pair $(q^\top, F)$ there exists a UB, piecewise smooth vector $\theta_* = [\theta_{1*}^\top, \theta_{2*}^\top]^\top$, $\theta_{1*}, \theta_{2*} : \mathbf{R}_+ \mapsto \mathbf{R}^n$ with*

3.4. PARAMETRIC MODELS OF TV I/O OPERATORS

possible discontinuities at $\{t_j\}_j$ such that the plant is described by the state-space model

$$\begin{aligned}
\dot{x}_F &= F x_F + \theta_{1*}(t) u_p + \theta_{2*}(t) y_p + \tilde{A}_F(t) x_F + \tilde{b}_F(t) u_p \\
x_F(t_j^+) &= \bar{P}(t_j) x_F(t_j^-) \\
y_p &= q^\top x_F + \tilde{c}_F^\top(t) x_F
\end{aligned} \qquad (3.17)$$

with the same internal stability properties as the original plant and such that:

- $\tilde{A}_F, \tilde{b}_F, \tilde{c}_F$ are UB, piecewise continuous matrices for which there exist constants $C, K > 0$ such that

$$\int_{t_0}^{t_0+T} |\cdot|^2 \leq C + K \mu' T$$

for all $t_0, T \geq 0$ and μ' as in Assumption 3.4.

- $\bar{P}$ is a UB piecewise smooth matrix with UB inverse and derivative everywhere except at $t = t_j$, $j = 1, 2, \ldots$. Furthermore, as the size of jumps of the nominal plant parameters and their derivatives approaches zero, $|\bar{P}(t_j) - I|$ and $|\theta_*(t_j^+) - \theta_*(t_j^-)|$ approach zero.

where t^+, t^- are used to denote right and left limits respectively. It follows that y_p can be expressed as

$$\begin{aligned}
y_p(t) &= \{G(s)[u_p \theta_{1*}]\}(t) + \{G(s)[y_p \theta_{2*}]\}(t) \\
&\quad + \{G(s)[\tilde{A}_F x_F + \tilde{b}_F u_p]\}(t) + \tilde{c}_F^\top(t) x_F(t) \\
&\quad + \sum_{t_j \leq t} q^\top \Phi_F(t, t_j)[\bar{P}(t_j) - I] x_F(t_j^-) + q^\top \Phi_F(t, t_0) x_F(t_0)
\end{aligned}$$

where $G(s) = q^\top (sI - F)^{-1}$ and $\Phi_F(.,.)$ is the state transition matrix associated with F.

Further, for any $\delta > 0$ there exist $\nu_0, \mu_0' > 0$ such that, for any $\nu \in [0, \nu_0)$, $\mu' \in [0, \mu_0')$ and for as long as $(u_p)_t \in L_\infty$, $x_F/m_p^{1/p}$ is UB, where m_p is a normalization signal as in Lemma 2.56 with $u = [u_p, y_p]^\top$. $\triangledown\triangledown$

Proof: Since the nominal part of the plant is strongly observable in all intervals (t_j, t_{j+1}), uniformly in j, there exist (Lyapunov) similarity transformations $P_j(t)$ putting the nominal plant in its observable canonical form inside (t_j, t_{j+1}) and such that $P_j(t), P_j^{-1}(t)$ and $\dot{P}_j(t)$, $t \in (t_j, t_{j+1})$ are UB with respect to t and j. Hence, the plant is described by

$$\begin{aligned}
\dot{x}_o &= A_o(t) x_o + b_o(t) u_p + \tilde{A}_o(t) x_o + \tilde{b}_o(t) u_p \; ; \; t \in [t_j, t_{j+1}) \\
y_p &= c_o^\top x + \tilde{c}_o^\top(t) x
\end{aligned} \qquad (3.18)$$

with boundary conditions arising from the continuity of the state vector of the original state-space description of the plant[7]

$$x_o(t_j) = P_j^{-1}(t_j^+)P_{j-1}(t_j^-)x_o(t_j^-)$$

and $[A_o(t), b_o(t), c_o]$ being in the observable canonical form. Since P_j, P_j^{-1} are UB, Assumption 3.4 implies that the perturbation part in (3.18) also satisfies

$$\int_{t_0}^{t_0+T} |\tilde{r}|^2 \leq C + K_o \mu' T$$

for some $C, K_o > 0$, for all $t_0, T \geq 0$.

Further, there exists a possibly discontinuous vector κ_o such that $A_o(t) - \kappa_o(t)c_o$ is the left-companion matrix which is similar to F. Thus, with (q, F) being an n-dimensional completely observable pair, (3.18) can be written in the form of (3.17) after a (constant) similarity transformation, $x_o = P_F x_F$ where

$$\theta_{1*} = P_F^{-1} b_o \ ; \ \theta_{2*} = P_F^{-1} \kappa_o$$

$$\tilde{A}_F = P_F^{-1}[\tilde{A} - \kappa_o \tilde{c}_o^T] P_F \ ; \ \tilde{b}_F = P_F^{-1} \tilde{b}_o \ ; \ \tilde{c}_F^T = \tilde{c}_o^T P_F$$

and

$$\bar{P}(t) = P_F^{-1} P_j^{-1}(t^+) P_{j-1}(t^-) P_F \ ; \ t \in [t_j, t_{j+1})$$

It is now quite straightforward to integrate (3.17) and obtain the expression for y_p given above. Notice that as the discontinuities in the nominal plant parameters and their derivatives vanish, $P_j(t_j^+)$ approaches $P_{j-1}(t_j^-)$ and hence, the discontinuities in θ_* vanish and $\bar{P}$ approaches the identity matrix.

Finally, for the last part of the lemma, consider the state-space description of the plant (3.8). Since (c_o, A_o) is strongly observable inside every (t_j, t_{j+1}), uniformly in j, it follows that for any $\delta' > \delta$ there exists a piecewise smooth, UB κ' such that the STM $\Phi'(.,.)$ of

$$\dot{x} = [A_o(t) - \kappa'(t)c_o^T(t)]x$$

satisfies

$$|\Phi'(t,\tau)| \leq k e^{-\delta'(t-\tau)} \ ; \ \forall t \geq \tau \in (t_j, t_{j+1}), \ \forall j$$

where k is a positive constant. For example, such a κ' can be constructed as in the first part of the proof. Hence, choosing $\delta'' \in (\delta, \delta')$, Corollary 3.8 and Assumption 3.6 imply that there exists $\nu_0 > 0$ such that for all $\nu \in [0, \nu_0)$

---

[7] Notice that the state of the canonical form (3.18) may be discontinuous.

3.4. PARAMETRIC MODELS OF TV I/O OPERATORS

$\Phi'(.,.)$ is exponentially decaying with rate at most $-\delta''$. Further, rewriting (3.8) as

$$\dot{x} = [A_o(t) - \kappa'(t)c_o^T(t)]x + [\tilde{A}(t) - \kappa'(t)\tilde{c}^T(t)]x + [b(t) + \tilde{b}(t)]u_p + \kappa'(t)y_p$$
$$y_p = c_o^T(t)x + \tilde{c}^T(t)x \qquad (3.19)$$

and invoking Lemma 2.45 and Assumption 3.4 we have that there exists $\mu_0' > 0$ such that for all $\mu' \in [0, \mu_0')$, the system

$$\dot{x} = [A_o(t) - \kappa'(t)c_o^T(t)]x + [\tilde{A}(t) - \kappa'(t)\tilde{c}^T(t)]x$$

is ES with rate at most $-\delta$. Since the plant is described by a bounded state-space representation, $(y_p)_t$ is in L_∞ for as long as $(u_p)_t \in L_\infty$. Hence, from Lemma 2.56 and Corollary 2.52 we obtain that $x/m_p^{1/p}$ is UB which, in turn, implies that $x_F/m_p^{1/p}$ is UB since $x_F(t) = P_F^{-1}P_j^{-1}(t^+)x(t)$ inside $[t_j, t_{j+1})$ and $P_j^{-1}(t)$, $t \in (t_j, t_{j+1})$ is UB, uniformly in j. (Also compare with Corollary 2.57). □

Lemmas 3.10 and 3.11 show that plants satisfying Assumptions 3.4–3.6 or plants with I/O operator $D_p^{-1}(s,t)N_p(s,t)$ can be described by a parametric model of the form

$$y_p(t) = \{G(s)[u_p\theta_{1*}]\}(t) + \{G(s)[y_p\theta_{2*}]\}(t) + \tilde{\eta}(t) \qquad (3.20)$$

where $\tilde{\eta}$ is a term appearing in the non-smooth parameter case and whose magnitude depends on the perturbation parameters ν, μ'. The properties of $\tilde{\eta}$ are discussed in some more detail in Chapter 6. Although the plant parametrization (3.20) is not in the inner product form required for the application of standard gradient-based estimators, it can be readily modified invoking the 'Swapping' Lemma 2.59 or Corollary 2.60 to yield

$$y_p = \{G(s)[u_p], G(s)[y_p]\}\theta_* + \eta + \tilde{\eta}$$
$$\triangleq w^T\theta_* + \eta + \tilde{\eta} \qquad (3.21)$$

where, $w^T = \{G(s)[u_p], G(s)[y_p]\}$ is a vector of signals which can be constructed from purely I/O information. When θ_* is absolutely continuous, an expression for η is directly obtained from Lemma 2.59 as

$$\eta = G(s)\{G'(s)[u_p\Pi_1]\dot{\theta}_*\} + G(s)\{G'(s)[y_p\Pi_2]\dot{\theta}_*\} \qquad (3.22)$$

where $G'(s) = (sI - F)^{-1}$. When θ_* contains a jump function, an analogous expression for η, with an additional term describing the effect of the jumps, is obtained from Corollary 2.60.

In a typical parameter estimation problem, the vector θ_* is unknown but u_p, y_p are continuously measured and therefore their filtered values w can be constructed and used to estimate θ_*. Of course, since θ_* is unknown, $\eta, \tilde{\eta}$ are also unknown and cannot be constructed from available measurements. Therefore, $\eta, \tilde{\eta}$ must be treated as 'noise' or modeling error when the parametrization (3.21) is used to estimate θ_* and they must be small in some sense in order for the estimation to be successful. As shown in Chapter 6, θ_* and consequently the unknown I/O operator of the plant in the P_L form can be identified within a small mean-square error in an I/O sense, provided that

- the absolutely continuous part of θ_*, say θ_*^s, is slowly varying in the mean,

- the average number of discontinuities in an interval, ν, is small and

- the parameter μ', characterizing the perturbation part of the plant, is small.

Since in typical applications the plant parameters contain very few discontinuity points inside long time intervals, the most critical parameter of the three is expected to be the speed of variation of θ_*^s. It is therefore desirable to avoid the estimation of any a priori known fast-varying components of θ_*^s, for example, by decomposing the parameter variations into a 'structured' and an 'unstructured' part; such a decomposition of θ_*^s is discussed in the following subsection.

3.4.1 Structured Parameter Variations

Parametric models of the form (3.21) have been widely used for the on-line estimation of the unknown parameters θ_* —resulting in the identification of the unknown plant I/O operator— in the LTI as well as the LTV case [G.S.84, N.A.89, S.B.89, M.G.88, Kre.86, A.J.83]. Naturally, the performance of a parameter estimator based on the linear model (3.21) depends heavily on the size of the 'noise' terms $\eta, \tilde{\eta}$ as well as the speed of variations of the unknown parameters θ_*. It is therefore of interest to exploit any available a priori knowledge about the form or structure of variations of the unknown parameters in order to decrease the size of the effective perturbation and, in particular, the speed of variation of the absolutely continuous part of the unknown parameters. This idea is illustrated in the following example.

3.12 Example: Let us assume that, for the plant of Lemma 3.10, the time variations of $\theta_*(t)$ are of the form

$$\theta_*(t) = \hat{\theta}_* \sin w_0 t \tag{3.23}$$

3.4. PARAMETRIC MODELS OF TV I/O OPERATORS

where the frequency w_0 is constant and known and $\hat{\theta}_*$ is an unknown but constant vector. If we parametrize the plant according to (3.21), we obtain

$$y_p = w^\top \theta_* + \eta$$

where η, given by (3.22), is an unknown signal. It now follows that if we were to use the last equation as a parametric model to estimate θ_*, we should require w_0 to be small in order for the estimation to be successful. Note that in this approach, we effectively treat θ_* as a constant, making no use of the a priori available knowledge about its structure of variation. On the other hand, using the available a priori information on the time-dependence of θ_* in (3.20) we obtain

$$y_p = G(s)[u_p \hat{\theta}_{1*} \sin w_0 t] + G(s)[y_p \hat{\theta}_{2*} \sin w_0 t]$$

which, by virtue of the Swapping Lemma, becomes

$$\begin{aligned} y_p &= G(s)[u_p \sin w_0 t]\hat{\theta}_{1*} + G(s)[y_p \sin w_0 t]\hat{\theta}_{2*} \\ &= w^\top \hat{\theta}_* \end{aligned} \quad (3.24)$$

where $w^\top = \{G(s)[u_p \sin w_0 t], G(s)[y_p \sin w_0 t]\}$ is a signal which can be constructed from available measurements and $\hat{\theta}_* = [\hat{\theta}_{1*}, \hat{\theta}_{2*}]$ is unknown but constant. Equation (3.24) is now of the same form as the parametrizations obtained in the LTI case without modeling errors or noise. It is therefore possible, using standard estimation techniques, to estimate $\hat{\theta}_*$ within an asymptotically converging to zero error (in an I/O sense). Once the estimate of $\hat{\theta}_*$ is available, we can obtain the estimate of $\theta_*(t)$ from (3.23). Note that, in contrast to the previous case where the parametric model (3.21) was used for estimation, the parametric model (3.24) results in zero estimation error for any value of w_0 (i.e., slow or fast parameter variations). (For details see Chapter 6, Theorem 6.6.) ▽▽

Let us now generalize the previous example by assuming that the plant parameters and consequently $\theta_*(t)$ are decomposed as

$$\theta_*(t) = \Pi(t)\hat{\theta}_*(t) \quad (3.25)$$

where $\hat{\theta}_*(t)$ is the unknown, or 'unstructured' part of $\theta_*(t)$ of dimension $l \in \mathbb{N}$ and $\Pi(t)$ is a known matrix with piecewise smooth UB elements. The flexibility of (3.25) in describing fully or partially structured or even unstructured parameter variations is demonstrated by the following simple examples.

1. *'Fully structured variations'*: Assume that $\theta_*(t) = A_0 + \sin(wt)A_1$ where w is known and A_0, A_1 are unknown constants. Then,

$$\Pi(t) = [I, \sin(wt)I] \quad ; \quad \hat{\theta}_* = [A_0^\top, A_1^\top]^\top$$

2. *'Partially structured variations':* Assume that $\theta_*(t) = A_0 + \sin[(w + \epsilon)t]A_1 + f(t)A_2$ where w is known, $f(t)$ is an unknown function and ϵ, A_0, A_1, A_2 are unknown constants. Then,

$$\Pi(t) = [I, \sin(wt)I, \cos(wt)I]$$
$$\hat{\theta}_*(t) = [A_0^\top + f(t)A_2^\top, \cos(\epsilon t)A_1^\top, \sin(\epsilon t)A_1^\top]^\top$$

3. *'Unstructured variations':* Assume that $\theta_*(t) = A_0 f(t)$ where both $A_0, f(t)$ are unknown. Then,

$$\Pi(t) = I \; ; \; \hat{\theta}_*(t) = \theta_*(t)$$

3.13 Remark: Notice that equation (3.25) can be augmented by an additional term $E^*(t)$ to describe the unknown fast, but small in amplitude, part ('jitter') of the variations of $\theta_*(t)$, i.e.,

$$\theta_*(t) = \Pi(t)\hat{\theta}_*(t) + E^*(t)$$

where $\|E^*(t)\|$ is small or small in the mean. This description may be useful in applications since, by treating $E^*(t)$ as parameter noise, we can avoid the use of high-dimensional $\hat{\theta}_*(t)$ in modeling the time-variations of $\theta_*(t)$. However, for reasons of clarity and ease of exposition, the discussion of this case is omitted as it presents no additional difficulty in the subsequent analysis.

▽▽

By incorporating the knowledge of $\Pi(t)$ in the regressor vector, we may estimate $\hat{\theta}_*(t)$ first and then obtain the estimate of $\theta_*(t)$ as $\theta(t) = \Pi(t)\hat{\theta}(t)$ where $\hat{\theta}(t)$ is the estimate of $\hat{\theta}_*(t)$ at time t. The benefit of this approach becomes clear when we consider the case where the parameters to be estimated, contain fast but known TV elements. Due to the finite speed of adaptation, a general adaptive law is not expected to yield a small prediction error $(w^\top(t)\theta(t) - y(t))$ when θ_* is fast TV. However, if a structured decomposition of the parameters succeeds in achieving a slowly varying $\hat{\theta}_*$, then, estimating the slowly TV component only, it is possible to ensure the smallness of the prediction error, despite the fact that the original parameters may be fast TV.

On the other hand, some caution should be exercised when the structured parameter variations approach is used. One of the shortcomings of this approach is that, in general, the relation between the actual system parameters and the vector θ_* is highly nonlinear, making the derivation of the exact structure of θ_* from the structure of the system parameters difficult. For example, consider an LTV system with I/O operator $D_p^{-1}(s,t)N_p(s,t)$ to be identified.

3.4. PARAMETRIC MODELS OF TV I/O OPERATORS

From Lemma 3.10, it follows that the variations of θ_* would have the same structure as the coefficients of $D_p(s,t)$ and $N_p(s,t)$. These coefficients, however, do not necessarily represent physical quantities and, in general, would be nonlinearly related to parameters with physical meaning through a similarity transformation. An additional issue of concern is that although, in principle, the speed of the structured part can be arbitrary, practical considerations put an upper bound on the derivatives of Π for which the results make sense. The reason is that, in general, the sensitivity of the prediction error bounds on the practically unavoidable uncertainties in Π increases as the derivatives of Π become larger. Consequently, if we allow Π to vary arbitrarily fast, we must also require that it is known with a practically unreasonable degree of accuracy. Thus, it should be emphasized that the intended purpose of the structured parameter variations approach is to reduce, rather than eliminate, the effective size of the perturbation introduced by the TV nature of the estimated parameters.

Chapter 4

Model Reference Control

4.1 Introduction

A class of feedback control strategies that has attracted considerable interest in the area of control systems and particularly in adaptive control is Model Reference Control (MRC). The principal idea behind MRC is to design the control law so as to achieve some prescribed tracking performance characteristics. In a typical MRC approach, a reference model is selected describing the desired I/O characteristics of the closed-loop plant, from the reference or command input to the plant output. The control law is then designed so that the I/O operator of the closed-loop plant, from the reference input to the plant output, matches the I/O operator of the reference model.

In the case of LTI plants, the MRC approach effectively amounts to a pole and zero placement design whereby feedback is used to place the plant poles at the desired locations while the plant zeros are cancelled and replaced by the desired zeros. For such a design to be meaningful and feasible, both the plant and the reference model must meet certain conditions. First, from an internal stability point of view, any cancellations must occur in the left half-plane, something that is often referred to as the *minimum phase plant* condition (or assumption). Second, from a realization/implementation point of view, the controller I/O operator should be at least proper, something that translates into a matching condition between the relative degrees of the plant and the reference model.

In this chapter we extend the MRC assumptions and design techniques, which are well understood in the LTI case, to the more intricate case of LTV plants. In particular, we study the design of MRC for LTV plants with the objective of forcing the LTV closed-loop plant to have the same (or approx-

4.2. PROBLEM STATEMENT

imately the same) I/O operator as a, typically LTI, reference model. We begin with Section 4.2 where we state the control problem for a class of LTV plants with smooth parameters, defined by a set of assumptions which are an extension of the LTI MRC assumptions. In Sections 4.3 and 4.4 we design and analyze MRC schemes meeting the MRC objective. The special case of slowly TV plants is treated in Section 4.5 where we study the applicability of pointwise LTI techniques in the design of (approximate) MRC's. In Section 4.6 we consider a more general class of plants with non-smooth and discontinuous parameters and establish that the MRC design of Sections 4.3 and 4.4 can be extended to this case as well, at the expense of some performance deterioration. We conclude this chapter with Section 4.6 where we present simple examples and simulations illustrating the design and performance of MRC schemes in the LTV case.

4.2 Problem Statement

Consider a SISO LTV plant described by the state-space equations

$$\begin{aligned} \dot{x}_p &= A(t)x_p + b(t)u_p \\ y_p &= c^\mathsf{T}(t)x_p \end{aligned} \quad (4.1)$$

and satisfying Assumptions 3.1–3.3.

The *MRC objective* is defined as follows:

Determine a control input u_p such that the closed-loop plant is internally stable and the plant output y_p tracks the output y_m of the LTI reference model[1]

$$y_m = W_m(s)[r] = k_m D_m^{-1}(s) N_m(s)[r] \quad (4.2)$$

for any UB, piecewise continuous reference input signal r.

In order to design a control law that meets the MRC objective, we need to impose certain additional conditions on the plant and the reference model. For reasons of convenience, these conditions are stated on the PDO factorization of the plant, rather than its state space descripion. Note, however, that such a statement is not restrictive since under Assumptions 3.1–3.2 the plant admits a left or right PDO factorization.

Our first condition concerns the high-frequency gain of the plant, as defined in Section 3.2, which should be bounded away from zero. Inherent in this

---
[1]The selection of the reference model as an LTI one is done for the simplicity of the analysis and design convenience; LTV reference models, satisfying similar assumptions, can be accommodated just as well.

condition is also the requirement that the relative degree of the plant should be constant.

4.1 Assumption: *The sign of the high-frequency gain, $k_p(t)$, is constant and $k_p(t)$ is smooth, UB and bounded away from zero. Without further loss of generality we assume that, there exists a constant c such that*

$$k_p(t) \geq c > 0$$

$\forall t \geq t_0$. ∎

By virtue of Assumptions 3.1–3.2 and 4.1, the I/O operator of the plant (4.1) admits PDO factorizations in the right form (P_R), i.e.,

$$y_p = k_p(t)N_p(s,t)D_p^{-1}(s,t)[u_p] \qquad (4.3)$$

or the left form (P_L), i.e.,

$$y_p = D_p^{-1}(s,t)N_p(s,t)k_p(t)[u_p] \qquad (4.4)$$

where $D_p(s,t), N_p(s,t)$ are monic PDO's with UB coefficients and of constant degree, denoted by n, m respectively. Furthermore, in (4.3) $D_p(s,t), N_p(s,t)$ are strongly right coprime while in (4.4) $D_p(s,t), N_p(s,t)$ are strongly left coprime PDO's in $[t_0, \infty)$.

Our second condition on the plant concerns the stability properties of $N_p^{-1}(s,t)$.

4.2 Assumption: $N_p^{-1}(s,t)$ *is an ES PIO with rate bounded from above by $-\alpha$, for some $\alpha > 0$* ∎

In other words, the state transition matrix associated with the differential equation $N_p(s,t)[x] = 0$, say $\Phi_N(.,.)$, is assumed to satisfy $\|\Phi_N(t,\tau)\| \leq ke^{-a(t-\tau)}$, for some positive constants k, a and all $t \geq \tau \geq t_0$. The differential equation $N_p(s,t)[x] = 0$ describes what is often referred to as the zero dynamics of the plant. Note that Assumption 4.2 is the LTV generalization of the minimum phase condition in the LTI MRC case.

Further, we assume that the reference model is selected to satisfy:

4.3 Assumption: $D_m(s)$ *and $N_m(s)$ are monic and Hurwitz (i.e., their inverses are ES PIO's) with $\deg[N_m(s)] \leq \deg[D_m(s)] - 1$.* ∎

4.4 Assumption: $W_m(s)$ *is designed so that $\deg[D_m(s)] \leq \deg[D_p(s,t)]$, $\deg[D_m(s)] - \deg[N_m(s)] = \deg[D_p(s,t)] - \deg[N_p(s,t)] (= n - m)$ and $k_m > 0$, i.e., the LTV plant and the reference model have the same (constant) relative degree $n^* \stackrel{\Delta}{=} n - m$.* ∎

4.3. TV MRC DESIGN

In the subsequent sections and chapters, we refer to Assumptions 4.1–4.4 as the MRC assumptions.

The requirement that the plant output y_p tracks the output of the reference model y_m for any UB reference input r implies that the control input u_p should be chosen such that the closed-loop I/O operator $r \mapsto y_p$ is equal to the I/O operator of the reference model. In the mathematical framework of Chapter 2, this problem can be cast in an elegant way as the solution of a Diophantine equation. The design and I/O properties of such an MRC law are presented in the following section.

4.3 TV MRC Design

Employing the techniques of [DLMS.80] we note that for the P_R plant, (4.3), a stabilizing controller is described by an I/O operator with a left factorization

$$u_p = N_2^{-1}(s,t)N_1(s,t)[y_p] \qquad (4.5)$$

for some PDO's $N_1(s,t), N_2(s,t)$. Furthermore, a controller with I/O operator

$$u_p = \frac{k_m}{k_p(t)} N_2^{-1}(s,t)N_1(s,t)[y_p] \qquad (4.6)$$

can also be used to stabilize the plant (4.4) in the P_L-form by selecting

$$N_2(s,t) = \frac{1}{k_m} \tilde{N}_2(s,t)N_p(s,t)k_m.$$

With this selection, the PDO $N_p(s,t)k_p(t)$ is directly cancelled by the controller PIO $N_2^{-1}(s,t)$; such a cancellation is permitted since $N_p^{-1}(s,t)$ is assumed to be exponentially stable and $k_p(t)$ is bounded away from zero. Thus, the plant I/O operator becomes effectively $D_p^{-1}(s,t)$, which belongs to the class of plants described by a P_R-form and can therefore be stabilized by a controller with a left factorization. In both cases, we must avoid the appearance of the controller PDO $N_1(s,t)$ as a PDO in the closed-loop I/O operator. This is achieved by writing the controller I/O operator in a proper stable factorization form as

$$u_p = N_2^{-1}(s,t)D(s)D^{-1}(s)N_1(s,t)[y_p]$$

where $D^{-1}(s)$ is ES and $\deg[N_1(s,t)] \leq \deg[D(s)] \leq \deg[N_2(s,t)]$ and then implementing $N_2^{-1}(s,t)D(s)$ in the forward path and $D^{-1}(s)N_1(s,t)$ in the feedback path.

The above discussion motivates the design of a MRC law as given by the following lemma.

4.5 Lemma: Consider the LTV plant (4.3) or (4.4) and suppose that Assumptions 4.1–4.4 hold. Further, consider the control input defined by

$$u_p = c_0(t)N_2^{-1}(s,t)D(s)\left[r + D^{-1}(s)N_1(s,t)y_p\right] \qquad (4.7)$$

where $D(s)$ is a monic, Hurwitz PDO[2] of degree $n-1$ and such that $N_m(s)$ is a right factor of $D(s)$, i.e., $D(s)$ can be expressed as $D_z(s)N_m(s)$ for some Hurwitz $D_z(s)$; $N_i(s,t)$, $i=1,2$ are PDO's of degree $n-1$ with $N_2(s,t)$ monic and $c_0(t)$ is a scalar function of time. Then, the controller parameters, i.e., $c_0(t)$ and the coefficients of $N_i(s,t)$, can be selected so that the closed-loop I/O operator $S_{ry} : r \mapsto y_p$ is BIBO stable and equal to the reference model I/O operator $W_m(s)$; furthermore, the controller parameters are smooth, UB functions of time and can be calculated by solving the algebraic design equations

$$N_2(s,t)c_0^{-1}(t)D_p(s,t) - N_1(s,t)k_p(t)N_p(s,t) = D_z(s)D_m(s)c_0(t)^{-1}N_p(s,t)$$

for a plant with I/O operator in the P_R-form (4.3), or

$$\tilde{N}_2(s,t)D_p(s,t) - k_m N_1(s,t) = k_m D_z(s)D_m(s)k_m^{-1}$$

$$N_2(s,t) = k_m^{-1}\tilde{N}_2(s,t)N_p(s,t)k_m$$

for a plant with I/O operator in the P_L-form (4.4). ▽▽

Proof: In Appendix IV.

Lemma 4.5 establishes the existence and provides a design procedure of a MRC law with smooth UB parameters which achieves the equality of the closed-loop plant and reference model I/O operators. Another important consideration in the design of control systems is the BIBO stability and I/O properties of the closed loop with respect to external inputs entering the loop at any break point between the controller and the plant, as shown in Fig. 4.1. Such external inputs are often used to model the effect of input disturbances (d_u), output disturbances (d_y), sensor noise (d_n) as well as effects of initial conditions. The properties of the I/O operators from any of these external inputs to any closed-loop signal, often referred to as *sensitivity* operators, are discussed in the following lemma.

4.6 Lemma: Consider the closed-loop system shown in Fig. 4.1 for which $c_0(t)$, $N_1(s,t)$, $N_2(s,t)$ are designed as in Lemma 4.5. Then the I/O depen-

---

[2] That is, $D^{-1}(s)$ is an ES PIO.

4.3. TV MRC DESIGN

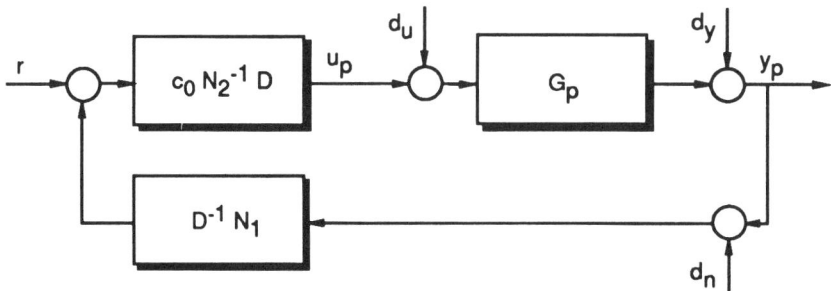

Figure 4.1: The TV MRC closed-loop plant.

dence of u_p, y_p on the external signals r, d_u, d_y, d_n is described by

$$\begin{bmatrix} u_p \\ y_p \end{bmatrix} = \begin{bmatrix} S_{ru} & S_{uu} & S_{yu} & S_{nu} \\ S_{ry} & S_{uy} & S_{yy} & S_{ny} \end{bmatrix} \begin{bmatrix} r \\ d_u \\ d_y \\ d_n \end{bmatrix} \quad (4.8)$$

where, omitting the PDO/PIO arguments for simplicity, the various sensitivity operators are given by:

1. For plants in the P_R-form, (4.3), $S_{ru} = [G_p^R]^{-1} W_m$, $S_{ry} = W_m$ and

$$\begin{aligned} S_{ru} &= D_p D_c^{-1} D & ; & \quad S_{ry} &= k_p N_p D_c^{-1} D \\ S_{uu} &= -1 + D_p D_c^{-1} N_2 c_0 & ; & \quad S_{uy} &= k_p N_p D_c^{-1} N_2 c_0^{-1} \\ S_{yu} &= S_{nu} = D_p D_c^{-1} N_1 & ; & \quad S_{ny} &= S_{yy} - 1 = k_p N_p D_c^{-1} N_1 \\ D_c &= N_2 c_0^{-1} D_p - N_1 k_p N_p &=& \quad D_z D_m c_0^{-1} N_p \end{aligned} \quad (4.9)$$

2. For plants in the P_L-form, (4.4), $S_{ru} = [G_p^L]^{-1} W_m$, $S_{ry} = W_m$ and

$$\begin{aligned} S_{ru} &= k_p^{-1} N_p^{-1} D_p D_c^{-1} k_m D & ; & \quad S_{ry} &= D_c^{-1} k_m D \\ S_{uu} &= -1 + k_p^{-1} N_p^{-1} D_p D_c^{-1} \tilde{N}_2 N_p k_p c_0 & ; & \quad S_{uy} &= D_c^{-1} \tilde{N}_2 N_p k_p \\ S_{yu} &= S_{nu} = k_p^{-1} N_p^{-1} D_p D_c^{-1} k_m N_1 & ; & \quad S_{ny} &= S_{yy} - 1 = D_c^{-1} k_m N_1 \\ D_c &= \tilde{N}_2 D_p - k_m N_1 = k_m D_z D_m k_m^{-1} & ; & \quad N_2 &= k_m^{-1} \tilde{N}_2 N_p k_m \end{aligned} \quad (4.10)$$

Furthermore, there exists $\delta_* > 0$ which in general depends on the stability margin of $D_m(s)$, $D(s)$ and $N_p(s,t)$ such that for any $\delta \in [0, \delta_*)$, and any initial time t_0, the various sensitivity operators are $L_p(\delta)$-stable, $p \in [1, \infty]$, uniformly in t_0; also, for the strictly proper sensitivity operators, the corresponding $g_{p,\delta}$ gains exist and are finite, uniformly in t_0. ▽▽

Proof: In Appendix IV.

At this point it must be emphasized that, although the MRC objective is originally motivated as a tracking problem, its practical application should also take into account the effect of possible disturbances and modeling errors on the plant output. As shown in Fig. 4.1, these disturbance effects are modeled by the external inputs d_u, d_y, d_n and their contribution to the total closed-loop response is completely determined by the respective sensitivity operators. These effects can be reduced, for example, by an appropriate selection of the reference model and the auxiliary filter to partially shape the appropriate sensitivity operators, for which $W_m(s)$ and $D(s)$ act as 'tuning' parameters (eqns. (4.9) and (4.10)). Tracking specifications can then be met by prefiltering the reference signal. Some additional sensitivity-shaping techniques are briefly discussed below where we consider the design of higher order TV MRC's which provide some additional degrees of freedom. We refer to such controllers as *over-parametrized TV MRC's*.

4.3.1 Design of Over-Parametrized TV MRC's

A direct consequence of Corollary 2.16 is that the TV MRC solution given by Lemma 4.5 is unique. Its uniqueness, however, relies heavily on our choice to design a TV MRC for which $\deg[N_2(s,t)] = \deg[N_1(s,t)] = n - 1$. This choice is by no means necessary and, in fact, a variety of higher order TV MRC's can be produced, all of them satisfying the basic MRC objective as stated above. Despite this obvious disadvantage due to the increased order and complexity, such a TV MRC may have some important advantages in applications, resulting from an additional flexibility of shaping its sensitivity operators to attenuate external disturbances and improve its robustness properties with respect to modeling errors. In the following we briefly discuss this issue by means of two examples.

4.7 Example: *Design of a Strictly Proper TV MRC:*

In this example we consider the design of a TV MRC whose I/O operator $N_2^{-1}(s,t)N_1(s,t)$: $y_p \mapsto u_p$ has relative degree $\ell \geq 1$. Such a design can be obtained as a direct extension of Lemma 4.5 by choosing

- $\deg[N_2(s,t)] = \deg[D(s)] = n + \ell - 1$;

- $\deg[N_1(s,t)] = n - 1$

where, as usual, $n = \deg[D_p(s,t)]$. While it is straightforward to verify that the results of Lemmas 4.5 and 4.6 are still valid —taking of course into account the different degrees of $N_2(s,t)$ and $D(s)$— Corollary 2.16 shows that this TV MRC solution is unique for any fixed $\ell \geq 0$.

4.3. TV MRC DESIGN

Among the advantages of a strictly proper TV MRC is an improved attenuation of high-frequency sensor noise, as it can be seen from the expressions for the sensitivity operator S_{ny} in (4.9) and (4.10). Furthermore, such a design is of particular interest in the adaptive control case, where it contributes to the simplification of the analysis and possibly improves the closed-loop robustness properties. ▽▽

4.8 Example: *Shaping the Sensitivity Operators:*

Let us consider an LTV plant in the P_R-form (4.3)[3] for which a TV MRC has been designed as in Lemma 4.5. Following the analysis of [DLMS.80], this basic TV MRC can now be used to derive a class of higher-order controllers which satisfy the same MRC objective and, in addition, allow some flexibility in shaping the characteristics of a closed-loop sensitivity operator.

Omitting the arguments for simplicity, let V, W, D_0, N_0 denote PDO's with UB coefficients such that

- $\deg[V] > \deg[W]$;

- V is a monic PDO and V^{-1} is ES;

- D_0 is monic and $N_0 D_p + D_0 k_p N_p = 0$ in $\mathbf{R}_+$ (the existence of D_0, N_0 is guaranteed by Assumptions 3.1–3.3 which imply the strong right coprimeness of $D_p, k_p N_p$);

Further, consider the control law

$$u_p = c_0 N_x^{-1} V D[r + (VD)^{-1} N_y y_p]$$

where

$$N_x = V N_2 + W N_0 c_0 \quad ; \quad N_y = V N_1 - W D_0$$

where N_2, N_1 and D are as in Lemma 4.5. It is quite straightforward to show that the results of Lemma 4.6 are valid for this control law as well, provided that in the expressions (4.9) we replace N_2, N_1, D, D_z by N_x, N_y, VD, VD_z respectively. Hence, $S_{ry} = W_m(s)$ and since V^{-1} is an ES PIO, this control law also satisfies the MRC objective. It does, however, introduce additional degrees of freedom in the TV MRC solution (namely the arbitrary PDO W) which can be used to alter the properties of the sensitivity operators, other than S_{ry} and S_{ru}.

Suppose, for example, that in addition to the TV MRC objective we would like to reject output disturbances d_y for which an internal model is available. That is, d_y satisfies the differential equation $L(s)[d_y] = 0$; typical examples are

---

[3] Similar techniques can be used for P_L-plants, (4.4), as well.

constant disturbances ($d_y = const.$, $L(s) = s$) or sinusoids ($d_y = \sin(w_0 t + \varphi)$, $L(s) = s^2 + w_0^2$). For this purpose, we may select W to have degree $\deg[L] - 1$ and satisfy a Diophantine equation of the form

$$XL + WD_0 = VN_1 + VD_z D_m / k_m$$

where X is some PDO of appropriate degree. It follows that this equation has a solution for W and X with UB coefficients provided that D_0 and L are strongly right coprime PDO's. Under this condition, the sensitivity operator S_{yy} in (4.9) takes the form

$$S_{yy} = k_m D_m^{-1} D_z^{-1} V^{-1} X L.$$

Thus, the contribution of d_y on the output, given by $S_{yy}[d_y]$, is exponentially decaying to zero.

The above procedure is nothing more than the TV version of the so-called Internal Model Principle (IMP) design which is frequently used to shape the closed-loop sensitivities in the LTI case. Also notice that in this example, the IMP design was facilitated by the controller structure and the assumption that N_p^{-1} is ES which may not be available in a general controller design. (For additional comments, see also the next chapter where an IMP design for P_L plants is considered.) ▽▽

In our development and study of the TV MRC problem so far we have dealt with only the I/O operator properties of the closed-loop plant, having tacitly assumed that all initial conditions are equal to zero. To account for arbitrary initial conditions in the closed-loop response we first need to specify the structure of the controller realization and then invoke Lemma 2.35 to establish the internal stability of the closed-loop plant. This problem is discussed in the following section.

4.4 Realization of the TV MRC and Internal Stability of the Closed-Loop Plant

The final issues to be resolved in the MRC problem for LTV plants, as posed in the previous section, are the state-space realization of the TV MRC compensators and the internal stability of the closed-loop plant. As mentioned above, the overall TV MRC scheme consists of two compensators, a cascade and a feedback one, with respective I/O operators $c_0(t) N_2^{-1}(s,t) D(s)$ and $D^{-1}(s) N_1(s,t)$. The realization of the two I/O operators in state-space follows the guidelines of the Examples 2.37 and 2.23. Note that the PDO's

4.4. REALIZATION AND INTERNAL STABILITY

$N_i(s,t)$, obtained as the solution of the respective Diophantine equations are in the left form and must be converted to the right form before realized in state-space (see Example 2.23). The state-space realization principles of the TV MRC scheme are summarized by the following corollary.

4.9 Corollary: *To realize in state-space the TV MRC scheme of Lemma 4.5 the plant output y_p and input u_p are used to generate a $(2n-1)$-dimensional auxiliary vector ω as follows:*

$$\dot{\omega}_1 = F\omega_1 + \theta_1 u_p \; ; \; \dot{\omega}_2 = F\omega_2 + \theta_2 y_p \; ; \; \omega_3 = \theta_3 y_p \quad (4.11)$$

$\omega = [\omega_1^T, \omega_2^T, \omega_3]^T \; ; \; \theta = [\theta_1^T, \theta_2^T, \theta_3]^T$ is a $(2n-1)$-dimensional parameter vector and $F \in \mathbf{R}^{(n-1)\times(n-1)}$ is a stable matrix with $\det(sI - F) = D(s)$. The input to the plant is then taken as

$$u_p = c_0(t)\left[g^T\omega + r\right] \quad (4.12)$$

where $g = [q^T, q^T, 1]^T$ is a constant vector such that (q^T, F) is an observable pair and c_0 is a scalar parameter. Then, there exists a control parameter vector $[\theta_*^T(t), c_{0*}(t)]^T$ such that the control law (4.11), (4.12) satisfies the TV MRC objective. Further, $[\theta_*^T(t), c_{0*}(t)]^T$ is UB and at least once differentiable with UB derivative, provided that the plant parameters are UB and possess a sufficiently large but finite number of UB derivatives. ▽▽

Proof: In Appendix IV.

Given the above realization of the controller, we are now in a position to describe the internal stability properties of the closed-loop plant. This result is a direct consequence of Lemmas 2.35 and 4.6 and establishes the well posedness of our solution to the MRC problem. That is, under the MRC assumptions, the TV MRC meets the MRC objective and guarantees the ES stability of the closed-loop plant for all uniform realizations of the plant and its BIBS/BIBO stability with respect to all exogenous signals and initial conditions. This result is made precise by the following theorem.

4.10 Theorem: *Under the conditions given in Lemma 4.5 and Corollary 4.9, the closed-loop plant is ES[4] and, therefore, BIBS stable for any external UB input. Furthermore, there exist constants $c, a > 0$ such that for all $t_0 \geq 0$ and any bounded initial conditions set at t_0, the ZIR of the closed-loop plant is bounded from above by $c\exp[-a(t-t_0)]$; c depends on the bound of the initial conditions and a depends on the location of the roots of $D(s), D_m(s)$ and the rate of exponential stability of $N_p^{-1}(s,t)$.* ▽▽

---
[4] Note that an LTV system is ES if the corresponding state transition matrix satisfies $\|\Phi(t,\tau)\| \leq ke^{-a(t-\tau)}$, for some positive constants k, a and all $t \geq \tau \geq t_0$.

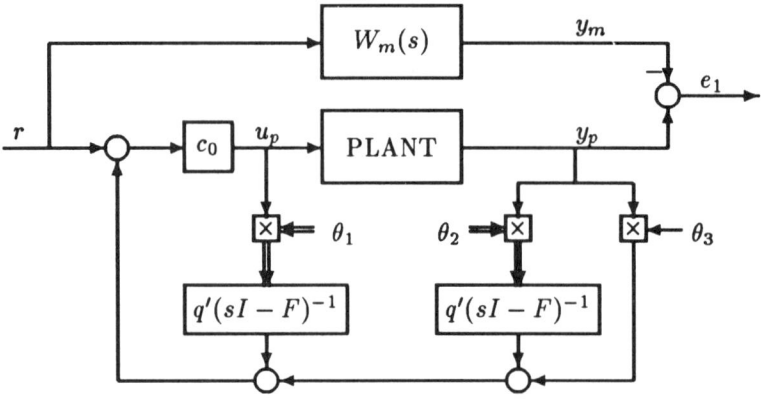

Figure 4.2: The MRC structure for LTV plants (TV MRC).

Proof: In Appendix IV.

Needless to say, the results of the above theorem are also valid for the over-parametrized TV MRC's presented in the previous section, provided that these controllers are realized according to Corollary 4.9; of course, some slight differences appear due to the increased order of the filters F and, for a strictly proper TV MRC, the absence of a direct throughput in the operator $N_2^{-1}(s,t)N_1(s,t) : y_p \mapsto u_p$, i.e., $\theta_3 = 0$.

4.11 Remark: It should be mentioned that a technical, but important, difference between the TV MRC for plants P_L and plants P_R is that the former involves direct cancellation of the plant PDO, while in the latter the plant PDO is cancelled by the closed-loop PIO, after an appropriate solution of the Diophantine equation. The implication of this observation is that the TV MRC state-space structure can be used in both TV MRC or TV PPC design of P_R plants by simply altering the choice of the desired closed-loop PIO. The same is not true for plants P_L for which a TV PPC design requires a different controller structure. ▽▽

Following the results of Lemma 4.5 and Corollary 4.9, the block diagram of the closed-loop plant with the TV MRC compensator is as shown in Fig. 4.2. We note, however, that the structure of this TV MRC scheme is essentially different from the standard one, shown in Fig. 4.3, which has been developed and widely used for LTI plants [N.V.78]. The difference between the two controller structures is due to the TV nature of the plant, for which the desired controller parameters are also TV, and it is discussed below.

4.4. REALIZATION AND INTERNAL STABILITY

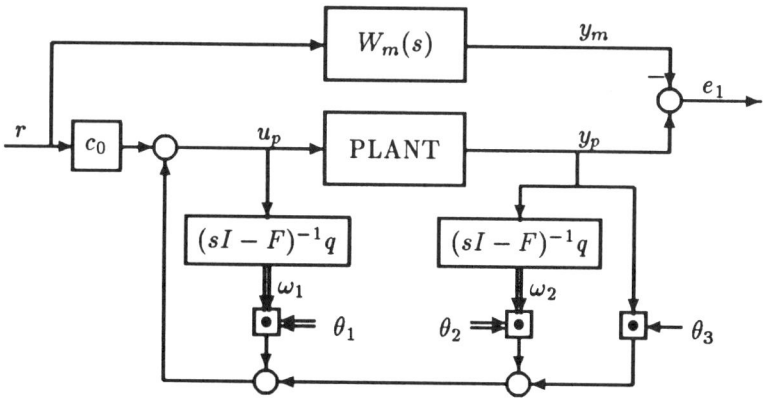

Figure 4.3: The standard (PW) MRC structure for LTI plants.

Using the standard MRC, the control input is generated by

$$\dot{\omega}_1 = F\omega_1 + qu_p \quad ; \quad \dot{\omega}_2 = F\omega_2 + qy_p \tag{4.13}$$

$$u_p = \theta^T \omega + c_0 r \tag{4.14}$$

where $\omega = [\omega_1^T, \omega_2^T, y_p]^T$; $\theta = [\theta_1^T, \theta_2^T, \theta_3]^T$ and F, q are as in Corollary 4.9. After some straightforward manipulations it follows that u_p can be written as

$$\begin{aligned} u_p &= \bar{M}_2(s,t)D^{-1}(s)[u_p] + \bar{M}_1(s,t)D^{-1}(s)[y_p] + \theta_3(t)y_p + c_0(t)r \\ &= D(s)\bar{M}_2^{-1}(s,t)\left[c_0(t)r + \bar{M}_1(s,t)D^{-1}(s)[y_p]\right] \end{aligned} \tag{4.15}$$

where $D(s) = \det(sI - F)$ and $\bar{M}_i(s,t)$ are PDO's such that $\bar{M}_i(s,t)D^{-1}(s) = \theta_i(sI - F)^{-1}q$. From the last equation, it becomes apparent that the cancellation of $D(s)$ and $D^{-1}(s)$ from the I/O operator $y_p \mapsto u_p$ of the controller is not possible in general, unless the PDO's $M_i(s,t)$ are TI. Hence, the matching condition cannot be expressed as a PDO equation and this controller does not satisfy the MRC objective in the general LTV case. Moreover, additional problems arise due to the location of $c_0(t)$, when $N_m(s)$ is not identically one. If, however, we assume that the plant is slowly TV, then we can perform the MRC design in a pointwise fashion that is, as if the plant were LTI at every time instant. We refer to the resulting MRC scheme as the pointwise MRC (PW MRC). The properties of the PW MRC as well as those of the TV MRC in the special but practically interesting case of slowly TV plants are discussed in the following section.

4.5 Slowly TV Plants

Our motivation to consider the case of slowly TV plants lies in the fact that in many practical applications the speed of the plant parameter variations is 'small' in some sense, e.g., the derivatives of the plant parameters are small for all $t \in \mathbf{R}_+$ or they are small in the mean. For reasons of clarity and ease of exposition, we first discuss the case where the plant parameter variations are slow, uniformly in time. The generalization of the results for slowly-in-the-mean TV plants or piecewise smooth plant parameters (e.g., 'jump' parameter variations), requires somewhat more involved arguments and is considered separately later in this chapter.

The intuitive idea behind the analysis of the slowly TV case is to consider a MRC law for the frozen plant (i.e., a pointwise design) for which the resulting frozen closed-loop plant is ES and then use Lemma 2.42 to establish stability for the TV closed-loop. In other words, since ES systems are robust with respect to 'small size' state dependent perturbations, the main difficulty is to establish the appropriate conditions which guarantee that the size of the closed-loop perturbation caused by the variation of the plant parameters is small in some sense. We begin with the analytically simpler case of LTV plants whose parameter variations are slow, uniformly in time. That is, we consider the LTV plant (4.1) which satisfies Assumptions 3.1–3.3 and, therefore, admits an I/O representation of the form (4.3) or (4.4). In addition, we assume that the plant parameters, denoted by the vector Θ_p, satisfy

4.12 Assumption: $\|\frac{d^i}{dt^i}\Theta_p(t)\| \leq \mu$, $\forall t \in \mathbf{R}_+$, $i = 1, 2, \ldots$ *for some 'small' parameter* $\mu \geq 0$. ∎

In Assumption 4.12 the parameter μ acts as a measure of the speed of variation of the plant parameters, in terms of the maximum magnitude of their derivatives. When $\mu = 0$ the plant parameters are constant and the plant is LTI. When μ is small, the plant parameters change slowly with time and the properties of the LTV plant can be approximated by the properties of the corresponding sequence of frozen LTI plants. For example, for sufficiently small μ, Assumption 3.2 is implied by

4.13 Assumption: *The PW (frozen) controllability and observability matrices of the triple $[A(t), b(t), c(t)]$ are strongly nonsingular.* ∎

Note that, given a plant in the form (4.1), Assumption 4.13 is easier to check than its TV counterpart 3.2, since the construction of the PW controllability and observability matrices is identical to the LTI case and does not involve the derivatives of the parameters. Similarly, the strong coprimeness of PDO's

4.5. SLOWLY TV PLANTS

can be simply checked by examining the coprimeness of the corresponding families of frozen polynomials (see Section 2.7, Lemma 2.41). The latter can serve to check the validity of Assumption 4.13 by examining the strong PW coprimeness of the polynomials in the frozen I/O representation of the plant, i.e., the numerator and denominator of the PW plant transfer function.

It should be pointed out that Assumption 4.12 is quite strong in the sense that a possibly large —but finite— number of derivatives of the plant parameters are required to be small. Later in this chapter we discuss how this assumption can be replaced by much weaker versions, e.g., $\|\dot{\Theta}_p(t)\|$ is small almost everywhere or small on the mean-square sense.

Let us now consider the MRC problem for the LTV plant (4.1) satisfying Assumptions 4.12, 4.13, in addition to 3.1, 3.3. Further, regarding the MRC Assumptions, since the plant is slowly TV, we may replace Assumption 4.2 by its pointwise counterpart:

4.14 Assumption: *The roots $\lambda_i(\tau)$ of the polynomial $N_{p\,\tau}(s) \in \{N_{p\,t}(s)\}_t$ satisfy*

$$Re[\lambda_i(t)] \leq -\alpha$$

$\forall t \geq 0$, *for some $\alpha > 0$.* ■

Notice that Assumption 4.14 is considerably easier to check than 4.2. The former is simply a condition on the roots of a family of polynomials while the latter requires the calculation of the state transition matrix associated with the differential equation $N_p(s,t)[x] = 0$. Again, for sufficiently small μ, if Assumption 4.14 holds then so does 4.2.

4.15 Corollary: *Consider the LTV plant 4.1 satisfying Assumptions 4.12, 4.13, 3.1, 3.3, 4.1, 4.14, 4.3, 4.4; then there exists $\mu_0 > 0$ such that the results of Lemma 4.5 hold for any $\mu \in [0, \mu_0)$.* ▽▽

Proof: Straightforward from Lemma 4.5, since there exists $\mu_0 > 0$ such that $\forall \mu \in [0, \mu_0)$, Assumptions 4.13, 4.14 imply 3.2, 4.2 (see Lemmas 2.33, 2.41 and 2.42). □□

4.16 Remark: It is possible to relax the strong controllability assumption to a 'strong stabilizability' one, without affecting the tracking performance of the TV MRC for P_L plants, or with an $O(\mu)$ error for P_R plants. This direction, however, is not pursued here as it imposes certain restrictive conditions on the internal structure of the plant (e.g., existence of Lyapunov transformations, structure of uncontrollable modes) in order to assure the internal stability of the closed-loop plant. ▽▽

The stability and tracking properties of the standard PW MRC, designed pointwise in time, can now be derived from those of the TV MRC and are given by the following theorem.

4.17 Theorem: *Consider a slowly TV plant satisfying Assumptions 4.12, 4.13, 3.1, 3.3, 4.1, 4.14, 4.3, 4.4. Then there exists $\mu_1 > 0$ such that the PW MRC (4.13), (4.14) guarantees that the closed-loop plant is ES and, therefore, BIBS stable for any $\mu \in [0, \mu_1)$. Furthermore, the plant output y_p satisfies*

$$y_p = W_m(s)[r] + L_1(s,t)[u_p] + L_2(s,t)[y_p]$$

where $L_1(s,t)$, $L_2(s,t)$ are strictly proper, ES I/O operators with UB parameters and rate that depends on $D^{-1}(s)$, $D_m^{-1}(s)$. In addition, there exists $\delta_* > 0$ such that for any fixed $\delta \in [0, \delta_*)$, $\gamma_{p,\delta}(L_i)$, $g_{p,\delta}(L_i) \leq O(\mu)$ where $p \in [1, \infty]$, $i = 1, 2$ and δ_* depends on the stability margin of $D^{-1}(s)$, $D_m^{-1}(s)$.
▽▽

Proof: In Appendix IV.

The significance of the above theorem is that for slowly TV systems, a controller can be designed and realized in a PW sense. In other words, for the purposes of a control system design, the plant can be assumed to be LTI at every time instant, something that simplifies a great deal all the necessary computations, at the expense of an $O(\mu)$-small deterioration in performance and stability margins (see also [S.A.91] for a more general and quantitative version of this result).

4.6 Non-Smooth Parameter Variations

Let us now consider the design of a MRC for the more general LTV plant

$$\begin{aligned}\dot{x} &= A_o(t)x + b_o(t)u_p + \tilde{A}(t)x + \tilde{b}(t)u_p \\ y_p &= c_o^T(t)x + \tilde{c}^T(t)x\end{aligned} \quad (4.16)$$

whose nominal and perturbation part satisfy Assumptions 3.4–3.6.

By virtue of Assumption 3.5, the nominal part of the plant admits an I/O operator with a PDO factorization as in (4.3) or (4.4), inside every interval (t_j, t_{j+1}). That is, for the nominal part of the plant (4.16)

$$\begin{aligned}\dot{x}_o &= A_o(t)x_o + b_o(t)u_p \\ y_p &= c_o^T(t)x_o\end{aligned}$$

we may write an I/O operator in the form (4.3) or (4.4), inside every interval (t_j, t_{j+1}). Further, let us assume that, for the nominal part of the plant, the

4.6. NON-SMOOTH PARAMETER VARIATIONS

MRC assumptions (4.1–4.4) are satisfied inside every interval, uniformly in j.[5] Then, the controller design procedures developed in Sections 4.3, 4.4 are applicable and we can design a TV MRC in a piecewise sense for each interval (t_j, t_{j+1}). Thus, the control input is determined from (4.7) or its state-space counterpart (4.12) in a piecewise sense.[6] In this section we apply this MRC input to the general LTV plant (4.16) and analyze the closed-loop stability properties.

Note that we should not expect the outcome of such a design procedure to meet the MRC objective exactly. A simple way to demonstrate this fact is to consider a plant with an output vector containing an infinite number of discontinuities. Since the plant output is a discontinuous function of time at an infinite number of points, it cannot be forced to track the continuous output of the reference model, no matter what control input is used (it is assumed, of course, that delta distributions are not admissible as control inputs).

Another issue of concern is that, even if the perturbation matrices $\tilde{A}, \tilde{b}, \tilde{c}$ are identically equal to zero, the piecewise I/O description of the plant is not complete since it does not include the necessary boundary conditions at each discontinuity point. Consequently, a controller which is designed based on such a description is not necessarily a stabilizing one. However, from Theorem 4.10 we have that the TV MRC guarantees the closed-loop exponential stability, whenever the pertinent assumptions are satisfied. In view of Corollary 3.8, a TV MRC could still preserve the closed-loop stability and achieve 'good' tracking in a mean-square sense, provided that on the average the discontinuities are not too frequent and the perturbation part is sufficiently small. The latter can be visualized by considering a plant with parameter discontinuities separated by large time intervals in the time scale of the closed-loop states. In this case, after a parameter discontinuity occurs the closed-loop plant output may depart from its reference trajectory and then converge to it exponentially fast. Thus, if the interval between two successive discontinuities is long enough, the plant output follows the reference trajectory for most of the time. This intuitive idea is made precise in the following theorem where, as in Chapter 3, ν is used to denote the average frequency of parameter discontinuities and μ' to denote the size of possible state perturbations caused, for example, by smooth approximations of non-differentiable parameters.

---

[5] For Assumption 4.2 in particular, uniformity in j means that there exist positive constants k, a, independent of j, such that the state transition matrix associated with the differential equation $N_p(s,t)[x] = 0$, say $\Phi_N(.,.)$, satisfies $\|\Phi_N(t,\tau)\| \leq ke^{-a(t-\tau)}$, for all $t \geq \tau \in (t_j, t_{j+1})$ and all j.

[6] At the points of discontinuity t_j the control input and the initial conditions of the filters in (4.12) can be arbitrary but UB. For example, a reasonable choice is the respective left limits.

4.18 Theorem: Consider the LTV plant (4.16) whose nominal and perturbation parts satisfy Assumptions 3.4–3.6. Further, suppose that the nominal part of the plant $[A_o, b_o, c_o]$ satisfies the MRC Assumptions 4.1–4.4 inside every interval (t_j, t_{j+1}), uniformly in j and the TV MRC control input is designed based on the nominal plant for each interval (t_j, t_{j+1}). Then there exist $\nu_0 > 0$, $\mu'_0 > 0$ such that $\forall \nu \in [0, \nu_0)$, $\forall \mu' \in [0, \mu'_0)$, the closed-loop plant is ES. Furthermore, there exist positive constants K, K', C such that

$$\int_{t_0}^{t_0+T} |y_p(t) - y_m(t)|^2 \, dt \leq C + K\nu T + K'\mu'T$$

for all $t_0, T \geq 0$. ▽▽

Proof: With the results of the previous sections and Corollary 3.8, the proof of the theorem is quite straightforward. Using a uniform realization of the plant and the appropriate controller realization, we obtain a state space representation for the closed-loop plant which is decomposed into a nominal and a perturbation part.

$$\dot{x}_c = A_c(t)x_c + b_c(t)r + \tilde{A}_c(t)x_c + \tilde{b}_c(t)r$$
$$y_p = c_c^T(t)x_c + \tilde{c}_c^T(t)x_c$$

where the subscript 'c' denotes the closed-loop states and nominal parameters and '~' denotes the closed-loop perturbation part, having the same properties as $\tilde{A}(t), \tilde{b}(t), \tilde{c}(t)$.

From Theorem 4.10 we have that the nominal part is ES inside all intervals (t_j, t_{j+1}) —with rate depending on $D_m^{-1}(s)$, $D^{-1}(s)$, $N_p^{-1}(s,t)$— and therefore, by Corollary 3.8, the nominal closed-loop plant is ES for sufficiently small ν. Hence, invoking Lemma 2.45, the overall closed-loop plant is ES $\forall \mu' \in [0, \mu'_0)$ and $\forall \nu \in [0, \nu_0)$, for some $\mu'_0 > 0$, $\nu_0 > 0$. Finally, the expression for $y - y_m$ is obtained by integrating the solutions of the respective differential equations:

$$y_m(t) = c_m(t)\Phi_m(t, t_j)x_m(t_j) + c_m(t)\int_{t_j}^{t} \Phi_m(t, \tau)b_m(\tau)r(\tau)\, d\tau \quad (4.17)$$

$$y_p(t) = [c_c(t) + \tilde{c}_c(t)]^T \Phi_c(t, t_j)x_c(t_j) \quad (4.18)$$
$$+ [c_c(t) + \tilde{c}_c(t)] \int_{t_j}^{t} \Phi_c(t, \tau)\left[\tilde{A}_c(\tau)x_c(\tau) + \tilde{b}_c(\tau)r(\tau)\right] d\tau$$
$$+ c_c(t) \int_{t_j}^{t} \Phi_c(t, \tau)b_c(\tau)r(\tau)\, d\tau + \tilde{c}_c(t) \int_{t_j}^{t} \Phi_c(t, \tau)b_c(\tau)r(\tau)\, d\tau$$

4.6. NON-SMOOTH PARAMETER VARIATIONS

where $\Phi_c(\cdot,\cdot)$ is the nominal state transition matrix associated with $A_c(t)$ and the subscript 'm' denotes the reference model. The result now follows from the equality of the I/O operators of $r \mapsto y_m$ and that of the nominal part of $r \mapsto y_p$ inside each interval (t_j, t_{j+1}), i.e.,

$$c_m(t) \int_{t_j}^{t} \Phi_m(t,\tau) b_m(\tau) r(\tau)\, d\tau = c_c(t) \int_{t_j}^{t} \Phi_c(t,\tau) b_c(\tau) r(\tau)\, d\tau$$

and the boundedness of x_c. Notice that when discontinuities appear in the plant description, the closed-loop I/O operator is equal to the desired one only inside the intervals between discontinuities and y_p may be discontinuous.
□□

An interesting observation is that y_p, given by (4.18), depends now on the complete closed-loop state vector which, in turn, depends on the zero dynamics of the plant. The appearance of the plant zero dynamics on the output has some important consequences in the selection of normalizing signals for the adaptive control case, considered in Chapter 7.

Finally, it should be mentioned that a somewhat more general (and more complicated) version of the above theorem can be established in the case where strong controllability/observability of the nominal plant is lost inside some time intervals of small-in-the-mean length. Such a result follows from Lemma 3.7 using similar arguments as in Theorem 4.18. We omit the details, however, since in our formulation such a situation can be treated by an appropriate selection of the nominal part of the plant. For example, during these intervals we may choose the nominal part as a fixed LTI system satisfying all the pertinent assumptions and incorporate the difference in the perturbation part.

4.6.1 Slowly TV Plants Revisited

In Section 4.5 we established the properties of a PW-designed MRC in the case of slowly TV plants under the quite restrictive Assumption 4.12. With this assumption we required that all the derivatives of the plant parameters, needed in our calculations, should be sufficiently small. It is intuitive, however, that since the frozen closed loop with a PW MRC is ES, the smallness of the first derivative of the plant parameters should suffice to guarantee the closed-loop stability. (Of course, the argument applies as well to the TV MRC structure with parameters designed for the frozen plant.) This observation is quantified in the following corollary.

4.19 Corollary: *Consider a slowly TV plant satisfying the MRC assumptions of Theorem 4.17 except that the plant parameters are only required to* be Lipschitz continuous UB functions of time and Assumption 4.12 is replaced by

4.20 Assumption: $\|\dot{\Theta}_p\|_\infty \leq \mu$. ∎

Then there exists $\mu_1 > 0$ such that

1. a PW-designed MRC guarantees that the closed-loop plant is ES and, therefore, BIBS stable for any $\mu \in [0, \mu_1)$;
2. the plant output y_p satisfies

$$y_p = W_m(s)[r] + L_1[x_c] + L_2[r]$$

where L_1, L_2 are strictly proper ES operators and x_c is the state vector of the closed-loop plant;

3. for any fixed $\mu'_1 \in [0, \mu_1)$ there exists $\delta_* > 0$ which depends on the rate of exponential stability of $D^{-1}(s)$, $D_m^{-1}(s)$, $N_p^{-1}(s,t)$ and the value of μ'_1 such that for any $\delta \in [0, \delta_*)$, $\gamma_{p,\delta}(L_i)$, $g_{p,\delta}(L_i) \leq O(\mu)$ where $\mu \in [0, \mu'_1]$, $i = 1, 2$ and $p \in [1, \infty]$. ▽▽

Proof: In Appendix IV.

Notice that, as in Theorem 4.18, the complete state vector of the closed-loop plant appears in the output which may now be affected by the zero dynamics of the plant. This is an important qualitative difference from the smooth-parameter case where the stability properties of the output perturbation operators were independent of the rate of exponential stability of $N_p^{-1}(s,t)$ (Theorem 4.17).

4.7 Examples

In the following examples we demonstrate the similarities and differences in the design and tracking performance of the TV and the PW MRC schemes.

In both examples and the corresponding simulations we consider the LTV plant

$$\frac{d^2}{dt^2}y_p + a_1 \frac{d}{dt}y_p + a_2 y_p = u_p \qquad (4.19)$$

where a_1, a_2 are TV parameters. The MRC objective is to make the plant output y_p track the output of the LTI reference model

$$[s^2 + 3s + 2]y_m = r \qquad (4.20)$$

where r is the reference input signal.

4.7. EXAMPLES

4.21 Example: *PW MRC Design for LTV Plants.* The standard PW MRC law, used for LTI plants, is shown in Fig. 4.3 and is summarized below for the plant (4.19).

$$\dot{\omega}_1 = -\omega_1 + u_p,\ \dot{\omega}_2 = -\omega_2 + y_p;\ u_p = \theta_1\omega_1 + \theta_2\omega_2 + \theta_3 y_p + r \quad (4.21)$$

where $\theta_1, \theta_2, \theta_3$ are the scalar controller parameters to be chosen for model-plant I/O matching. Using the properties of the PDO's and PIO's the closed-loop plant may be written as

$$[(s+1-\theta_1)(s+1)^{-1}(s^2 + a_1 s + a_2)(s+1) - \theta_2 - \theta_3(s+1)](s+1)^{-1} y_p = r \quad (4.22)$$

For I/O matching, i.e., $y_p = y_m$, $\forall r$ we should find $\theta_1, \theta_2, \theta_3$ such that

$$(s+1-\theta_1)(s+1)^{-1}(s^2 + a_1 s + a_2)(s+1) - \theta_2 - \theta_3(s+1) = (s+1)(s^2 + 3s + 2) \quad (4.23)$$

As shown in [N.V.78],[7] in the special case of LTI plants where a_1, a_2 are constants, there exist constants θ_{*i} for which (4.23) is satisfied. In the TV case, however, a_1, a_2 are functions of time (the argument 't' is dropped for simplicity). Since the PDO's with TV parameters do not commute with respect to multiplication, i.e., $(s+1)^{-1}(s^2 + a_1 s + a_2)(s+1) \neq s^2 + a_1 s + a_2$ in general, (4.23) cannot be solved for $\theta_1, \theta_2, \theta_3$ directly as it is done in the LTI case. Despite this difficulty, expressions for $\theta_1, \theta_2, \theta_3$ can be obtained by solving (4.23) pointwise in time, i.e., by solving

$$(s+1-\theta_1) \star (s^2 + a_1 s + a_2) - \theta_2 - \theta_3(s+1) = (s+1)(s^2 + 3s + 2) \quad (4.24)$$

for $\theta_1, \theta_2, \theta_3$, where $P(s,t) \star Q(s,t)$ denotes the pointwise multiplication of two PDO's.[8] The solution of (4.24) is then given as

$$\begin{aligned}
\bar{\theta}_{*1} &= a_1 - 3 \\
\bar{\theta}_{*2} &= a_1^2 - 4a_1 - a_1 a_2 + 3a_2 + 3 \\
\bar{\theta}_{*3} &= 4a_1 + a_2 - a_1^2 - 5
\end{aligned} \quad (4.25)$$

Using (4.25) in (4.22) the output y_p of the plant is expressed as

$$y_p = (s^2 + 3s + 2)^{-1} r + L(s,t) r \quad (4.26)$$

and the mismatch operator $L(s,t)$ is of the form

$$\begin{aligned}
L(s,t) &= -[(s^2 + 3s + 2)(s+1) + X(s,t)]^{-1} X(s,t)(s^2 + 3s + 2)^{-1} \\
X(s,t) &= (a_1 - 3)[\dot{a}_1 + (s+1)^{-1}(\dot{a}_2 - \dot{a}_1 - \ddot{a}_1)]
\end{aligned} \quad (4.27)$$

---
[7] Also follows from Lemma 4.5 and Corollary 4.9 when the plant is LTI.
[8] That is, $s \star a(t) = a(t) \star s = a(t)s$.

In general, the solution (4.25) does not satisfy (4.23), unless $\dot{a}_1 = \dot{a}_2 = 0$, $\forall t \geq 0$, i.e., the plant parameters are time-invariant. That is, due to the time variation of the plant parameters, (4.23) can only be solved approximately and the plant I/O operator cannot be made exactly equal to the I/O operator of the reference model with the standard MRC structure. By using an approximate solution of (4.23) (e.g. the pointwise one) we can guarantee stability and small tracking error provided that the plant parameters vary slowly with time. Notice that for slowly TV plants the mismatch operator $L(s,t)$ is stable; this can be seen by writing a, not necessarily minimal, state-space representation of $L(s,t)$ whose a state matrix has diagonal blocks the state matrices corresponding to $(s^2 + 3s + 2)(s+1)$ and $(s+1)$ and at least one of the off-diagonal blocks being $O(\dot{a}_1, \dot{a}_2, \ddot{a}_1)$. ▽▽

4.22 Example: *TV MRC Design for LTV Plants.* Using the TV MRC structure, shown in Fig. 4.2, the control law is given as:

$$\dot{\omega}_1 = -\omega_1 + \theta_1 u_p , \quad \dot{\omega}_2 = -\omega_2 + \theta_2 y_p ; \quad u_p = \omega_1 + \omega_2 + \theta_3 y_p + r \quad (4.28)$$

Thus, the closed-loop plant can be written as

$$(s+1)^{-1}\left[(s+1-\theta_1)[s^2 + a_1 s + a_2] - [\theta_2 + (s+1)\theta_3]\right] y_p = r \quad (4.29)$$

For model-plant following we should determine θ_1, θ_2, θ_3 such that $y_p = y_m$, $\forall r$, i.e.,

$$(s+1-\theta_1)[s^2 + a_1 s + a_2] - [\theta_2 + (s+1)\theta_3] = (s+1)(s^2 + 3s + 2) \quad (4.30)$$

Comparing (4.30) with (4.23) it is clear that in the former no PIO appears either in the left- or the right-hand side and therefore no commutativity problem arises. From (4.30) we obtain

$$(1 - \theta_1 + a_1 - 4)s^2 + [(1-\theta_1)a_1 + \dot{a}_1 + a_2 - \theta_3 - 5]s$$
$$+ (1-\theta_1)a_2 + \dot{a}_2 - \theta_2 - \dot{\theta}_3 - 2 - \theta_3 = 0$$

That is, for

$$\theta_1 = \theta_{*1} = a_1 - 3$$
$$\theta_2 = \theta_{*2} = a_1^2 - 4a_1 - a_1 a_2 + 3a_2 + 3 - 5\dot{a}_1 + 2a_1 \dot{a}_1 - \ddot{a}_1 \quad (4.31)$$
$$\theta_3 = \theta_{*3} = 4a_1 + a_2 - a_1^2 - 5 + \dot{a}_1$$

equation (4.30) is satisfied and the I/O operator of the closed-loop plant is equal to that of the reference model. We note that, in this case, the controller

4.7. EXAMPLES

parameters $\bar{\theta}_{*1}, \bar{\theta}_{*2}, \bar{\theta}_{*3}$ are well defined, bounded, smooth functions of time for any bounded, smooth functions a_1, a_2 (i.e., a_1, a_2 may be fast TV).

Further, to demonstrate the results of Theorem 4.17, let us consider again the PW MRC design of the previous example. The control law can then be written as

$$u_p = \bar{\theta}_{*1}(s+1)^{-1}[u_p] + \bar{\theta}_{*2}(s+1)^{-1}[y_p] + \bar{\theta}_{*3}y_p + r$$

or, operating on both sides by $(s+1)$

$$(s+1)[u_p] = \bar{\theta}_{*1}[u_p] + \bar{\theta}_{*2}[y_p] + (s+1)[\bar{\theta}_{*3}y_p] + (s+1)[r] + X_1$$

where $X_1 = \dot{\bar{\theta}}_{*1}(s+1)^{-1}[u_p] + \dot{\bar{\theta}}_{*2}(s+1)^{-1}[y_p]$. Furthermore, letting $\tilde{\theta}_i = \bar{\theta}_{*i} - \theta_{*i}$, the PW MRC law becomes

$$(s+1)[u_p] = \theta_{*1}u_p + \theta_{*2}y_p + (s+1)[\theta_{*3}y_p] + (s+1)[r] + X_1 + X_2$$

where $X_2 = \tilde{\theta}_1 u_p + \tilde{\theta}_2 y_p + (s+1)[\tilde{\theta}_3 y_p]$ and from (4.25) and (4.31), the $\tilde{\theta}_i$'s depend only on the derivatives of the plant parameters. It is now straightforward to verify that

$$y_p = W_m(s)[r] + L_1(s,t)[u_p] + L_2(s,t)[y_p]$$

where

$$L_1(s,t) = W_m(s)(s+1)^{-1}\{\dot{\bar{\theta}}_{*1}(s+1)^{-1} + \tilde{\theta}_1\}$$
$$L_2(s,t) = W_m(s)(s+1)^{-1}\{\dot{\bar{\theta}}_{*2}(s+1)^{-1} + \tilde{\theta}_2\} + W_m(s)\tilde{\theta}_3$$

It is now apparent that the perturbation operators $L_1(s,t), L_2(s,t)$ are strictly proper, ES and their $L_p(\delta)$ gains, $\delta < 1$, are $O[\dot{a}_1, \dot{a}_2, \ddot{a}_1]$. ▽▽

4.23 Simulations: Let us now simulate the response of the plant (4.19) with the PW and TV MRC for $r = 10\sin t$, $a_1 = -6$ and $a_2 = 2\sin\mu t$. The controller parameters are computed using (4.25) for the PW MRC structure (4.21) and (4.31) for the TV MRC structure (4.28). When $\mu = 0.1$ the PW MRC results in a bounded but nonzero tracking error while for the TV MRC the tracking error converges to zero (Fig. 4.4).

Increasing the value of μ to one, however, the tracking error for the PW MRC grows unbounded with time, as shown in Fig. 4.5.a, but the TV MRC still results in a tracking error that converges to zero, as shown in Fig. 4.5.b.

The unbounded closed-loop response with the PW MRC is due to the larger value of μ which results in an unstable mismatch operator $L(s,t)$, given by (4.27). We should note that for this example the solution θ_* for the TV MRC structure is the same as the pointwise solution $\bar{\theta}_*$ for the PW MRC

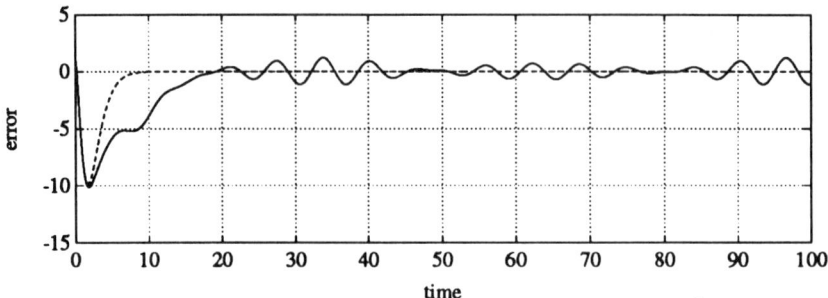

Figure 4.4: MRC tracking error response. Known, slowly TV plant parameters; $\mu = 0.1$. a. (—) PW MRC law; b. (- - -) TV MRC law.

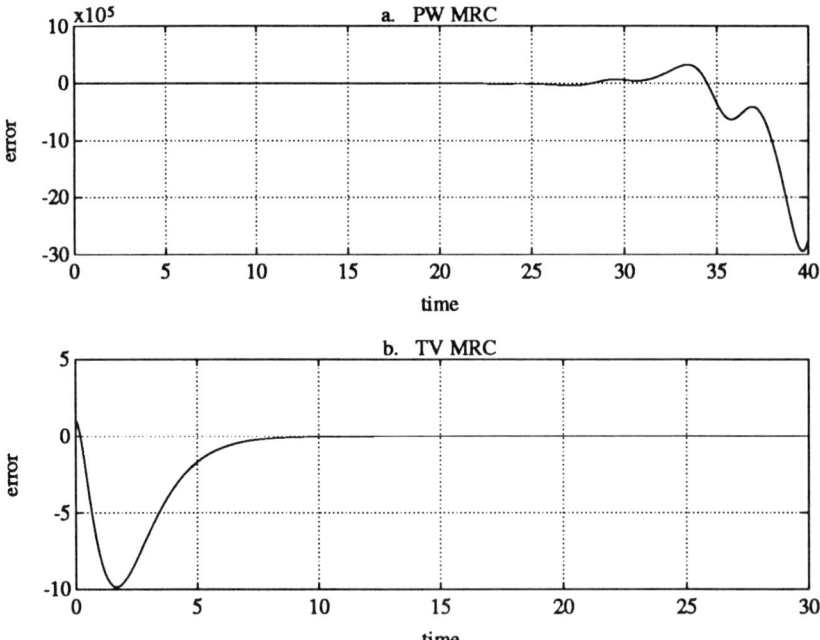

Figure 4.5: MRC tracking error response. Known, fast TV plant parameters; $\mu = 1$. a. PW MRC law: Unbounded response due to fast parameter variations; b. TV MRC law.

APPENDIX IV

structure, i.e., $\theta_* = \bar{\theta}_*$ when $a_1 = constant$. This demonstrates that the exact model-plant matching achieved by the TV MRC structure is a characteristic of the new structure and not only the specific choice of θ_*. ▽▽

Proof of Lemma 4.5:

Plant P_R: From (3.4) and (4.7) we obtain that the PIO's involved in the description of the closed-loop system are $D^{-1}(s)$, which is due to the internal cancellations in the control law and is ES by design and $D_c^{-1}(s,t)$, where

$$D_c(s,t) = N_2(s,t)c_0^{-1}(t)D_p(s,t) - N_1(s,t)k_p(t)N_p(s,t)$$

and which should be made ES by an appropriate selection of the controller parameters. Further, after some straightforward calculations, it follows that

$$y_p = k_p(t)N_p(s,t)D_c^{-1}(s,t)D(s)r \stackrel{\Delta}{=} S_{ry}(s,t)r \qquad (4.32)$$

where $S_{ry}(s,t) : r \mapsto y_p$ is the closed-loop plant I/O operator. To satisfy the model following objective we need to find $c_0(t)$ and the coefficients of $N_1(s,t)$, $N_2(s,t)$ such that $D_c^{-1}(s,t)$ is ES and $S_{ry}(s,t) = W_m(s)$. Substituting in (4.32) we get

$$N_2(s,t)c_0^{-1}(t)D_p(s,t) - N_1(s,t)k_p(t)N_p(s,t) = D_z(s)D_m(s)k_m^{-1}k_p(t)N_p(s,t) \qquad (4.33)$$

which also implies that the closed-loop PIO is ES, since $N_p^{-1}(s,t)$ is ES. For the PDO of the right- and left-hand side of (4.33) to have the same leading coefficient $c_0(t)$ should be selected as

$$c_0(t) = c_{0*}(t) \stackrel{\Delta}{=} k_p^{-1}(t)k_m \qquad (4.34)$$

Thus, invoking Corollary 2.16, the Diophantine equation (4.33) can be solved for $N_i(s,t)$ with smooth, UB coefficients. Furthermore, since $N_p^{-1}(s,t)k_p^{-1}(t)$, $D_m^{-1}(s)$, $D_z^{-1}(s)$ and $D(s)$ are all ES PIO's (note that $k_p(t)$ is bounded away from zero), it follows that the controller (4.7), with the so-selected parameters, also guarantees the BIBO stability of $S_{ry}(s,t)$.

Plant P_L: From (3.5) and (4.7) we obtain

$$\{N_2(s,t)c_0^{-1}(t)k_p^{-1}(t)N_p^{-1}(s,t)D_p(s,t) - N_1(s,t)\} y_p = D(s)r \qquad (4.35)$$

Letting

$$c_0(t) = c_{0*}(t) \stackrel{\Delta}{=} k_p^{-1}(t)k_m \ ; \ N_2(s,t) = k_m^{-1}\tilde{N}_2(s,t)N_p(s,t)k_m \qquad (4.36)$$

where $\tilde{N}_2(s,t)$ is a monic PDO of degree $n-m-1$ to be determined, the PIO's involved in the description of the closed loop are $D^{-1}(s)$, $k_p^{-1}(t)N_p^{-1}(s,t)$, due to direct cancellations and $D_c^{-1}(s,t)$ where

$$D_c(s,t) = \tilde{N}_2(s,t)D_p(s,t) - k_m N_1(s,t)$$

Further, the I/O operator $S_{ry}(s,t) : r \mapsto y_p$ is expressed as

$$S_{ry}(s,t) = D_c^{-1}(s,t)k_m D(s)r \tag{4.37}$$

Note that $k_p^{-1}(t)N_p^{-1}(s,t)$ has been cancelled directly by $N_2(s,t)$, an operation that is permitted since $N_p^{-1}(s,t)$ is ES and $k_p(t)$, $k_p^{-1}(t)$ are both UB. Next, in order to meet the control objective $\tilde{N}_2(s,t)$ and $N_1(s,t)$ are selected so that $S_{ry}(s,t) = W_m(s)$, i.e.,

$$\tilde{N}_2(s,t)D_p(s,t) - k_m N_1(s,t) = k_m D_z(s) D_m(s) k_m^{-1} \tag{4.38}$$

As before, the Diophantine equation (4.38) can be solved for $N_1(s,t)$ and $\tilde{N}_2(s,t)$ since both PDO's in the left- and right-hand side of (4.38) are monic and of the same degree $(2n-m-1)$ and the PDO's $D_p(s,t)$ and 1 are always right coprime. Hence, by Corollary 2.16, (4.38), (4.36) can be solved for $N_i(s,t)$ with smooth, UB coefficients. Furthermore, S_{ry} is BIBO stable since all the PIO's involved in its description are ES. □□

Proof of Lemma 4.6:

The part of the lemma regarding the expressions for the sensitivity operators is actually quite straightforward and follows from linearity and the definitions of N_2, N_1 and c_0 as the PDO's satisfying the Diophantine equations given in Lemma 4.5. For the rest, we must first verify that the various expressions make sense as an LTV system description, i.e., that we can write a state space representation for the given I/O operator.

For plants in the P_R-form the operators to be realized are of the form $PD_c^{-1}Q$ where D_c^{-1} is ES and $\deg[D_c] \geq \deg[P] + \deg[Q]$. From example 2.23, the state space realization of the operator $D_c^{-1}Q$ has an input matrix with the top $\deg[D_c] - \deg[Q] - 1$ elements being identically zero (if $\deg[D_c] = \deg[Q]$ then $D_c^{-1}Q$ has a throughput term and $\deg[P] = 0$). Consequently, the output of $D_c^{-1}Q$ can be differentiated at least $\deg[D_c] - \deg[Q] \geq \deg[P]$ times without requiring differentiation of its input. Hence, the output of the operator $PD_c^{-1}Q$ can be obtained as a linear combination of the states of $D_c^{-1}Q$ with weights depending on the coefficients of P and D_c as well as the derivatives of the latter. Furthermore, from the boundedness and smoothness

APPENDIX IV. 115

of the coefficients of the various PDO's, it is apparent that the overall STM is ES with the rate of D_c^{-1}.

Similar arguments apply in the case of plants in the P_L-form. In this case, however, we must also realize an operator of the form $D_1^{-1}QD_2^{-1}$ where D_1^{-1}, D_2^{-1} are ES and $\deg[Q] \leq \deg[D_1]+\deg[D_2]$. The easiest way to perform such a realization comes from the fact that the set of PDO's with smooth coefficients is an associative ring (non-commutative, though). Hence, we can apply the Euclidean algorithm to write $Q = Q_1Q_2 + R$ where $\deg[Q_1] \leq \deg[D_1]$, $\deg[Q_2] \leq \deg[D_2]$ and $\deg[R] \leq \deg[Q_1]$. Hence, we can realize $D_1^{-1}QD_2^{-1}$ as a cascade combination of $Q_2D_2^{-1}$ and $D_1^{-1}Q_1$ plus a cascade combination of D_2^{-1} and $D_1^{-1}R$. Again, the overall STM is ES with rate which depends on D_1^{-1} and D_2^{-1}.

It is also interesting to observe that the form of the various PDO's does not affect the result since their parameters are smooth and UB and they can be converted to the appropriate form. That is, denoting by D_R, D_L the right and left form of the same PDO, if the ODE $D_R x = 0$ is ES then $D_L x = 0$ is also ES. An alternative way of verifying this is by considering the state space descriptions of D_R^{-1} (controllable canonical form) and D_L^{-1} (observable canonical form). It follows trivially that they are both completely controllable and observable and thus uniform realizations of the same impulse response and topologically equivalent.

Finally, the boundedness and smoothness of the coefficients of the various PDO's together with the ES property of the realizations of the sensitivity operators implies that the associated impulse response $h(t,\tau)$ of the strictly proper part is UB and there exist constants $k, a > 0$ such that $\|h(t,\tau)\| \leq k\exp[-a(t-\tau)]$ for all $t \geq \tau \geq 0$ where a depends on the PIO's in the I/O description of the operator. Thus, $L_p(\delta)$ stability follows from Corollary 2.52 with $\delta_* = a$; the same corollary also shows that for the strictly proper sensitivity operators, the corresponding $g_{p,\delta}$ gains also exist and are finite.

□□

Proof of Corollary 4.9:

From (4.11), (4.12) we have

$$u_p = c_0(t)\left\{D^{-1}(s)\bar{N}_2(s,t)[u_p] + D^{-1}(s)\bar{N}_1(s,t)[y_p] + \theta_3(t)y_p + r\right\} \quad (4.39)$$

where

$$\begin{aligned}\bar{N}_2(s,t) &= [Q_1(s),\ldots,Q_{n-1}(s)]\theta_1(t)\\ \bar{N}_1(s,t) &= [Q_1(s),\ldots,Q_{n-1}(s)]\theta_2(t)\end{aligned} \quad (4.40)$$

$D(s) = det(sI - F)$ and $Q_j(s)$, $j \in \overline{n-1}$ are TI PDO's such that $q^{\mathsf{T}}(sI - F)^{-1} = D^{-1}(s)[Q_1(s), \ldots, Q_{n-1}(s)]$. From (4.39) and (4.7) $\bar{N}_i(s,t)$ should satisfy

$$\bar{N}_2(s,t) = [D(s) - N_2(s,t)]c_0^{-1}(t) \; ; \; \bar{N}_1(s,t) = N_1(s,t) - D(s)\theta_3(t) \quad (4.41)$$

Note that $\deg[D(s) - N_2(s,t)] = n-2 = \deg[\bar{N}_2(s,t)]$ and that $\deg[N_1(s,t) - D(s)\theta_3(t)] = \deg[\bar{N}_1(s,t)] = n-2$ implies that $\theta_{*3}(t)$ is equal to the leading coefficient of $N_1(s,t)$. Thus, from (4.40) and (4.41), $\theta_i(t)$ should satisfy

$$[Q_1(s), \ldots, Q_{n-1}(s)]\theta_i(t) = \bar{N}_i(s,t) \;,\; i = 1, 2 \quad (4.42)$$

Further, by expressing the PDO's $\bar{N}_i(s,t)$ in the right form, (4.42) yields

$$\theta_{*i}(t) = Q^{-1}[\bar{n}_{i1}(t), \bar{n}_{i2}(t), \cdots, \bar{n}_{i\,n-1}(t)]^{\mathsf{T}} \quad (4.43)$$

where $Q = [\underline{q}_1^{\mathsf{T}}, \ldots, \underline{q}_{n-1}^{\mathsf{T}}]^{\mathsf{T}}$, $\underline{q}_j$ are vectors of the coefficients of s^{n-1-j} of $Q_j(s)$ and $\bar{n}_{ij}(t)$ are the coefficients of s^{n-1-j} of $\bar{N}_i(s,t)$ in the right PDO form, $i = 1, 2$, $j = 1, \ldots, n-1$, obtained from (4.41). Note that Q^{-1} exists due to the observability of (q^{T}, F) [Kai.80]. Finally, the boundedness and differentiability properties of $\theta_*(t)$ follow by inspection, directly from (4.43) and the respective Diophantine equation (4.33) or (4.38). □□

Proof of Theorem 4.10:

Observe first that the control law can be put in the form Σ_c of Lemma 2.35 with $u_1 = u = u_p$ and the plant is uniformly realizable. It remains to show that for $r = 0$ and any bounded initial conditions at t_0, $\forall t_0 \geq 0$, the signals u_p, y_p decay exponentially fast, uniformly in t_0.

Initial conditions in the auxiliary filters at $t = t_0$ can be introduced directly as a contribution of a term $\Phi_F(t,t_0)w_0$ in the filter states where $\Phi_F(\cdot,\cdot)$ is the STM of F (a matrix exponential if F is TI). It is straightforward to verify that the effect of all such terms can be expressed as an exponentially decaying external input, denoted by w_e, entering the closed-loop system at the reference input node (see Fig. 4.1); consequently, their contribution to u_p, y_p is characterized by the sensitivity operators S_{ru} and S_{ry}. This approach, however, is not used for initial conditions associated with the plant states since the latter is not necessarily ES.

To account for the plant initial conditions we first note that bounded initial conditions of any uniform realization correspond to bounded initial conditions of a canonical form. Further, since the plant is uniformly completely

APPENDIX IV.

controllable, for any bounded initial conditions $x(t_0)$ there exists $d_c > 0$ and a bounded $\hat{u}(t)$; $t \in [t_0, t_0 + d_c]$ such that [S.A.68]

$$\int_{t_0}^{t_0+d_c} \Phi(t,\tau) B(\tau) \hat{u}(\tau)\, d\tau = \Phi(t_0 + d_c, t_0) x(t_0)$$

where $\|\hat{u}\| \leq a_1(d_c, \|x(t_0)\|)$ and $a_1(\cdot)$ is a constant determined by its arguments. Let us now denote by $x(t)$ the system response with input u_p and initial conditions $x(t_0)$ and by $x_*(t)$ the response with input $u_p + \hat{u}$, for any u_p. Since $\hat{u} = 0$, $\forall t > t_0 + d_c$, and employing the linearity assumption, we obtain that $x(t) = x_*(t)$, $\forall t > t_0 + d_c$. Also, since the plant is in a bounded realization, $\|x(t) - x_*(t)\|$ is UB.

In other words, under uniform complete controllability, the fictitious input $\hat{u}$, emulates the effect of arbitrary initial conditions exactly after some finite time, before which all signals are bounded. The net result of the procedure is that the initial conditions of the plant are translated to external inputs which can be manipulated using I/O techniques. These external inputs are $\hat{u}$, entering at the node of input disturbances (d_u in Fig. 4.1), and $\hat{y}$ which is due to the difference between $x(t)$ and $x_*(t)$ and is a bounded, finite function;[9] $\hat{y}$ enters the loop at the node of output disturbances (d_y in Fig. 4.1).

From Lemma 4.6, it follows that the plant input and output satisfy

$$u_p = S_{ru}[\hat{u}] + S_{yu}[\hat{y}] + S_{ru}[w_e] \tag{4.44}$$

$$y_p = S_{yy}[\hat{y}] + W_m[w_e] + S_{uy}[\hat{u}] \tag{4.45}$$

where the various sensitivity operators are given in Lemma 4.6 for either P_R or P_L plants. We observe now that w_e is exponentially decaying with the rate of $D^{-1}(s)$ and therefore $w_e \in L_\infty(\delta)$ where $\delta \in [0, \delta_*)$ and δ_* is a positive constant as in Lemma 4.6. Moreover, since $\hat{u}, \hat{y}$ are UB functions of bounded support, $\hat{u}, \hat{y} \in L_\infty(\delta)$. Using Lemma 4.6, the various sensitivity operators are $L_\infty(\delta)$ stable and therefore, $u_p, y_p \in L_\infty(\delta)$ and there exist constants $k, a > 0$, $a < \delta_*$ and k depending on d_c and $x(t_0)$, w_0, such that

$$\|[u_p(t), y_p(t)]\| \leq k \exp[-a(t-t_0)]$$

for any initial time $t_0 \geq 0$. Thus, invoking Lemma 2.35 the proof follows. Notice that since ES implies BIBS stability, the closed-loop system is internally stable for any external UB input and any initial conditions, uniformly in t_0. Moreover, this property is shared by any other uniform realization of the plant I/O operator. □□

---

[9]That is, it is a function of finite (compact) support, vanishing outside a finite interval.

Proof of Theorem 4.17:

The proof of the theorem relies on the use of swapping techniques to express the PW MRC the control input as

$$u_p = c_0(t)N_2^{-1}(s,t)N_1(s,t)[y_p] \qquad (4.46)$$
$$+ c_0(t)N_2^{-1}(s,t)D(s)\left[r + \hat{L}_1(s,t)[u_p] + \hat{L}_2(s,t)[y_p]\right]$$

where $N_i(s,t)$ are the PDO's corresponding to the TV MRC and $\hat{L}_i(s,t)$ are proper, stable perturbation operators whose gains are $O(\mu)$. Of course, for the TV MRC design to make sense, μ should be in $[0, \mu_0)$ (see Corollary 4.15). To demonstrate the application of these techniques we derive (4.46) in two ways: one using an I/O approach and property P4 of PDO's and one using a state-space approach and Lemma 2.59.

- *(I/O approach:)* Let $\bar{M}_i(s,t)$ denote the controller PDO's and $\bar{\theta}_3(t)$ the pure gain feedback, corresponding to the standard PW MRC, which are obtained from a pointwise design. Using Property P4 of the PDO's from Chapter 2, the plant input can be written as

$$u_p = c_0(t)r + D^{-1}(s)D(s)\bar{M}_2(s,t)D^{-1}(s)[u_p]$$
$$+ D^{-1}(s)D(s)\bar{M}_1(s,t)D^{-1}(s)[y_p] + \bar{\theta}_3(t)y_p$$

and, consequently,

$$D(s)c_0^{-1}(t)[u_p] = c_0^{-1}(t)\bar{M}_2(s,t)[u_p] + c_0^{-1}(t)\bar{M}_1(s,t)[y_p]$$
$$+ D(s)c_0^{-1}(t)\bar{\theta}_3(t)[y_p] + D(s)[r] \qquad (4.47)$$
$$+ X_2(s,t)D^{-1}(s)[u_p] + X_1(s,t)D^{-1}(s)[y_p]$$

where $c_0(t) = k_m/k_p(t)$ and

$$X_i(s,t) = D(s)c_0^{-1}(t)\bar{M}_i(s,t) - c_0^{-1}(t)\bar{M}_i(s,t)D(s) \; ; \quad i = 1,2$$

Thus, using the properties of PDO's, $\deg[X_i(s,t)] \leq 2n - 4$ and the coefficients of $X_i(s,t)$, say $\Theta_X(t)$, are of order of the derivatives of $c_0^{-1}(t)$ and $\Theta_M(t)$, the coefficients of $\bar{M}_i(s,t)$. Further, in a PW design $\bar{\theta}_3(t)$ and $\Theta_M(t)$ are obtained from the solution of a PW Diophantine equation which, in turn, can be expressed as $S_0(t)\bar{\Theta}_M(t) = a_0(t)$ where $S_0(t)$ is the PW Sylvester matrix of $D_p(s,t)$ and $k_p(t)N_p(s,t)$, and $S_0(t)$, $a_0(t)$ depend only on the plant parameters $\Theta_p(t)$ and not on their derivatives.[10] Since, under Assumption 4.13, $S_0(t)$ is strongly

---
[10] Note that in a pointwise approach there is no distinction between the P_R and P_L forms.

APPENDIX IV. 119

nonsingular for all t, the PW controller parameters are smooth, UB functions of time and therefore,

$$\|\Theta_X(t)\| \leq O(\mu). \tag{4.48}$$

Further, (4.47) can be rewritten as

$$\begin{aligned}
D(s)c_0^{-1}(t)u_p &= \bar{N}_2(s,t)u_p + \bar{N}_1(s,t)y_p + D(s)\bar{\theta}_{*3}(t)y_p + D(s)r \\
&\quad + X_2(s,t)D^{-1}(s)u_p + X_1(s,t)D^{-1}(s)y_p \\
&\quad + Y_2(s,t)u_p + Y_1(s,t)y_p
\end{aligned} \tag{4.49}$$

where $\bar{N}_i(s,t), \bar{\theta}_{*3}(t)$ correspond to the TV MRC design and

$$Y_2(s,t) = c_0^{-1}(t)\bar{M}_2(s,t) - \bar{N}_2(s,t)$$

$$Y_1(s,t) = \underbrace{\left\{c_0^{-1}(t)\bar{M}_1(s,t) - \bar{N}_1(s,t)\right\}}_{Y_{11}(s,t)} + \underbrace{\left\{D(s)[c_0^{-1}(t)\bar{\theta}_3(t) - \bar{\theta}_{*3}(t)]\right\}}_{Y_{12}(s,t)}$$

After a straightforward comparison of the Diophantine equations for the TV and PW MRC, it follows that $\deg[Y_2(s,t)] \leq n-2$, $\deg[Y_{11}(s,t)] \leq n-2$, $\deg[Y_{12}(s,t)] \leq n-1$ and

$$\|\Theta_Y(t)\| \leq O(\mu) \tag{4.50}$$

from which we obtain (4.46) with the input perturbation operators being at least proper, ES with rate depending on $D(s)$ and given by

$$\hat{L}_1(s,t) = D^{-1}(s)X_2(s,t)D^{-1}(s) + D^{-1}(s)Y_2(s,t)$$

$$\hat{L}_2(s,t) = D^{-1}(s)X_1(s,t)D^{-1}(s) + D^{-1}(s)Y_1(s,t)$$

- *(State-Space approach:)* Let $c_0(t)$ and $\bar{\theta}_i(t)$, $i = 1, 2, 3$, denote the controller parameters of the PW MRC. Then, the control input is given by

$$u_p = G(s)[u_p]\bar{\theta}_1(t) + G(s)[y_p]\bar{\theta}_2(t) + \bar{\theta}_3(t)y_p + c_0(t)r$$

where $G(s)$ denotes the operator

$$G(s): u \mapsto q^T(sI - F^T)^{-1}[u]$$

and F is the state matrix and q is the input vector of the state-space representation of the auxiliary filters. Letting $G'(s)$ denote the operator

$$G'(s): u \mapsto (sI - F^T)^{-1}[u]$$

and using Lemma 2.59 the term $v_1 = G(s)[u_p]\bar{\theta}_1(t)$ can be written as

$$v_1 = c_0(t)G(s)[u_p\bar{\theta}_1/c_0] + c_0(t)G(s)\left[G'(s)[u_p](\frac{d}{dt}[\bar{\theta}_1/c_0])\right]$$

It now follows from the PW MRC assumptions that $\|\frac{d}{dt}[\bar{\theta}_1/c_0]\|_\infty = O(\mu)$ (similarly for $\bar{\theta}_2$ and y_p).

Further, let $\theta_i(t)$ denote the controller parameters of the TV MRC realized with a state matrix F^T and output vector q^T, i.e., the control law for the TV MRC would be given by

$$u'_p = c_0(t)\left\{G(s)[u'_p\theta_1] + G(s)[y'_p\theta_2] + \theta_3(t)y'_p + r\right\}$$

Next, define $\tilde{\theta}_i = \theta_i - \bar{\theta}_i/c_0$. From the respective Diophantine equations for the PW and TV MRC and Assumption 4.12 it follows that $\|\tilde{\theta}_i\|_\infty = O(\mu)$.

Thus, the PW MRC control input can be expressed as

$$\begin{aligned} u_p &= c_0(t)\{G(s)[u_p\theta_1] + G(s)[y_p\theta_2] + \theta_3(t)y_p + r\} \\ &\quad + \hat{L}_1(s,t)[u_p] + \hat{L}_2(s,t)[y_p] \end{aligned}$$

from which (4.46) follows with the input perturbation operators $\hat{L}_i(s,t)$ being at least proper, ES with rate depending on F and given by

$$\hat{L}_1(s,t)[u_p] = c_0(t)G(s)\left[G'(s)[u_p](\frac{d}{dt}[\bar{\theta}_1/c_0])\right] - c_0(t)G(s)[u_p\tilde{\theta}_1]$$

$$\begin{aligned} \hat{L}_2[y_p] &= c_0(t)G(s)\left[G'(s)[y_p](\frac{d}{dt}[\bar{\theta}_2/c_0])\right] \\ &\quad - c_0(t)G(s)[y_p\tilde{\theta}_2] - c_0(t)\tilde{\theta}_3(t)y_p \end{aligned} \quad (4.51)$$

In other words, the closed loop of the LTV plant with the PW MRC can be effectively described as the TV MRC loop perturbed by two dynamic operators $\hat{L}_1$ and $\hat{L}_2$, with respective inputs u_p and y_p and outputs entering the closed-loop system at the node of the reference input (see Fig. 4.1). Furthermore, it is quite straightforward to verify using the results of Chapter 2 that, for any fixed $\delta \in [0, \delta_*)$ (δ_* as in the Theorem) and $p \in [1, \infty]$, $\gamma_{p,\delta}[\hat{L}_i(s,t)]$ and $g_{p,\delta}[\hat{L}_1(s,t)]$ are $O(\mu)$.

Thus, the PW MRC loop without any external inputs is described by

$$\begin{bmatrix} u_p \\ y_p \end{bmatrix} = \begin{bmatrix} S_{ru} \\ S_{ry} \end{bmatrix} v \; ; \; v = [\hat{L}_1(s,t), \hat{L}_2(s,t)] \begin{bmatrix} u_p \\ y_p \end{bmatrix}$$

and since $S_{ry} = W_m(s)$ we have that

$$y_p = W_m(s)[r] + L_1(s,t)[u_p] + L_2(s,t)[y_p]$$

APPENDIX IV.

where $L_i(s,t) = W_m(s)\hat{L}_i(s,t)$ with the properties stated in the Theorem.[11]

Further, for any $\delta \in [0, \delta_*)$ and $p \in [1, \infty]$ a sufficient condition for the closed-loop $L_p(\delta)$-stability is [D.V.75]

$$\gamma_{p,\delta}\begin{bmatrix} S_{ru} \\ S_{ry} \end{bmatrix} \gamma_{p,\delta}[\hat{L}_1(s,t), \hat{L}_2(s,t)] < 1 \qquad (4.52)$$

Since $\gamma_{p,\delta}[\hat{L}_i(s,t)] \leq O(\mu)$ it follows that there exists $\bar{\mu}_1 > 0$ ($\bar{\mu}_1 \leq \mu_0$), depending on the particular choice of p and δ, such that for any $\mu \in [0, \bar{\mu}_1)$ the above inequality is satisfied. Moreover, from Lemma 2.51, if the closed-loop plant (without initial conditions) is $L_p(\delta)$ stable, it is also $L_p(\hat{\delta})$ stable for any $\hat{\delta} \leq \delta$.

To complete the proof we need to show that there exists $\mu_1 > 0$ such that $\forall \mu \in [0, \mu_1)$ the closed-loop system is ES. Let $\delta \in (0, \delta_*)$ and choose, for simplicity, μ_1 such that (4.52) is satisfied for $p = \infty$ and all $\mu \in [0, \mu_1)$. Since the closed loop with the TV MRC is ES, in the presence of arbitrary but bounded initial conditions the truncations of u_p, y_p at t satisfy

$$\|[u_p, y_p]_t\|_{\infty, \delta} \leq \gamma_{\infty, \delta}(S_r)\gamma_{\infty, \delta}(\hat{L})\|[u_p, y_p]_t\|_{\infty, \delta} + \beta \sup_{t_0 \leq \tau \leq t} \left[e^{\delta \tau} e^{-a(\tau - t_0)} \right]$$

where β depends on the size of the initial conditions at t_0 (u_p and y_p are taken as 0 for $t < t_0$), $-a \leq -\delta_*$ is the exponential rate of decay of the closed loop with the TV MRC and

$$S_r = \begin{bmatrix} S_{ru} \\ S_{ry} \end{bmatrix} \; ; \; \hat{L} = [\hat{L}_1(s,t), \hat{L}_2(s,t)]$$

Hence, for any $\mu \in [0, \mu_1)$,

$$e^{\delta t}\|[u_p(t), y_p(t)]\| \leq \|[u_p, y_p]_t\|_{\infty, \delta} \leq \frac{\beta e^{\delta t_0}}{1 - \gamma_{\infty, \delta}(S_r)\gamma_{\infty, \delta}(\hat{L})} < \infty; \; \forall t \geq t_0 \geq 0$$

Thus, $\|[u_p(t), y_p(t)]\| \leq k e^{-\delta(t-t_0)}$ for some constant k independent of t_0. The proof now follows from Lemma 2.35 and the observations that the PW MRC satisfies the general controller structure requirements of that lemma and under the given assumptions the plant is uniformly completely observable for $\mu < \mu_1$.

It should be pointed out that in an attempt to reduce the conservatism of the result, we should find the supremum of μ_1 with respect to δ and the

---

[11] Notice that these expressions are slightly different than the ones obtained in [T.I.87] in that the perturbation operators $L_i(s,t)$ do not depend on the zero-dynamics of the plant; this result is of value in the adaptive control case and its derivation relies heavily on the existence and properties of the TV MRC solution given by Lemma 4.5.

selection of the —possibly weighted— $L_p(\delta)$-norm. For the former, it is easily obtained that the supremum occurs as $\delta \to 0$ which, of course, implies that increased speeds of parameter variations tend to decrease the exponential stability margin. For the latter, although the selection of p is not straightforward due to the lack of explicit formulas for the $L_p(\delta)$-gains of TV operators, the result of the theorem is still valid with some modifications in the proof. That is, for sufficiently small μ such that (4.52) holds, the closed-loop system is $L_p(\delta)$-stable. Furthermore, for $t > t_0$, the derivatives of u_p, y_p can be expressed as

$$\dot{u}_p = sS_{ru}\left\{\hat{L}_1(s,t)[u_p] + \hat{L}_2(s,t)[y_p]\right\} + \varepsilon$$

$$\dot{y}_p = sS_{ry}\left\{\hat{L}_1(s,t)[u_p] + \hat{L}_2(s,t)[y_p]\right\} + \varepsilon$$

with ε denoting exponentially decaying terms due to the initial conditions. Since $sS_{ru}\hat{L}_2(s,t)$ is the only possibly non-proper term, we substitute the expression for y_p in $\dot{u}_p$ to obtain

$$\dot{u}_p = sS_{ru}\left\{\hat{L}_1(s,t) + \hat{L}_2(s,t)S_{ry}\hat{L}_1(s,t)[u_p] + \hat{L}_2(s,t)S_{ry}\hat{L}_2(s,t)[y_p]\right\} + \varepsilon$$

Thus, with the notation being obvious from the previous relationships,

$$\begin{bmatrix}\dot{u}_p \\ \dot{y}_p\end{bmatrix} = S_L\begin{bmatrix}u_p \\ y_p\end{bmatrix} + \varepsilon$$

where, now, S_L is a proper, $L_p(\delta)$-stable operator. Hence, since $\|[u_p, y_p]_t\|_{p,\delta}$ is UB, we obtain that $\|[\dot{u}_p, \dot{y}_p]_t\|_{p,\delta}$ is UB and the proof follows as an application of Lemma 2.55. □□

Proof of Corollary 4.19:

The first part of the theorem is a direct consequence of Lemma 2.42 and the ES property of the frozen closed-loop system with a PW-designed MRC.

To obtain the expression for the plant output, consider a smooth approximation of the plant parameters so that a sufficient number of derivatives are available for the purposes of the TV MRC design. The plant is thus brought in the form of (4.16) with the parameters of the perturbation part being $O[\mu/a]$-small in the uniform norm and a being arbitrary as in Lemma 2.64. From the same lemma, the ith derivative of a parameter approximate is $O[\mu(2a)^{i-1}]$; hence, by Assumptions 4.13 and 4.14 we obtain that for sufficiently small μ, we can select a so that the MRC assumptions are satisfied while $n_I = 1$ in Assumption 3.6 due to the continuity of the plant parameters. At this point and for simplicity we take μ_1 such that the closed loop is ES (from the first

APPENDIX IV.

part of the proof) and MRC assumptions hold. Hence, we can apply Theorem 4.18 from which the output of the closed-loop plant with the TV MRC designed for the approximate smooth nominal plant is given by (4.18). Thus, for the TV MRC, there are two output perturbation operators of the form

$$L'_1[x_c](t) = [c_c(t) + \tilde{c}_c(t)] \int_{t_0}^{t} \Phi_c(t,\tau)\tilde{A}_c(\tau)x_c(\tau)\,d\tau$$

$$L'_2[r](t) = [c_c(t) + \tilde{c}_c(t)] \int_{t_0}^{t} \Phi_c(t,\tau)\tilde{b}_c(\tau)r(\tau)\,d\tau + \tilde{c}_c(t) \int_{t_0}^{t} \Phi_c(t,\tau)b_c(\tau)r(\tau)\,d\tau$$

which satisfy the properties stated in the corollary. Finally, similar expressions are valid for a PW-designed MRC where the difference between the controller parameters of the TV MRC and its PW counterpart should also be taken into account. As in Theorem 4.17 this difference is $O[\mu]$, for a fixed smooth approximation, and therefore, it can be absorbed in the perturbation part of the closed-loop system leaving the rest of the proof qualitatively unaffected.

□□

Chapter 5

TV 'Pole-Placement' Control

5.1 Introduction

One of the main drawbacks of MRC, studied in the previous chapter, is the requirement that the zero dynamics of the plant are ES. Since this requirement may be quite restrictive in applications, it is desirable to develop alternative control strategies —based on a different control objective— which do not suffer the same limitations. In the LTI case, such a strategy is the Pole-Placement Control (PPC) whose objective is to place the poles of the closed-loop system at prescribed locations usually determined by the stability and regulation performance specifications. On the other hand, PPC is rather limited to LTI plants due to the absence of the 'pole' notion in the TV case. We therefore extend the definition of the (TI) PPC objective as to *design a controller such that the PIO's of the closed-loop I/O operator is equal to some prescribed, TV or TI, PIO's*. We refer to this extended objective as the TV PPC objective. For lack of a better name, we refer to a controller that meets the TV PPC objective for an LTV plant, as a TV PPC. When the plant is LTI, the TV PPC objective obviously reduces to the classical PPC objective and is met by a standard PPC scheme. It goes without saying that TI PPC's are a subclass of TV PPC's.

Existing PPC structures for LTV plants (e.g., [M.G.88, Kre.86, G.S.84]), have been derived by pointwise (PW) calculations, based on the 'frozen' plant approach, i.e., under the assumption that the plant is LTI at each time instant. Such controllers, termed as PW PPC, may yield acceptable closed-loop performance, if restricted to the case of slowly TV plants. Of course, a TV

PPC is not the same as a PW PPC, the latter being unable to satisfy the control objective or even guarantee stability in the general LTV case.

In this chapter we present the design and analysis of TV PPC schemes for LTV plants. We begin with Sections 5.2, 5.3 where we design and realize in state-space a TV PPC for LTV plants with smooth parameters and establish the closed-loop stability properties. In Section 5.4 we discuss the issue of incorporating some tracking performance features in a TV PPC through the use of internal models. In Section 5.5 we consider the case of non-smooth parameters and generalize the results of Sections 5.2 and 5.3. The special case of slowly TV plants and PW designs is analyzed in Section 5.6. Finally, we present some simple examples and simulations illustrating the design and properties of TV PPC's in Section 5.7.

5.2 TV PPC Design

Consider a SISO LTV plant described by the state-space equations

$$\begin{aligned} \dot{x}_p &= A(t)x_p + b(t)u_p \\ y_p &= c^\mathsf{T}(t)x_p \end{aligned} \quad (5.1)$$

and satisfying Assumptions 3.1–3.3.

As it was pointed out in Chapter 3, Assumptions 3.1–3.2 imply that the I/O operator of the plant (5.1) admits PDO factorizations in the right form (P_R), i.e.,

$$y_p = N_p(s,t)D_p^{-1}(s,t)[u_p] \quad (5.2)$$

or the left form (P_L), i.e.,

$$y_p = D_p^{-1}(s,t)N_p(s,t)[u_p] \quad (5.3)$$

where $D_p(s,t)$ is a monic PDO with UB coefficients and of constant degree, denoted by n and $N(s,t)$ is a PDO of degree $\leq n-1$ with UB coefficients. Furthermore, in (5.2) $D_p(s,t), N_p(s,t)$ are strongly right coprime while in (5.3) $D_p(s,t), N_p(s,t)$ are strongly left coprime PDO's in $[t_0,\infty)$.

The *TV PPC objective* is defined as follows:

Determine a control input u_p such that the closed-loop plant is internally stable and the closed-loop PIO[1] is equal to a prescribed ES PIO $A_*^{-1}(s,t)$ where $A_*(s,t)$ is a monic PDO of degree $2n-1$ with smooth, UB coefficients.

In this section we develop and analyze the I/O properties of controllers that meet the TV PPC objective. We start with the following lemma which

---
[1] Modulo, of course, ES PIO's which are due to internal cancellations in the controller.

establishes the existence of a TV PPC and provides the design equations for its construction in an I/O operator form.

5.1 Lemma: *Suppose that for the LTV plant (5.1) Assumptions 3.1–3.3 are satisfied. Then, there exist two $(n-1)$-degree PDO's $N_1(s,t)$, $N_2(s,t)$ with smooth, UB coefficients and $N_2(s,t)$ monic such that the closed-loop PIO of*

a. *the P_R plant (5.2) with the controller*

$$u_p = -N_2^{-1}(s,t)N_1(s,t)[y_p] \tag{5.4}$$

where $N_1(s,t), N_2(s,t)$ satisfy the Diophantine equation

$$N_2(s,t)D_p(s,t) + N_1(s,t)N_p(s,t) = A_*(s,t)$$

or,

b. *the P_L plant (5.3) with the controller*

$$u_p = -N_1(s,t)N_2^{-1}(s,t)[y_p] \tag{5.5}$$

where $N_1(s,t), N_2(s,t)$ satisfy the Diophantine equation

$$D_p(s,t)N_2(s,t) + N_p(s,t)N_1(s,t) = A_*(s,t)$$

is equal to the desired PIO $A_*^{-1}(s,t)$. ▽▽

Proof: Straightforward from the expressions for the I/O operators of the plant and the controller and Corollary 2.16. □□

The design of the controller I/O operator as given in the above lemma, with $A_*^{-1}(s,t)$ being ES, also ensures the BIBO stability of the closed-loop system with respect to external inputs. To make this statement more precise, consider the realization of the TV PPC according to Examples 2.38 and 2.39. That is, the control input from the TV PPC is obtained as

$$u_p = -C(s,t)H(s,t)[y_p] \tag{5.6}$$

where, for the P_R plant and the control law (5.4)

$$C(s,t) = N_2^{-1}(s,t)D(s) \quad ; \quad H(s,t) = D^{-1}(s)N_1(s,t)$$

and for the P_L plant and the control law (5.5)

$$C(s,t) = N_1(s,t)D^{-1}(s) \quad ; \quad H(s,t) = D(s)N_2^{-1}(s,t)$$

and $D^{-1}(s)$ is an ES PIO[2] of order $n-1$. This closed-loop configuration is depicted in Fig. 5.1 where the various external signals may represent command inputs (r or v), input disturbances (d_u), output disturbances (d_y), measurement noise (r) or even effects of initial conditions. The total closed-loop

---

[2] $D(s)$ may be TV with smooth UB coefficients.

5.2. TV PPC DESIGN

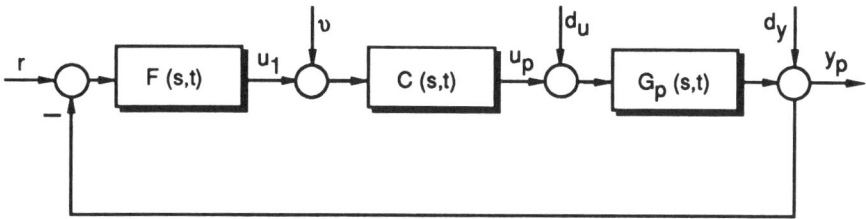

Figure 5.1: The TV PPC closed-loop system.

response to the various external inputs is simply obtained using superposition, with each input filtered by the appropriate sensitivity operator.

With reference to Fig. 5.1, the following lemma describes the BIBO and $L_p(\delta)$ stability properties of the closed-loop plant with the TV PPC and provides expressions of the various sensitivity operators.

5.2 Lemma: *The I/O description of the closed-loop plant with the TV PPC (5.6), shown in Fig. 5.1, is given by*

$$\begin{bmatrix} u_p \\ y_p \\ u_1 \end{bmatrix} = \begin{bmatrix} S_{ru} & S_{vu} & S_{uu} & S_{yu} \\ S_{ry} & S_{vy} & S_{uy} & S_{yy} \\ S_{r1} & S_{v1} & S_{u1} & S_{y1} \end{bmatrix} \begin{bmatrix} r \\ v \\ d_u \\ d_y \end{bmatrix} \quad (5.7)$$

where, omitting the PDO/PIO arguments for simplicity,

$$S_{yu} = -S_{ru} \quad ; \quad S_{ry} = 1 - S_{yy} \quad ; \quad S_{y1} = -S_{r1}$$

and

1: *for the plant (5.2) in the P_R-form*

$$\begin{aligned}
S_{ru} &= D_p A_*^{-1} N_1 & ; \quad S_{vu} &= D_p A_*^{-1} D & ; \quad S_{uu} &= 1 - D_p A_*^{-1} N_2 \\
S_{yy} &= 1 - N_p A_*^{-1} N_1 & ; \quad S_{vy} &= N_p A_*^{-1} D & ; \quad S_{uy} &= N_p A_*^{-1} N_2 \\
S_{r1} &= D^{-1} N_2 D_p A_*^{-1} N_1 & ; \quad S_{v1} &= -D^{-1} N_1 N_p A_*^{-1} D & & \\
& & ; \quad S_{u1} &= -D^{-1} N_1 N_p A_*^{-1} N_2 & &
\end{aligned}$$
(5.8)

2: *for the plant (5.3) in the P_L-form*

$$\begin{aligned}
S_{ru} &= N_1 A_*^{-1} D_p & ; \quad S_{vu} &= N_1 A_*^{-1} D_p N_2 D^{-1} & ; \quad S_{uu} &= -N_1 A_*^{-1} N_p \\
S_{yy} &= N_2 A_*^{-1} D_p & ; \quad S_{vy} &= N_2 A_*^{-1} N_p N_1 D^{-1} & ; \quad S_{uy} &= N_2 A_*^{-1} N_p \\
S_{r1} &= D A_*^{-1} D_p & ; \quad S_{v1} &= -D A_*^{-1} N_p N_1 D^{-1} & ; \quad S_{u1} &= -D A_*^{-1} N_p
\end{aligned}$$
(5.9)

Furthermore, there exists $\delta_* > 0$ which in general depends on $A_*(s,t)$ and $D(s)$ such that for any $\delta \in [0, \delta_*)$, and any initial time t_0, the various sensitivity operators are $L_p(\delta)$-stable, $p \in [1, \infty]$, uniformly in t_0; also, for the strictly proper sensitivity operators, the corresponding $g_{p,\delta}$ gains exist and are finite, uniformly in t_0. ▽▽

Proof: Straightforward, following similar arguments as in Lemma 4.6. □□

At this point, it should be noted that under Assumptions 3.1–3.3, a plant in the P_L-form can be expressed in the P_R-form and vice-versa. A controller, designed for P_R-plants, has the advantage that it can be implemented so that only fixed PDO's, together with $N_p(s,t)$, appear in the closed-loop I/O operator $v \mapsto y_p$. Such a property is important when the closed-loop response to command signals is considered. Consequently, it is desirable —and feasible— to design the TV PPC for plants in the P_L-form by first converting the plant in the P_R-form and then perform the TV PPC design for the P_R-plant. The apparent drawback of this procedure is an increase in the computational load which may be crucial when the controller calculations are performed on-line. This issue is further discussed in Chapter 8, where an indirect adaptive controller is designed by estimating the plant parameters (necessarily in the P_L form) and calculating the corresponding TV PPC parameters.

Further, overparametrized TV PPC designs can be obtained, as in the TV MRC case, with a higher order controller I/O operator $y_p \mapsto u_p$ (see Examples 4.7 and 4.8). Such a TV PPC may have strictly proper I/O operator and/or possess additional degrees of freedom to allow some partial shaping of the closed-loop sensitivity operators. For example, suppose $-N_1 N_2^{-1}$ is a TV PPC for a P_L plant $D_p^{-1} N_p$, and let V, W, D_0, N_0 denote PDO's with UB coefficients such that

- $\deg[V] > \deg[W]$;
- V is a monic PDO and V^{-1} is ES;
- D_0 is monic and $D_p N_0 + N_p D_0 = 0$ in $\mathbf{R}_+$.

Then the controller $-N_y N_x^{-1}$ with $N_y = N_1 V + D_0 W$ and $N_x = N_2 V + N_0 W$ is also a TV PPC. Taking into account the slight modifications in the degrees of the various PDO's and replacing N_2, N_1, D, A_* in Lemma 5.2 by N_x, N_y, DV, A_*V respectively, the same results are applicable in this case as well. However, notice that the improvement of the properties of the closed loop sensitivity operators via an IMP design may not be as simple as in the TV MRC case, due to the lack of the ES property of N_p^{-1} and, particularly for P_L plants, the different controller factorization.

5.3. REALIZATION AND INTERNAL STABILITY

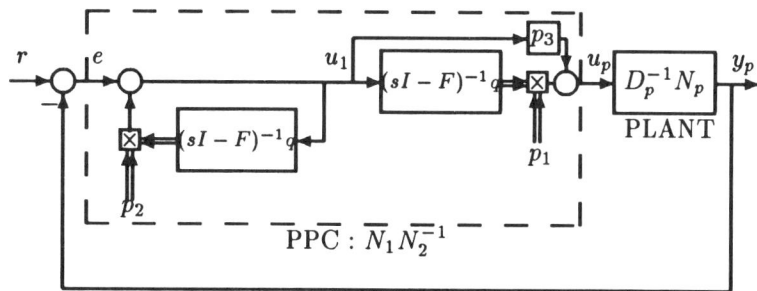

Figure 5.2: The TV PPC structure for LTV plants.

5.3 Realization of the TV PPC and Internal Stability of the Closed-Loop Plant

In order to establish the exponential and, therefore, internal stability of the closed-loop system, we consider a state-space realization of the TV PPC law of Lemma 5.1 according to the guidelines of Examples 2.39 and 2.38. In the case of plants in the P_L-form, the control law is generated by using an auxiliary filter as follows

$$\begin{aligned} \dot{\omega}_1 &= F\omega_1 + qu_1 \\ u_1 &= p_2^T(t)\omega_1 + (r - y_p) \\ u_p &= p_1^T(t)\omega_1 + p_3(t)u_1 \end{aligned} \tag{5.10}$$

where r is the reference (command) signal, $u_1 \in \mathbf{R}$ is an internal signal, F is a constant Hurwitz matrix of dimension $(n-1) \times (n-1)$ and (F, q) is a completely controllable pair. The vectors $p_1(t)$, $p_2(t)$ and the scalar $p_3(t)$ are the controller parameters which are to be selected such that the controller has the I/O operator $y_p \mapsto u_p$ specified in Lemma 5.1. The block diagram of the corresponding closed-loop system is shown in Fig. 5.2.

Alternatively, in an I/O operator-notation, the control law (5.10) is expressed as

$$\begin{aligned} u_p &= \{p_1^T(t)G(s) + p_3(t)\}[u_1] \\ u_1 &= p_2^T(t)G(s)[u_1] + (r - y_p) \\ G(s) &= (sI - F)^{-1}q \end{aligned} \tag{5.11}$$

The existence of parameters $p_i(t)$ such that (5.10) or (5.11) represent a TV PPC for a P_L-plant is established in the following corollary.

5.3 Corollary: Let $N_1(s,t)$, $N_2(s,t)$ be $(n-1)$-degree PDO's with smooth, UB coefficients and with $N_2(s,t)$ monic. Then, there exist smooth, UB parameters $p_i(t)$ such that the I/O operator $y_p \mapsto u_p$ of (5.11) is equal to $-N_1(s,t)N_2^{-1}(s,t)$. ▽▽

Proof: Immediate from Example 2.38. Notice that the parameters $p_i(t)$ depend on the coefficients of the left form of the PDO's $N_i(s,t)$. These PDO's are initially obtained in the right form as the solution of the corresponding left Diophantine equation (see Lemma 5.1). Consequently, the calculation of the controller parameters $p_i(t)$ involves an additional intermediate step converting right-form PDO's into left-form ones. □□

5.4 Remark: The TV PPC realization for plants in the P_R-form follows similarly by using the results of Example 2.39 to realize the I/O operators $N_2^{-1}(s,t)D(s)$ and $D^{-1}(s)N_1(s,t)$ as in the TV MRC case and is summarized below.[3]

$$\begin{aligned}
\dot{\omega}_1 &= F\omega_1 + \theta_1 u_p \\
\dot{\omega}_2 &= F\omega_2 + \theta_2 y_p \\
\omega_3 &= \theta_3 y_p \\
u_p &= g^T \omega + v
\end{aligned} \tag{5.12}$$

where $\omega = [\omega_1^T, \omega_2^T, \omega_3]^T$ is a $(2n-1)$-dimensional vector, $F \in \mathbf{R}^{(n-1)\times(n-1)}$ is a stable matrix with $\det(sI - F) = D(s)$ and $g = [q^T, q^T, 1]^T$ is a constant vector such that (q^T, F) is an observable pair.

Then, there exists a control parameter vector $[\theta_{1*}^T, \theta_{2*}^T, \theta_{3*}]$ such that the I/O operator $y_p \mapsto u_p$ of (5.12) is equal to that of the TV PPC (5.6) as given by Lemma 5.1 for a P_R-plant. (see Corollary 4.9 for details). ▽▽

Having specified the state-space realization of the controller, it is now possible to describe the exponential and BIBS stability properties of the closed-loop plant, as well as the effects of arbitrary initial conditions, for the TV PPC designed in Lemma 5.1 and realized according to Corollary 5.3 or Remark 5.4.

5.5 Theorem: The closed-loop plant (5.1) with the TV PPC is ES and, therefore, BIBS stable for any external UB input. Furthermore, for arbitrary initial conditions set at t_0, the ZIR of the closed-loop plant decays as $c\exp[-a(t-t_0)]$ where c, a are positive constants independent of t_0; c depends

---
[3] In this case, it is more convenient to consider v as the command input.

5.4. COMMAND TRACKING WITH THE TV PPC

on the size of the initial conditions and a depends on the rate of exponential stability of $A_*^{-1}(s,t)$ and $D^{-1}(s)$. ▽▽

Proof: The proof is a direct consequence of Lemmas 2.35 and 5.2 and is obtained along the same lines as Theorem 4.10. Notice that for P_L plants, the main difference is that the initial conditions of the filters enter as exponentially decaying disturbances at the 'r' and 'd_u' nodes of Fig. 5.1, while for P_R plants they enter at the 'v' node —same as in the TV MRC case. □□

5.6 Remark: Throughout the development of the TV PPC we have considered a general TV form of the desired PIO $A_*^{-1}(s,t)$ which allows many of the subsequent results to be established in a compact and unified way. In most practical applications, a TI $A_*(s)$ would be sufficient to meet the performance specifications and considerably easier to select and realize. In particular cases, however, it may be possible to exploit the additional flexibility, offered by a TV $A_*(s,t)$, to compensate —at least in part— for time variations in the closed-loop PDO's and improve the closed-loop performance.
▽▽

5.4 Command Tracking with the TV PPC

Although the design of a TV (or PW) PPC scheme is a very general one, including the MRC design as a special case,[4] it does not guarantee any particular tracking performance capabilities for the associated closed-loop system. The reason is that the PPC objective deals mainly with the properties of the closed-loop PIO and state transition matrix, while it is assumed that any tracking performance requirements have already been incorporated in the selection of the desired closed-loop PIO. In the previous chapter we discussed the case where the performance objective was defined in terms of a reference model. This objective, however, required the zero dynamics of the plant to be exponentially stable. It is therefore apparent that, in order to avoid imposing any such limitations in a PPC design, it is desirable to consider weaker objectives than the MRC when tracking performance requirements should be incorporated in the PPC.

A frequently studied tracking performance objective is to achieve exact tracking for a class of reference inputs whose internal model is specified a priori. In the LTI case the design of a PPC satisfying this objective is a rather straightforward procedure, involving the so-called internal model principle

---

[4] The TV MRC can be obtained from a TV PPC with $A_*(s,t)$ containing $N_p(s,t)$ as a factor and some minor modifications to adjust the high-frequency gain.

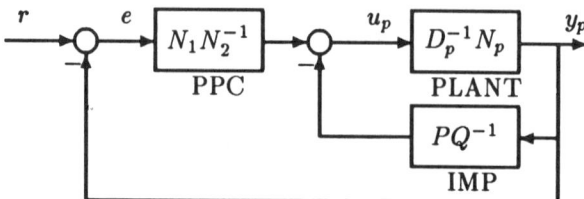

Figure 5.3: TV internal model principle/pole placement controller design.

(IMP) [Bng.77]. Briefly described, the essence of this procedure is to introduce certain zeros, corresponding to the internal model, in the sensitivity transfer function from the reference input to the tracking error. The same procedure can be extended to the LTV case, where the condition for the TV IMP/PPC design is a *skew* coprimeness of the internal model and the plant PDO. This condition is essentially similar to that of [Bng.77] in the multivariable LTI case and reduces to the well known coprimeness condition of the internal model and the plant numerator polynomials in the SISO LTI case. In the general LTV case, however, the skew coprimeness cannot be expressed in terms of algebraic equations, as the right or left coprimeness do, which complicates the design of a TV IMP/PPC.

Let us begin our discussion of the TV IMP/PPC design by considering the LTV plant (5.1) and its I/O operator in the P_L form (5.3). The TV IMP/PPC objective is defined as follows.

Determine the control input u_p so as to meet the TV PPC objective and force the output y_p to track reference signals r satisfying the differential equation

$$\Lambda(s)[r] = 0$$

where $\Lambda(s)$ is an a priori specified TI PDO.

In order to achieve exact tracking, the control input u_p is generated as the difference of the output of two compensators as shown Fig. 5.3, i.e.,

$$u_p = N_1(s,t)N_2^{-1}(s,t)[r - y_p] - P(s,t)Q^{-1}(s)[y_p] \qquad (5.13)$$

which is realized in state space using ES filters as in the TV PPC case of the previous section.

The first compensator has an input $r - y_p$ and is designed to stabilize the closed loop, while the second has an input y_p and is designed to introduce the internal model of r in the forward path. The properties and assumptions of such a design are summarized in the following corollary.

5.4. COMMAND TRACKING WITH THE TV PPC 133

5.7 Corollary:

a. *Consider the LTV plant (5.1) and let $N_R(s,t), D_R(s,t)$ denote the PDO's of the corresponding P_R form of the plant.[5] Also let $Q^{-1}(s)$ be an exponentially stable TI PIO of order $\deg[\Lambda(s)] - 1$ and such that $Q(s)$ and $N_R(s,t)$ are strongly left coprime in $\mathbf{R}_+$ and $P(s,t)$ be a PDO of degree $\deg[\Lambda(s)] - 1$ with smooth UB coefficients. Further, select $N_2(s,t)$, $N_1(s,t)$ to satisfy the design equations*

$$[D_p(s,t)Q(s) + N_p(s,t)P(s,t)]\tilde{N}_2(s,t) + N_p(s,t)N_1(s,t) = \bar{A}_*(s,t) \quad (5.14)$$

$$N_2(s,t) = Q(s)\tilde{N}_2(s,t)$$

where $\bar{A}_*^{-1}(s,t)$ is an ES PIO of order $2n-2+\deg[\Lambda(s)]$. Then, the control law (5.13), realized in state space according to Examples 2.39 and 2.38, guarantees the exponential stability of the closed-loop plant. The rate of exponential decay of the closed-loop state transition matrix depends on $\bar{A}_*(s,t)$, $Q(s)$ and the auxiliary filter used in the realization of $N_1(s,t)N_2^{-1}(s,t)$.

b. *Suppose that in (a.) the PDO $P(s,t)$ also satisfies*

5.8 Assumption:

$$X(s,t)\Lambda(s) - N_p(s,t)P(s,t) = D_p(s,t)Q(s) \quad (5.15)$$

for some PDO $X(s,t)$ with smooth, UB coefficients and degree $n - 1$. ∎

Then, in addition to the result in (a), the tracking error $e = r - y_p$ converges to zero exponentially fast. Such a controller is referred to as TV IMP/PPC.

▽▽

Proof: In Appendix V.

From the above corollary it is apparent that the main limitation in the design of a TV IMP/PPC lies in the selection of the PDO $P(s,t)$ as to satisfy Assumption 5.8 which is a dynamical equation with respect to the coefficients of the unknown PDO's $P(s,t)$, $X(s,t)$. Since for implementation purposes the controller parameters are required to be UB, we also need to assume the existence of bounded solutions of (5.15). Notice, however, that the exponential stability of the closed-loop plant is guaranteed even if $P(s,t)$ does not satisfy Assumption 5.8, provided of course that it has smooth and UB coefficients.

A general but quite restrictive sufficient condition for (5.15) to have a solution with UB coefficients, for any $D_p(s,t)$, $Q(s)$, is that for any UB smooth

---
[5] That is, $y_p = N_R(s,t)D_R^{-1}(s,t)[u_p]$.

function $f(t)$, there exist PDO's $Y_1(s,t)$, $Y_2(s,t)$ with UB coefficients such that
$$Y_1(s,t)\Lambda(s) + N_p(s,t)Y_2(s,t) = f(t) \tag{5.16}$$
which can be interpreted as skew coprimeness of the PDO's $\Lambda(s)$, $N_p(s,t)$. At this point, it is interesting to notice that considerably weaker conditions can be developed in several special cases of interest, as discussed in the following example.

5.9 Example: Consider the case where $\Lambda(s) = s$ (similar arguments hold for any $\Lambda(s) = s^k$). Then, $P(s,t) = \rho(t)$ and by expressing the PDO's in (5.15) in the left form, it follows that a UB solution for ρ and the coefficients of $X(s,t)$ exists, provided that the ODE
$$N_p(s,t)[\rho] = \alpha(t) \tag{5.17}$$
has a UB solution, where $\alpha(t)$ is the coefficient of s^0 of the PDO $D_p(s,t)Q(s)$ expressed in the left form. This is trivially true if $\alpha = 0$ (in other words, s is a right factor of $D_p(s,t)$) or if $N_p^{-1}(s,t)$ is an ES PIO. These conditions can be further relaxed in the case of periodic systems. For example, if $\alpha(t)$ is a periodic function with period T, then a UB solution of (5.17) can be found, by an appropriate selection of the initial conditions of $\rho(t_0), \dot{\rho}(t_0), \ldots$ provided that the state transition matrix $\Phi_N(\cdot,\cdot)$ of (5.17) satisfies $det[I - \Phi_N(t_0 + T, t_0)] \neq 0$. ▽▽

5.5 Non-Smooth Parameter Variations

In this section we extend the previous design procedures and results to the more general case where the LTV plant is described by
$$\begin{aligned} \dot{x} &= A_o(t)x + b_o(t)u_p + \tilde{A}(t)x + \tilde{b}(t)u_p \\ y_p &= c_o^T(t)x + \tilde{c}^T(t)x \end{aligned} \tag{5.18}$$
whose nominal and perturbation part satisfy Assumptions 3.4–3.6.

Due to the discontinuities in the plant parameters, a PDO/PIO factorization may not be available for the LTV plant (5.18) and therefore the TV PPC or TV IMP/PPC design procedures (Lemma 5.1 or Corollary 5.7) are not directly applicable in this case. Under Assumption 3.5, however, the nominal part of the plant possesses a PDO/PIO factorization inside every interval (t_j, t_{j+1}) satisfying Assumptions 3.1–3.3 in a piecewise sense. It is therefore possible to design a TV PPC or TV IMP/PPC[6] for the nominal

---
[6] Of course, for the TV IMP/PPC, Assumption 5.8 should also be satisfied in a piecewise sense.

plant $[A_o, b_o, c_o]$ inside each interval (t_j, t_{j+1}). Thus, the control input is determined from (5.4) or (5.5) or their state-space counterparts (5.12) or (5.10) in a piecewise sense while the parameter discontinuities and the perturbation part of the plant are effectively treated as modeling errors. According to Theorem 5.5, this control input guarantees that the nominal closed-loop is piecewise ES and therefore, invoking Corollary 3.8, ES for a sufficiently small modeling error. We make this idea precise in the following theorem and corollary where we establish the stability and performance properties of such TV PPC designs for the plant (5.18).

5.10 Theorem: *Consider the LTV plant (5.18) whose nominal and perturbation parts satisfy Assumptions 3.4–3.6 with the TV PPC determined from (5.4) or (5.5) in a piecewise sense. Then, there exist $\nu_0 > 0$, $\mu_0' > 0$ such that $\forall \nu \in [0, \nu_0)$, $\forall \mu' \in [0, \mu_0')$, the closed-loop system with the TV PPC (or TV IMP/PPC) is ES with rate depending on $A_*^{-1}(s,t)$ and the values of ν and μ'.* ▽▽

Proof: As in Theorem 4.18.

Notice that, in contrast to the MRC case, the exponential rate of decay of the closed-loop state transition matrix depends now on $A_*(s,t)$ and the values of ν and μ'. Since $A_*(s,t)$ is selected by the designer, the rate of exponential stability of the closed loop can be determined a priori, for sufficiently small ν and μ'.

Finally, for the TV IMP/PPC we can give a characterization of the tracking performance deterioration, due to the parameter discontinuities and the perturbation part of the plant, as stated by the following corollary.

5.11 Corollary: *Under the conditions of Theorem 5.10, suppose that Assumption 5.8 holds inside every interval (t_j, t_{j+1}), uniformly in j and the TV IMP/PPC (5.13) is used to generate the control input. Then, in addition to the results of Theorem 5.10, there exist positive constants K, K', C such that*
$$\int_{t_0}^{t_0+T} |y_p(t) - r(t)|^2 \, dt \leq C + K\nu T + K'\mu' T$$
for all $t_0, T \geq 0$. ▽▽

Proof: As in Theorem 4.18.

5.6 Slowly TV Plants

In the special case where the plant parameters vary slowly with time, the calculation of the controller parameters is considerably simplified by adopting a pointwise approach in the design of a PPC (PW PPC). This simplification is obtained at the expense of some deterioration in the closed-loop performance defined by the TV PPC objective.

In order to make this idea precise, let us consider the plant (5.1) satisfying Assumptions 3.1, 3.3 and, denoting the plant parameters by the vector Θ_p,

5.12 Assumption: $\|\frac{d^i}{dt^i}\Theta_p(t)\| \leq \mu$, $\forall t \in \mathbf{R}_+$, $i = 1, 2, \ldots$ *for some 'small' parameter $\mu \geq 0$.* ∎

5.13 Assumption: *The PW (frozen) controllability and observability matrices of the triple $[A(t), b(t), c(t)]$ are strongly nonsingular.* ∎

Under Assumption 5.12, the properties of the LTV plant can be approximated by the properties of the corresponding sequence of frozen LTI plants. For example, for sufficiently small μ, Assumption 3.2 is implied by the easier to check Assumption 5.13. In addition, the validity of Assumption 5.13 may also be checked by examining the strong PW coprimeness of the polynomials in the frozen I/O representation of the plant, i.e., the numerator and denominator of the PW plant transfer function.

Therefore, for sufficiently small μ, Assumptions 3.1, 3.3, 5.12, 5.13, imply that the LTV plant (5.1) admits an I/O representation of the form (5.2) or (5.3), i.e.,

$$y_p = N_p(s,t)D_p^{-1}(s,t)[u_p]$$

or

$$y_p = D_p^{-1}(s,t)N_p(s,t)[u_p]$$

where the coefficients of the PDO's $D_p(s,t), N_p(s,t)$ are slowly TV.[7] Furthermore, for sufficiently small μ, the PDO's $D_p(s,t), N_p(s,t)$ are strongly pointwise coprime and strongly left or right coprime. Consequently, considering the P_L-form of the plant, we may design a PW PPC by taking the controller I/O operator $y_p \mapsto u_p$ to be $-N_1(s,t)N_2^{-1}(s,t)$, i.e.,

$$u_p = -N_1(s,t)N_2^{-1}(s,t)[y_p] \qquad (5.19)$$

with $N_1(s,t), N_2(s,t)$ such that

$$D_p(s,t) \star N_2(s,t) + N_p(s,t) \star N_1(s,t) = A_*(s)$$

where '$\star$' denotes pointwise multiplication and $A_*^{-1}(s)$ is the desired, ES PIO[8] of order $2n-1$. The closed-loop stability properties with such a PW PPC are

---
[7] Note that, in general, the PDO's D_p, N_p of the P_R and P_L form are not the same but their coefficients may be different by $O(\mu)$.

[8] In this case we take $A_*(s)$ to be TI; if otherwise selected, $A_*(s,t)$ should be slowly TV since the speed of variation of its coefficients affects the range of μ for which closed-loop stability can be guaranteed.

5.6. SLOWLY TV PLANTS

given by the following theorem (similarly for a PW PPC corresponding to the P_R-form of the plant).

5.14 Theorem: *Consider the LTV plant (5.1) satisfying Assumptions 3.1, 3.3, 5.12, 5.13 and its I/O operator expressed in the P_L-form (5.3). Further, consider the PW PPC (5.19), realized in state-space by (5.10) and applied to the LTV plant (5.1). Then there exists a constant $\mu_2 > 0$ such that $\forall \mu \in [0, \mu_2)$,*

1. *the closed-loop plant is ES and therefore, BIBS and BIBO stable;*

2. *the closed-loop state transition matrix is exponentially decaying with rate that depends on $A_*^{-1}(s)$, $D^{-1}(s)$ and the value of μ;*

3. *the closed-loop PIO $D_c^{-1}(s,t)$ satisfies*

$$D_c(s,t) = A_*(s) + \Delta(s,t)$$

where $\Delta(s,t)$ is a PDO of degree at most $2n-2$, with smooth, UB coefficients $O(\mu)$. ▽▽

Proof: In Appendix V.

The results of Theorem 5.14 can be extended to a wider class of slowly TV plants and a wider class of control laws. For example, Assumption 5.12 may be relaxed as

5.15 Assumption: $\|\dot{\Theta}_p\|_\infty \leq \mu$. ∎

5.16 Theorem: *Suppose that for the LTV plant (5.1), Assumptions 5.15, 3.1, 5.13 and 3.3 are satisfied except that in 3.1 the plant parameters are only required to be Lipschitz continuous. Also suppose that an LTV controller, of constant order and with a UB parameter vector $\Theta_c(t)$ satisfying $\|\dot{\Theta}_c\|_\infty \leq \mu$,[9] is designed so that the frozen closed loop is ES with rate at most $-a_*$ for all $t \in \mathbf{R}_+$. Then there exists a constant $\mu_2 > 0$ such that $\forall \mu \in [0, \mu_2)$,*

1. *the closed-loop system is ES and, therefore, BIBS stable;*

2. *the exponential rate of decay the closed-loop state transition matrix depends on a_* and the value of μ.*

▽▽

---

[9] Such a condition holds if, for example, the controller parameter vector Θ_c is a Lipschitz continuous function of the plant parameter vector Θ_p.

Proof: Immediate from Lemma 2.42. Also notice that, in view of Theorem 5.10, it is quite straightforward to further relax the conditions of the theorem to hold in a piecewise sense and on the average. □□

5.17 Remark: Theorem 5.16 allows the construction of a large class of stabilizing controllers for slowly TV plants, including most of the controllers that can be designed using LTI techniques. For example, an LTI PPC, or a Linear Quadratic Gaussian Regulator with guaranteed stability margin (see [Kai.80]), satisfies the above conditions[10] and can be used as PW designs. Furthermore, under Assumption 4.14, a PW MRC also belongs to the same class as a special case of a PPC. Note that a more general and quantitative version of these results has been given in [S.A.91] where the properties of the closed-loop TV sensitivity operators are approximated by the properties of the corresponding operators of the frozen closed-loop plant. ▽▽

5.18 Remark: In the special case of an IMP/PPC design we may select the PDO $P(s,t)$, entering in equations (5.14) and (5.15), in a PW sense (PW IMP) and then perform a TV PPC design for the augmented plant. This approach requires less restrictive and easier to check assumptions while preserving the stabilizing properties of the TV PPC for arbitrarily fast TV plants (see Corollary 5.7). Moreover, it has the advantage that even if Assumption 5.8 is not satisfied, small tracking errors can be achieved for slowly TV plants.

For example, assuming that $N_p(s,t)$, $\Lambda(s)$ are strongly left coprime PDO's in $\mathbf{R}_+$,[11] we can solve

$$\Lambda(s)\hat{X}(s,t) - N_p(s,t)\hat{P}(s,t) = D_p(s,t)Q(s) \qquad (5.20)$$

instead of (5.15), by solving a system of linear algebraic equations. Next the TV PPC parameters are calculated as in Corollary 5.7 with $P(s,t), X(s,t)$ being replaced by $\hat{P}(s,t), \hat{X}(s,t)$ respectively. In this case the tracking error becomes

$$\begin{aligned} e &= N_2(s,t)\bar{A}_*^{-1}(s)\hat{X}(s,t)Q^{-1}(s)\Lambda(s)[r] \\ &\quad + N_2(s,t)\bar{A}_*^{-1}(s)[\Lambda(s)\hat{X}(s,t) - \hat{X}(s,t)\Lambda(s)]Q^{-1}(s)[r] \end{aligned} \qquad (5.21)$$

The coefficients of the PDO $[\Lambda(s)\hat{X}(s,t) - \hat{X}(s,t)\Lambda(s)]$ depend on the derivatives of the plant parameters and, hence, the contribution of the last term of

---

[10] Their order is constant and, under Assumption 5.13, their parameters are Lipschitz continuous functions of the plant parameters.

[11] Note that for slowly TV plants, the strong pointwise coprimeness of $N_p(s,t)$ and $\Lambda(s)$ guarantees the strong left coprimeness of the PDO's $N_p(s,t)$, $\Lambda(s)$.

5.7. EXAMPLES

(5.21), for a UB reference input, is $O(\mu)$-small if the plant is slowly TV. However, the PPC objective, i.e., $\bar{A}_*^{-1}(s)$ is the closed-loop PIO, is still satisfied for slow and fast TV plants.

A similar result can also be established by using the conventional approach for SISO LTI plants, whereby $\Lambda^{-1}(s)$ is included in the forward path of the loop and the PPC is designed for an augmented plant. Needless to say, in both cases, the exponential stability of the closed-loop system is guaranteed by the PPC design, for arbitrarily fast variations should the TV PPC be used or, for sufficiently slow variations with the PW PPC. Finally, with this approach, an $O(\mu)$ tracking error is also obtained for slowly TV plants with Lipschitz continuous parameters. The details of these statements, however, are omitted as completely analogous to Theorem 5.16. $\nabla\nabla$

5.7 Examples

The following examples demonstrate the design principles and properties of the TV and PW PPC for LTV plants. For simplicity we drop the argument t in the expressions of the various TV parameters.

5.19 Example: *TV and PW PPC Design.* Let us consider the second order plant

$$\frac{d^2}{dt^2} y_p + \frac{d}{dt}(a_1 y_p) + a_2 y_p = \frac{d}{dt} u_p + b u_p \qquad (5.22)$$

where a_1, a_2, b are the TV parameters of the plant and the control objective is to design a controller that makes the closed-loop PIO equal to $(s^3 + 6s^2 + 11s + 6)^{-1}$. According to the previously presented analysis, we realize a TV PPC law, choosing $F = -2$ and $q = 1$, as follows:

$$u_p = p_1(s+2)^{-1}[u_1] + p_3 u_1 \;;\; u_1 = p_2(s+2)^{-1}[u_1] + (r - y_p) \qquad (5.23)$$

where p_1, p_2, p_3 are the controller parameters to be determined. From (5.23) the plant input can also be written as

$$u_p = (s\psi_2 + \psi_3)(s + \psi_1)^{-1}[r - y_p] \;; \qquad (5.24)$$

$$\psi_1 = 2 - p_2 \;;\; \psi_2 = p_3 \;;\; \psi_3 = p_1 + 2p_3 - \dot{p}_3 \qquad (5.25)$$

Combining (5.22) with (5.24) we obtain the PIO of the closed loop

$$\left[(s^2 + s a_1 + a_2)(s + \psi_1) + (s + b)(s\psi_2 + \psi_3)\right]^{-1} \qquad (5.26)$$

which is to be made equal to $[s^3 + 6s^2 + 11s + 6]^{-1}$. Thus, we obtain the following set of equations that ψ_1, ψ_2, ψ_3 must satisfy:

$$\begin{pmatrix} 1 & 1 & 0 \\ a_1 & b & 1 \\ a_2 & -\dot{b} & b \end{pmatrix} \begin{pmatrix} \psi_1 \\ \psi_2 \\ \psi_3 \end{pmatrix} = \begin{pmatrix} 6 - a_1 \\ 11 - a_2 + \dot{a}_1 \\ 6 + \dot{a}_2 \end{pmatrix} \quad (5.27)$$

or, in a compact form with obvious notation,

$$S_L(t)\psi(t) = A(t) \quad (5.28)$$

Using Cramer's rule, and letting $A = (A_1, A_2, A_3)^\mathsf{T}$ we obtain the solution for ψ

$$\begin{aligned}
\psi_1 &= \frac{A_1(b^2 + \dot{b}) - A_2 b + A_3}{\det[S_L(t)]} \\
\psi_2 &= \frac{A_1(a_2 - a_1 b) + A_2 b - A_3}{\det[S_L(t)]} \\
\psi_3 &= \frac{-A_1(a_1 \dot{b} + b a_2) + A_2(a_2 + \dot{b}) + A_3(b - a_1)}{\det[S_L(t)]}
\end{aligned} \quad (5.29)$$

where, for the solution (5.29) to exist and be bounded $\det[S_L(t)] = b^2 + \dot{b} - a_1 b + a_2$ must be bounded away from zero, for all $t \geq 0$ (Left Coprimeness condition). From (5.25), we obtain the desired controller parameters as

$$p_1 = \psi_3 - 2\psi_2 + \dot{\psi}_2 \; ; \; p_2 = 2 - \psi_1 \; ; \; p_3 = \psi_2 \quad (5.30)$$

Thus, the response of the plant (5.22) with the controller (5.23) and the parameters (5.30), (5.29) is

$$y_p = r - (s + 2 - p_2)(s^3 + 6s^2 + 11s + 6)^{-1}(s^2 + sa_1 + a_2)[r] \quad (5.31)$$

and, hence, the control objective is satisfied exactly with no restrictions on the speed of variation of the plant parameters, other than being finite.

In contrast to the above solution, the design of a PW PPC proceeds by 'freezing' the values of a_1, a_2, b in (5.26) at each time instant [Kre.86, M.G.88], i.e., by taking their derivatives to be identically zero. Thus, the parameters ψ_1, ψ_2, ψ_3 are calculated by solving

$$\begin{pmatrix} 1 & 1 & 0 \\ a_1 & b & 1 \\ a_2 & 0 & b \end{pmatrix} \begin{pmatrix} \psi_1 \\ \psi_2 \\ \psi_3 \end{pmatrix} = \begin{pmatrix} 6 - a_1 \\ 11 - a_2 \\ 6 \end{pmatrix} \quad (5.32)$$

Furthermore, for a pointwise realization of the PPC we select

$$p_1 = \psi_3 - 2\psi_2 \; ; \; p_2 = 2 - \psi_1 \; ; \; p_3 = \psi_2$$

5.7. EXAMPLES

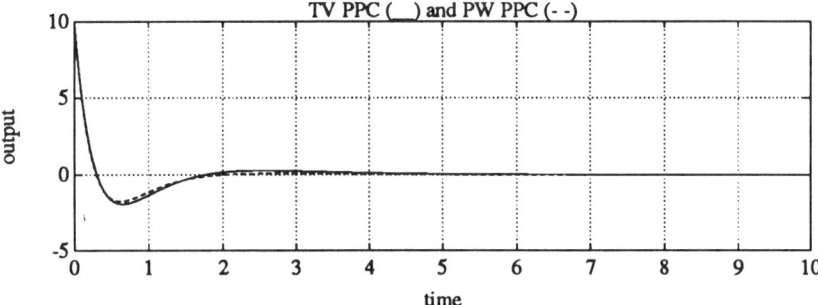

Figure 5.4: Slowly TV plant ($\mu = 0.5$): Exact asymptotic regulation with the TV and PW PPC.

A comparison with (5.25) shows that the effective controller parameters ψ_i are now ψ_1, ψ_2, $\psi_3 + \dot{\psi}_2$. Thus, using the PW PPC, the PIO (5.26) becomes

$$[s^3 + 6s^2 + 11s + 6 - s(\dot{a}_1 - \dot{\psi}_2) - \dot{a}_2 - \dot{b}\psi_2 + b\dot{\psi}_2]^{-1} \qquad (5.33)$$

From (5.33) it becomes apparent that, in general, this solution cannot satisfy the control objective exactly, unless the plant is TI. Furthermore, the closed-loop stability cannot be guaranteed unless the plant is slowly TV, i.e., $\dot{a}_1, \dot{a}_2, \dot{b}$ are small which, together with the assumed strong nonsingularity of the Sylvester matrix, implies that $\tilde{\psi}_i$ is also small. ▽▽

5.20 Simulations: Next, we assign some numerical values for the plant parameters a_1, a_2, b and calculate the corresponding PPC parameters. Let

$$a_1 = 20 + 12\sin\mu t \; ; \; a_2 = 6\cos\mu t \; ; \; b = -1 \qquad (5.34)$$

where μ is a positive constant which determines the speed of variation of a_1, a_2. Then $\det[S_L(t)] = 21 + 12\sin 2t + 6\cos 2t \geq 7.5836 \; \forall t$ and the TV PPC parameters are obtained, in a straightforward way, from (5.29) and (5.30). Moreover, the PW PPC parameters are similarly calculated from (5.32) by observing that, for this example, the PW and left TV Sylvester matrices are the same since $\dot{b} = 0$.

In Fig. 5.4 the response of the closed loop system during regulation ($r = 0$) is shown for the TV and PW PPC. Notice that due to the 'small' value of μ, the PW design is able to preserve the closed-loop stability.

On the other hand, while a larger value of μ does not affect the closed-loop stability for the TV PPC, it is likely to cause the failure of a PW PPC. Letting $\mu = 2$ and using the TV PPC, exact regulation is achieved, as shown in Fig. 5.5. Using the PW PPC, however, the regulation response of the

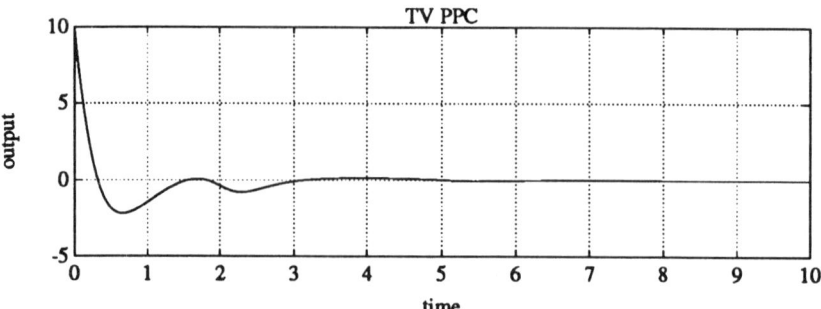

Figure 5.5: Fast TV plant ($\mu = 2$): Exact asymptotic regulation with the TV PPC.

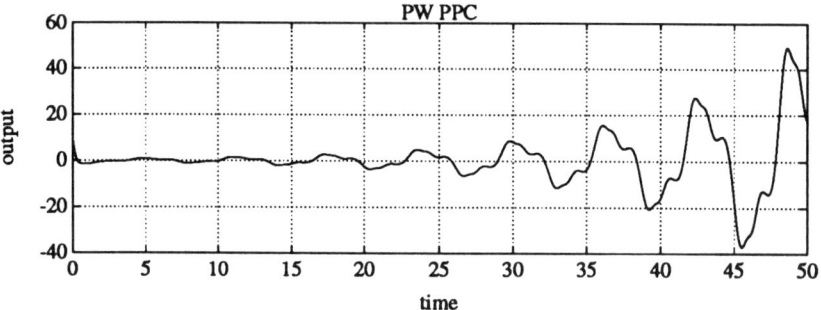

Figure 5.6: Fast TV plant ($\mu = 2$): Unbounded response obtained with the PW PPC during regulation.

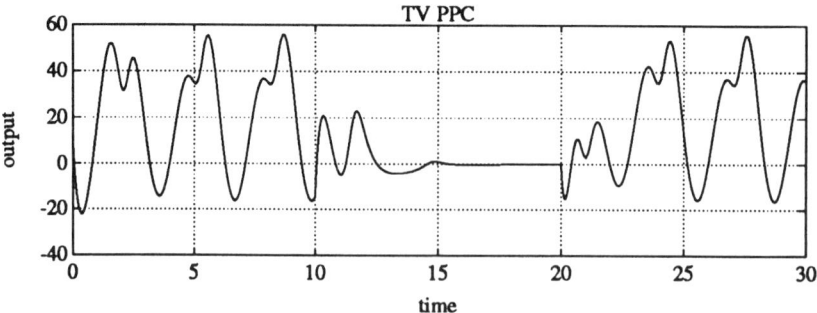

Figure 5.7: Fast TV plant ($\mu = 2$): Poor tracking of step reference inputs with the TV PPC.

5.7. EXAMPLES

plant becomes unbounded, as shown in Fig. 5.6. In fact, it can be shown (via Floquet analysis) that, for this example, the PIO (5.33) corresponding to the closed-loop with the PW PPC, is unstable. ▽▽

5.21 Example: *TV IMP/PPC Design.* Let us now consider the case where, in addition to the closed-loop stability, the control objective is that the output of the plant (5.22) should track constant reference signals. In general, the performance of a simple (TV or PW) PPC scheme with respect to such an objective, may be very poor. This is demonstrated in Fig. 5.7 where we use the TV PPC of the previous example (with $\mu = 2$) to track a square wave reference input with values alternating between 0 and 10.

In order to enhance the tracking performance of the TV PPC we employ the results presented in Section 5.4 to design a TV IMP/PPC. Thus, according to Corollary 5.7, we take $\Lambda^{-1}(s) = s^{-1}$ and design the plant input as

$$u_p = -\rho y_p + \bar{u}_p \; ; \; \bar{u}_p = (s\psi_1 + \psi_2)(s + \psi_3)^{-1}[r - y_p] \; ; \qquad (5.35)$$

where ρ is a time-varying gain and $\bar{u}_p$ is realized as in the TV PPC case (eqns. (5.23), (5.24)). We observe that in this case, we need not increase the order of the controller since $\deg[\Lambda(s)] = 1 \rightsquigarrow \deg[Q(s)] = 0$. From Corollary 5.7 we have that ρ should satisfy

$$s^2 + sa_1 + a_2 + (s+b)\rho = (s+x)s \qquad (5.36)$$

for some x. Performing the calculations in (5.36) we obtain

$$\dot{\rho} = -b\rho - (a_2 + \dot{a}_1) \; ; \; x = a_1 + \rho \qquad (5.37)$$

At this point we assume that we can find a bounded function of time, ρ, s.t. (5.37) is satisfied. Note that in the LTI case ($\dot{\rho} = 0$) this assumption reduces to $b \neq 0$, i.e., the plant zero should not be a zero of the internal model $\Lambda(s)$. The rest of the TV IMP/PPC design is done as in the PPC case with $[s + (a_1 + \rho)]s$ in the place of $[s^2 + sa_1 + a_2]$. The final TV IMP/PPC design guarantees the closed-loop internal stability as well as the exact tracking of constant (step) reference inputs. ▽▽

5.22 Simulations: Using the numerical values of Example 5.19, with $\mu = 2$, we obtain

$$\rho = -12\sin 2t + 6\cos 2t \; ; \; x = 20 + 6\cos 2t \qquad (5.38)$$

and the plant PIO, for which the TV PPC should be designed, is now $[s^2 + s(20 + 6\cos 2t) + (12\sin 2t)]^{-1}$ and the calculation of the gains p_1, p_2, p_3 is

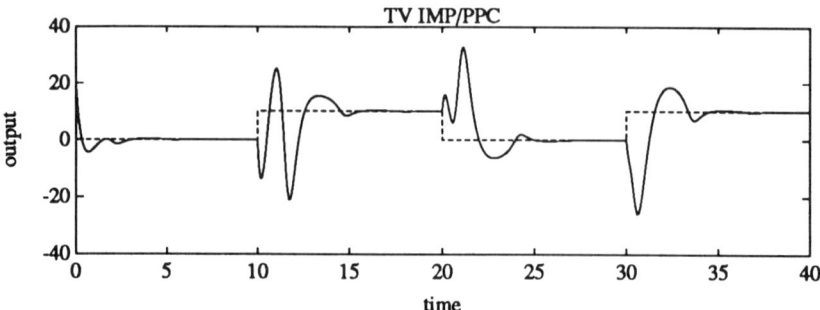

Figure 5.8: Fast TV plant ($\mu = 2$): Exact asymptotic tracking of step reference inputs with the TV IMP/PPC.

performed in the same fashion as in Example 5.19, by making the following substitutions:

$$a_1 \longleftarrow 20 + 6\cos 2t$$
$$a_2 \longleftarrow 12\sin 2t \tag{5.39}$$

The exact asymptotic tracking of step reference inputs is demonstrated in Fig. 5.8, where the output of the closed-loop plant is required to follow a square wave reference input. ▽▽

5.23 Example: *TV PPC + PW IMP Design.* Finally, to illustrate the modified IMP design mentioned in Section 5.6, let us consider Example 5.21 again and take $u_p = -\rho y_p + \bar{u}_p$. From (5.22) we obtain

$$[s^2 + sa_1 + a_2 + (s+b)\rho][y_p] = (s+b)[\bar{u}_p]$$

We may now choose ρ to introduce the internal model as a factor of the plant PIO in a PW sense. For example, assuming $b \neq 0$, we may take $\rho = -a_2/b$ (other choices are also possible) which yields the following description for the modified plant

$$[s^2 + s(a_1 + \rho)][y_p] = (s+b)[\bar{u}_p]$$

With this choice, $L(s) = s$ becomes a left factor of the plant PIO. Next, the control law $\bar{u}_p$ is designed as a TV PPC for the modified plant, i.e., for the plant $[s^2 + s(a_1 + \rho)]^{-1}[s + b]$. The resulting controller guarantees the closed-loop stability irrespective of the speed of the plant parameter variations and $O(\mu)$ tracking error to step reference inputs.

This is demonstrated in Figs. 5.9, 5.10 and 5.11. In the first of these figures $\mu = 2$, i.e., the plant parameters are fast TV; the closed-loop is shown

5.7. EXAMPLES

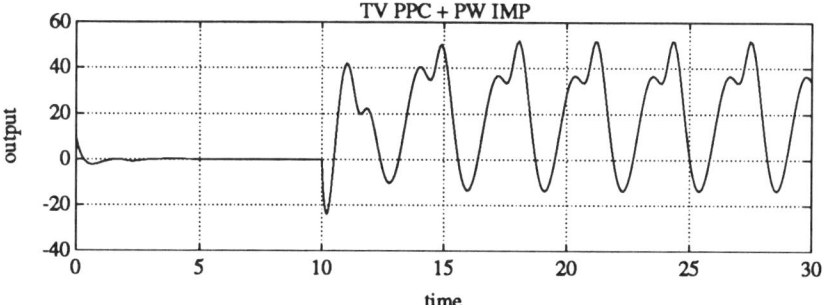

Figure 5.9: Fast TV plant ($\mu = 2$): Bounded response but poor tracking of step reference inputs with the TV PPC + PW IMP.

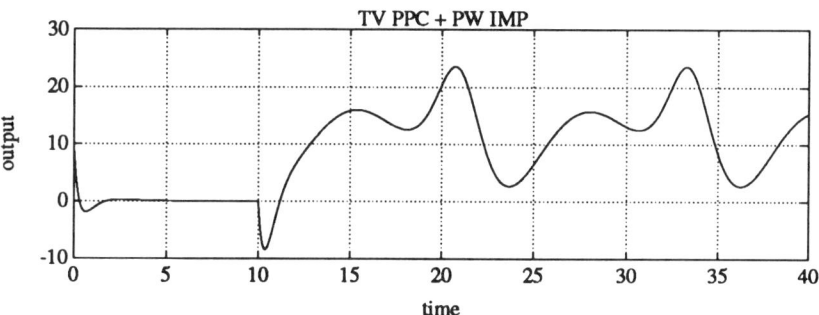

Figure 5.10: Slowly TV plant ($\mu = 0.5$): Somewhat improved tracking of step reference inputs with the TV PPC + PW IMP.

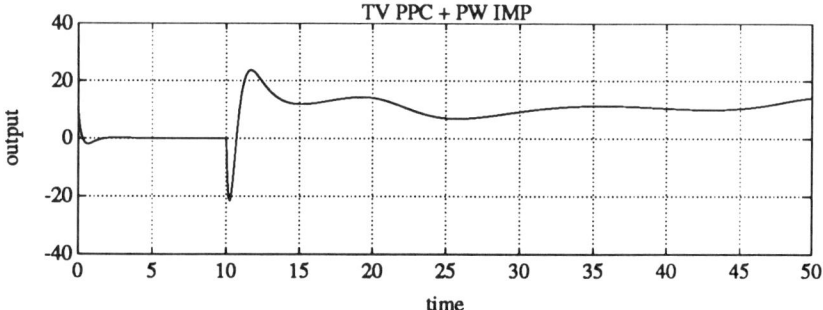

Figure 5.11: Slowly TV plant ($\mu = 0.2$): Considerably improved tracking of step reference inputs with the TV PPC + PW IMP.

146   CHAPTER 5. TV 'POLE-PLACEMENT' CONTROL

to have bounded response but tracking is very poor.[12] In the last two figures, $\mu = 0.5, 0.2$ respectively; the tracking of a step input improves considerably as the speed of the plant parameters decreases. ▽▽

APPENDIX V

Proof of Corollary 5.7:

For the first part of the Corollary we must show that $D_p Q$ and N_p are strongly left coprime PDO's in $\mathbf{R}_+$. From Definition 2.12, it suffices to show that there exist PDO's X, Y, X_0, Y_0 with smooth UB coefficients, such that

$$D_p Q X + N_p Y = 1 \; ; \; D_p Q X_0 + N_p Y_0 = 0.$$

Since D_p and N_p are strongly left coprime in $\mathbf{R}_+$, the PDO's D_R, N_R, corresponding to the right coprime factorization of the plant, exist and have smooth UB coefficients and satisfy $D_p N_R - N_p D_R = 0$. Further, since Q and N_R are strongly left coprime in $\mathbf{R}_+$, there exist PDO's X_q, Y_q, X_R, Y_R with smooth UB coefficients such that

$$Q X_q + N_R Y_q = 1 \; ; \; Q X_R + N_R Y_R = 0.$$

Hence, $D_p Q X_R + D_p N_R Y_R = D_p Q X_R + N_p D_R Y_R = 0$. Next, let X_1, Y_1 be the PDO's satisfying $D_p X_1 + N_p Y_1 = 1$ which, by the strong coprimeness of D_p and N_p, exist and have smooth UB coefficients. Since Q and N_R are strongly left coprime in $\mathbf{R}_+$, there exist PDO's X, W with smooth UB coefficients, such that $Q X - N_R W = X_1$. Define the PDO $Y = Y_1 - D_R W$. It follows that $D_p N_R W - N_p D_R W = 0$ and $D_p(X_1 + N_R W) + N_p(Y_1 - D_R W) = 1$ from which we obtain $D_p Q X + N_p Y = 1$. Letting $X_0 = X_R$ and $Y_0 = D_R Y_R$ it follows from Definition 2.12 that the PDO's $D_p Q$ and N_p are strongly left coprime in $\mathbf{R}_+$.

The above result guarantees the existence of two PDO's $\tilde{N}_1, \tilde{N}_2$ with smooth UB coefficients such that $D_p Q \tilde{N}_2 + N_p \tilde{N}_1 = \bar{A}_*$. Letting $N_1 = \tilde{N}_1 - P \tilde{N}_2$, we obtain equation (5.14). The rest of the proof of part (a.) follows from Lemma 2.35 using similar arguments as in Theorems 4.10 and 5.5.

For part (b.) and after some straightforward calculations we obtain the

---

[12] Step reference input; 0-10 transition at $t = 10$.

APPENDIX V.   147

following equation for the tracking error $e = r - y_p$

$$e = N_2(s,t) \left[X(s,t)\Lambda(s)\tilde{N}_2(s,t) + N_p(s,t)N_1(s,t) \right]^{-1} X(s,t)\Lambda(s)Q^{-1}(s)[r] \tag{5.40}$$

where, from Assumption 5.8,

$$X(s,t)\Lambda(s) = D_p(s,t)Q(s) + N_p(s,t)P(s,t).$$

Since $\Lambda(s)Q^{-1}(s)[r] = Q^{-1}(s)\Lambda(s)[r] = Q^{-1}(s)[0]$ we get

$$e = N_2(s)\bar{A}_*^{-1}(s)X(s)Q^{-1}(s)[0] \tag{5.41}$$

Consequently, by part (a.) the closed-loop system is ES and the tracking error decays to zero exponentially fast.   □□

Proof of Theorem 5.14:

Note that, by Lemma 2.41, either Assumption 3.2 or its pointwise version 5.13 can be used, since the latter implies the former for $\mu \in [0, \mu_{20})$ and some $\mu_{20} > 0$. For the PW PPC, the coefficients of the controller PDO's $M_i(s,t)$ are obtained by a pointwise solution of the corresponding Diophantine equation (see Lemma 5.1) i.e.,

$$D_p(s,t) \star M_2(s,t) + N_p(s,t) \star M_1(s,t) = A_*(s)$$

where $(\star)$ denotes a pointwise operation. Hence, from property P4 of the PDO's, it follows that $M_i(s,t)$ satisfy

$$D_p(s,t)M_2(s,t) + N_p(s,t)M_1(s,t) = A_*(s) + \Delta_1(s,t) \tag{5.42}$$

where $\Delta_1(s,t)$ is a PDO of degree $2n-2$, with coefficients $O(\mu)$. Furthermore, since the controller parameters $p_i(t)$ are found by pointwise calculations from $M_i(s,t)$, it follows that the realized control law corresponds to PDO's $\hat{M}_i(s,t)$ such that the PDO's $\hat{M}_i(s,t) - M_i(s,t)$ have coefficients $O(\mu)$ and $\hat{M}_2(s,t) - M_2(s,t)$ is of degree $n-2$. Thus, the actual closed-loop system has PIO

$$D_c^{-1}(s,t) = [D_p(s,t)\hat{M}_2(s,t) + N_p(s,t)\hat{M}_1(s,t)]^{-1} = [A_*(s) + \Delta(s,t)]^{-1} \tag{5.43}$$

where, again, $\Delta(s,t)$ is a PDO of degree $2n-2$, with coefficients $O(\mu)$. Hence, by Lemma 2.45, the closed-loop system is BIBO stable $\forall \mu \in [0, \mu_2)$, for some $\mu_2 > 0$ (and $\mu_2 \leq \mu_{20}$). Further, the internal stability of the plant and the exponential stability of the closed-loop system follow by using similar arguments as in the proof of Theorem 5.5 where, now, D_c is given by (5.43).
                                                                                                     □□

Chapter 6

On-Line Parametric Identification

6.1 Introduction

In the previous chapters we established some general controller design techniques which under their respective assumptions and given a complete knowledge of the plant parameters produce a stabilizing controller. In practice, however, such a knowledge is more often than not unavailable, introducing some uncertainty in the dynamical description of the plant. This uncertainty can be in the form of a general dynamical operator (dynamic uncertainty) and/or in the form of parametric uncertainty, that is an error in the parameters of the state-space representation of the (nominal) plant. For both types of uncertainty, a controller designed so as to make the nominal closed-loop plant exponentially stable is also able to guarantee stability in the presence of sufficiently 'small' amounts of uncertainty. For example, such a result may easily be established by employing the small-gain theorem [D.V.75] for dynamic uncertainty whose operator has small gain or Lemma 2.45 for small parametric uncertainty. In the context of TV plants, however, the requirement that the parametric uncertainty is small may be too restrictive and difficult to satisfy. Consider for example a parameter varying as $\sin(w_0 t)$; any small perturbation of the frequency w_0 is sufficient to destroy the knowledge of the parameter within a small or small-in-the-mean-square error.

On the other hand, it is intuitively possible to use a parameter estimation scheme to identify the parameters of the nominal plant on-line, thus reducing a large parametric uncertainty to a level that can be tolerated by the controller. The implementation of this deceivingly simple idea is the subject

6.2. ESTIMATION OF TV PARAMETERS

of the rest of this book. The main theoretical problem, introduced by such an implementation, is the severe nonlinear coupling between the parameter estimator and the control law. This raises questions not only about the stability/boundedness of the closed-loop system but about the existence of solutions of the differential equation describing the closed-loop as well.

In order to provide an answer to these questions, we first need to establish some fundamental properties of parameter estimation algorithms operating in a closed-loop environment. In such an environment the various signals cannot be assumed to be bounded a priori —or, for that matter, even exist at all. Moreover, any convenient arguments on the convergence of the estimated parameters are also absent since the closed-loop signals are not in the disposal of the designer. The derivation of the properties of parameter estimators under such weak conditions is the subject of this chapter. We begin with Section 6.2 where we consider a general affine in the (TV) parameters model and analyze some basic algorithms for the estimation of the TV parameters. The results of Section 6.2 are subsequently employed in Section 6.3 where we discuss the parametric identification of the I/O operator of a general LTV plant. Finally, we present a simple example to illustrate the main ideas of this chapter.

6.2 Affine Parametric Models and Estimation of TV Parameters

Let us consider the parametric model,

$$y(t) = w^\mathsf{T}(t)\theta_*(t) + \eta(t) \tag{6.1}$$

where $y : [t_0, t_0 + T] \mapsto \mathbf{R}$, $w : [t_0, t_0 + T] \mapsto \mathbf{R}^n$ are signals available for measurement, $\theta_* : \mathbf{R}_+ \mapsto \mathbf{R}^n$ is a vector of unknown parameters and $\eta : [t_0, t_0 + T] \mapsto \mathbf{R}$ is an unknown signal which is typically due to modelling error effects or noise in (6.1).

Parametric models of the form (6.1) have been extensively studied in the identification of LTI plants, where θ_* is constant (e.g., see [S.B.89, G.S.84]). As we have shown in Chapter 2, the parametric model (6.1) is also applicable in the case of LTV plants, e.g., see equation (3.21) where θ_* is a vector of the PDO coefficients of the plant I/O operator in the P_L form and η is a swapping term depending on $\dot{\theta}_*$. In this section, our objective is to develop and study parameter estimators or adaptive laws to estimate $\theta_*(t)$ in (6.1). For this purpose, we assume that

6.1 Assumption: θ_* *is UB and piecewise Lipschitz continuous on* $\mathbf{R}_+$ *with piecewise UB derivative; we use* n_I *to denote the number of points of*

discontinuity of θ_* —always of the first kind— in an interval $I \subseteq [t_0, t_0 + T]$; ■

6.2 Assumption: *w, η (and therefore y) are UB and piecewise continuous on $[t_0, t_0 + T]$; in this context 'UB' also denotes that the bounds are independent of T and t_0;* ■

6.3 Assumption: *a bounded, convex set $\mathcal{M}(t)$ with smooth boundary and such that $\theta_*(t) \in \mathcal{M}(t)$, $\forall t \in \mathbf{R}_+$, is known a priori. For simplicity, we assume that the set $\mathcal{M}(t)$ is a ball in $\mathbf{R}^n$, i.e.,*

$$\mathcal{M}(t) = \mathcal{M} = \{\theta \in \mathbf{R}^n : |\theta - \theta_c| \leq M_0\}$$

where the center θ_c and the radius M_0 are constant. ■

Assumption 6.2 requires w, η and y to be UB on $[t_0, t_0 + T]$ with bounds that are independent of t_0 and T. In most identification problems of LTV plants parametrized by (6.1), such an assumption is satisfied by considering either stable LTV plants or plants which are stabilized by a fixed (non-adaptive) controller. On the other hand, if Assumption 6.2 fails, it is often possible to rewrite (6.1) as

$$\bar{y}(t) = \bar{w}^\mathsf{T}(t)\theta_*(t) + \bar{\eta}(t)$$

where $\bar{x}$ denotes the *normalized signal x*, i.e., $\bar{x}(t) = x(t)/m(t)$ and $m(t)$ is a suitable normalization signal selected so that $\bar{w}, \bar{\eta}$ and $\bar{y}$ satisfy Assumption 6.2. This situation typically arises in an adaptive control setup where the closed-loop signals may not be assumed to be UB a priori. Normalization is therefore a useful tool which allows us to extend any results developed for the parametric model (6.1) to such cases. A more detailed discussion on the design of a normalization signal is given later in this and the next chapter.

In Assumption 6.3, the parameter M_0 used in the definition of the set $\mathcal{M}$ is a measure of the size of the parametric uncertainty in $\theta_*(t)$. With a little additional effort both M_0 and θ_c can be allowed to vary with time. For example, under a similar formulation, it is possible to admit generalized ellipsoids as parametric uncertainty sets, i.e.,

$$\mathcal{M}(t) = \{\theta \in \mathbf{R}^n : |M(t)[\theta - \theta_c(t)]| \leq 1\}$$

where $M(t)$ is a known, strongly nonsingular, positive definite matrix $\forall t \in \mathbf{R}_+$, $\theta_c(t)$ is a known vector and M, θ_c are Lipschitz continuous and UB on $\mathbf{R}_+$. At this stage, we do not consider such a generalization, as it unnecessarily

6.2. ESTIMATION OF TV PARAMETERS

complicates the presentation without offering any significant improvement of the results; it may be useful, however, in subsequent studies.

Furthermore, the vector norm in the definition of the set $\mathcal{M}$ can be chosen as the most convenient one for the particular problem although norms whose unit balls have non-smooth boundaries introduce some additional difficulties in the estimator construction. In our discussion we avoid this rather minor issue by considering only two vector norms, namely the $|\cdot|_2$ and the $|\cdot|_\infty$ which are a 'natural' selection in most applications. For example, $|\cdot|_2$ would be less conservative if the uncertainty in $\theta_*(t)$ is specified in terms of the radius of a sphere while $|\cdot|_\infty$ would be more appropriate if it is specified in terms of the minimum and maximum value of its components. Notice that in this case, the estimation problem can be treated component-wise with interval constraints and thus satisfy the smooth boundary assumption; an example of this is presented later.

We may now develop an adaptive law to estimate θ_* in (6.1) according to the following procedure:

Let $\theta(t)$ be the estimate of $\theta_*(t)$ at time t. Then the estimated value of y at time t, based on the estimate $\theta(t)$, is

$$\hat{y}(t) = w^\top(t)\theta(t) \tag{6.2}$$

The *estimation error*

$$\epsilon_1 = \hat{y} - y = w^\top \theta - y \tag{6.3}$$

is therefore a measure of the 'quality' of estimation in the sense of (6.1), that is, how well is the partially unknown parametric model (6.1) approximated by (6.2). Substituting (6.1) and (6.2) in (6.3), it follows that the estimation error is expressed as

$$\epsilon_1 = w^\top \phi - \eta$$

where $\phi \triangleq \theta - \theta_*$ is the parameter error. In other words, ϵ_1 contains information about the parameter error, in an inner product form, corrupted by the noise term η. An adaptive law to update the estimate θ is designed by using the gradient projection method to minimize the cost

$$J(\theta) = \epsilon_1^2 = (w^\top \theta - y)^2$$

with respect to θ and subject to the constraint $\theta \in \mathcal{M}$. Such an update law has the form

$$\dot{\theta} = \mathcal{P}(-\gamma \epsilon_1 w) \ ; \ \ \theta(t_0) \in \mathcal{M} \tag{6.4}$$

where $\gamma > 0$ is a constant gain referred to as the *adaptation gain* and, for simplicity, is taken as scalar and $\mathcal{P}$ denotes a projection operator, designed to guarantee $\theta \in \mathcal{M}$.

Essentially, the operator $\mathcal{P}$ is the identity when θ is in the interior of $\mathcal{M}$ and projects $-\gamma\epsilon_1 w$ on the tangent hyperplane at $\theta(t)$ when the latter is on the boundary of $\mathcal{M}$ and the vectorfield points towards the exterior of $\mathcal{M}$ (e.g., see [Ega.79, G.S.84, S.B.89]). Some additional provisions are taken so that the vectorfield in (6.4) is (at least locally) Lipschitz continuous, in order to avoid any problems with the existence and uniqueness of θ, as well as problems in the numerical simulations of the adaptive law. For this purpose, we may slightly increase the size of the set $\mathcal{M}$, i.e., define a boundary region of some small but nonzero thickness, where we make a smooth transition between projected and unprojected vectorfields. The thickness of the boundary region is denoted throughout by ϵ_* which is treated as a small design parameter. In the following examples we illustrate such a design of the projection operator in the two, most frequently encountered, cases where the set $\mathcal{M}$ is specified in terms of an L_∞ and an L_2 vector norm. It should be mentioned that the existence and uniqueness (in the sense of Fillipov) of solutions of differential equations of the form (6.4) have been established in [P.I.91] for a general projection operator. It is therefore theoretically possible to select $\epsilon_* = 0$. In practice, however, such a choice may cause numerical problems due to the discontinuous vectorfield in the differential equation with projection. For this reason, in the following we assume that $\epsilon_* > 0$.

6.4 Example: When the set $\mathcal{M}$ is defined in terms of the minimum and maximum values of the components of θ_*, say θ_{i*min} and θ_{i*max} respectively, the projection operator $\mathcal{P}$ is simply constructed as a multiplier of the form

$$\mathcal{P} = \mathrm{diag}[\sigma_{pi}]$$

$$\sigma_{pi} = \begin{cases} \max\left\{0, \min\left[1, 1 + \frac{\theta_i - \theta_{i*min}}{\epsilon_*}\right]\right\} & \text{when } \epsilon_1 w_i > 0 \\ \max\left\{0, \min\left[1, 1 + \frac{\theta_{i*max} - \theta_i}{\epsilon_*}\right]\right\} & \text{when } \epsilon_1 w_i < 0 \\ 1 & \text{otherwise} \end{cases} \quad (6.5)$$

where ϵ_* is an arbitrary, small positive constant whose purpose is to ensure the Lipschitz continuity (in θ) of $\sigma_{pi}\epsilon_1 w_i$, something that can be shown using Assumptions 6.1-6.3. Pictorially, the functions σ_{pi} are shown in Fig. 6.1.

▽▽

6.5 Example: A similar construction of the projection operator $\mathcal{P}$ is also possible when the set $\mathcal{M}$ is defined in terms of the Euclidean norm of θ_*. In this case the set $\mathcal{M}$ is of the general form

$$\mathcal{M} = \{\theta \in \mathbf{R}^n : |\theta - \theta_c|_{2,M} \leq 1\}$$

where $|\theta|_{2,M}$ denotes the weighted Euclidean norm $\left(\theta^\top M \theta\right)^{1/2}$ and M is a

6.2. ESTIMATION OF TV PARAMETERS

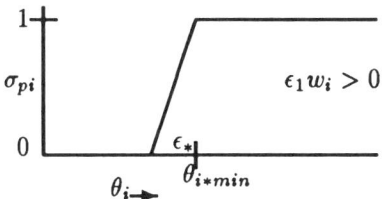
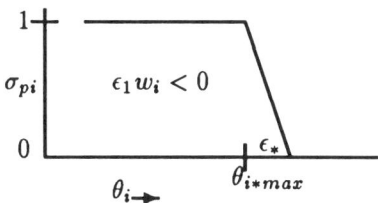

Figure 6.1: The functions σ_{pi}.

symmetric positive definite matrix. In other words, $\mathcal{M}$ is a generalized ellipsoid in $\mathbf{R}^n$ centered at θ_c.

For such a set, one simple form of a projection operator can be given in terms of a multiplier

$$\mathcal{P} = I - \sigma_p \theta_\perp \theta_\perp^T$$

where

$$\theta_\perp = \frac{M(\theta - \theta_c)}{|M(\theta - \theta_c)|_2}$$

is the normal vector, pointing outwards, on the surface $|\theta - \theta_c|_{2,M} = constant$,[1]

$$\sigma_p = \begin{cases} \max\left\{0, \min\left[1, \frac{|\theta-\theta_c|_{2,M}-1}{\epsilon_*}\right]\right\} & \text{when } \epsilon_1 w^T \theta_\perp < 0 \\ 0 & \text{otherwise} \end{cases} \quad (6.6)$$

and ϵ_* is again an arbitrary, small positive constant. $\triangledown\triangledown$

The properties of the adaptive law (6.4) are given by the following theorem.

6.6 Theorem: *Consider the parametric model (6.1), satisfying Assumptions 6.1–6.3, and the adaptive law (6.4). Then,*

a. θ, $\dot{\theta}$ are UB on $[t_0, t_0 + T]$ (θ is within distance ϵ_* from $\mathcal{M}$);

b. there exist constants C_0, C_1, K_ϕ, K_J, independent of T, t_0, such that for any interval $I = [t_I, t_I + T_I] \subseteq [t_0, t_0 + T]$,

1. $\displaystyle\int_{t_I}^{t_I+T_I} \epsilon_1^2(t)\, dt \leq C_0 + \int_{t_I}^{t_I+T_I} \eta^2(t)\, dt + \frac{K_\phi}{\gamma}\int_{t_I}^{t_I+T_I} |\dot{\theta}_*^s(t)|_i\, dt + \frac{K_J}{\gamma} n_I$

2. $\displaystyle\int_{t_I}^{t_I+T_I} |\dot{\theta}(t)|^2\, dt \leq \gamma^2 C_1 \int_{t_I}^{t_I+T_I} \epsilon_1^2(t)\, dt$

---

[1] The constant used here is the current value of $|\theta - \theta_c|_{2,M}$, i.e., $\theta_\perp$ is the normal vector on the boundary of a scaled ellipsoid, similar to $\mathcal{M}$, that passes through the point θ. Note that a more efficient but more complicated type of projection would be to take $\theta_\perp$ to be the vector of minimum distance between θ and $\mathcal{M}$.

where θ_*^s denotes the Lipschitz continuous part of θ_*, i.e., $\theta_* = \theta_*^s + \theta_*^J$, for some piecewise constant (jump) function θ_*^J and $|\cdot|_i$ is the linear functional norm induced by the norm employed in the definition of the set $\mathcal{M}$. ▽▽

Proof: In Appendix VI.

At this point, it is worthwhile to make some observations on the performance of the adaptive law (6.4), as given by Theorem 6.6 and its relation with the various design parameters.

- The adaptive law (6.4) identifies the I/O properties of θ_*, viewed as an operator defined by (6.1), in the mean-square sense. In this sense, the mean-square value of the estimation error is a measure of the quality of identification. Thus, the identification is 'successful' from an I/O point of view, provided that θ_* varies slowly with time and the noise term η is small.

- The constants used in the theorem depend on the size of parametric uncertainty (M_0) and the gain of adaptation (γ) as follows:

$$C_0 = O(M_0^2/\gamma) \ ; \ K_\phi = O(M_0) \ ; \ C_1 = \|w\|_\infty^2$$

while K_j is of the order of the maximum size of jumps. These relations imply that if the adaptation gain is large enough compared with the speed of variation of θ_*, the mean-square value of the estimation error is of the order of the mean-square value of the noise term η.

- In the special case where $\eta = 0$ and $\theta_* = constant$, the estimation error ϵ_1 is square integrable. It follows that if ϵ_1 is also uniformly continuous, then as $T \to \infty$, $\epsilon_1 \to 0$. This does not imply, however, that for small η and $\dot{\theta}_*$, ϵ_1 remains uniformly small as $T \to \infty$. Instead, properties (2) and (3) allow for the possibility of 'burst' phenomena, whereby the estimation error may attain $O(M_0)$-large values during short time periods. Such phenomena can be avoided in the case of constant parameters by using some a priori information on the L_∞ bound of η and a 'dead-zone' modification in the adaptive law (e.g., see [P.N.82, MGHM.88, G.S.84]). It is not clear though, whether this technique can be transferred to the TV case with similar results.

- Even in the TI case, Theorem 6.6 does not provide any conclusions on the 'strong' convergence of the operator θ to θ_*, i.e., the convergence of $\|\theta - \theta_*\|$ to zero. Although such a convergence is desirable, it does not seem to be possible unless θ_* is constant and the vector w is *persistently exciting* [And.77]. When the unknown parameter vector θ_* is slowly TV this result can be extended to the parametric model (6.1) within an error that depends on the speed of variation of θ_* [M.G.87].

Analogous results can be obtained with estimators employing the so-called

6.2. ESTIMATION OF TV PARAMETERS

σ-modification [I.K.83]. With this modification, a term $-\sigma(\theta - \theta_c)$ is introduced in the vectorfield of the estimator, providing the necessary component to ensure the boundedness of the parameter estimates, for example

$$\dot{\theta} = -\gamma\epsilon_1 w - \sigma(\theta - \theta_c) \tag{6.7}$$

where σ is a positive constant.

6.7 Theorem: *Consider the parametric model (6.1), satisfying Assumptions 6.1–6.2 and the adaptive law (6.7). Then,*

a. θ, $\dot{\theta}$ are UB on $[t_0, t_0 + T]$ and $\theta(t)$ converges exponentially fast to a residual set

$$\left\{ \theta : |\theta(t) - \theta_c|_2 \leq \sqrt{\frac{\gamma}{2}} \mathcal{E}_{-\sigma}(t) \| (y - w^T \theta_c)_t \|_{2,\sigma} \right\}$$

b. there exist constants C_0, C_1, K_J, independent of T, t_0, such that for any interval $I = [t_I, t_I + T_I] \subseteq [t_0, t_0 + T]$,

1. $\displaystyle\int_{t_I}^{t_I+T_I} \epsilon_1^2(t)\,dt \leq C_0 + \int_{t_I}^{t_I+T_I} \eta^2(t)\,dt \ldots$

   $\displaystyle + \frac{1}{2\gamma}\int_{t_I}^{t_I+T_I} |\sqrt{\sigma}(\theta_*(t) - \theta_c) + \frac{1}{\sqrt{\sigma}}\dot{\theta}_*^s(t)|_2^2\,dt + \frac{K_J}{\gamma}n_I$

2. $\displaystyle\left[\int_{t_I}^{t_I+T_I} |\dot{\theta}(t)|_2^2\,dt\right]^{1/2} \leq \gamma\left[C_1 \int_{t_I}^{t_I+T_I} \epsilon_1^2(t)\,dt\right]^{1/2} \ldots$

   $\displaystyle + \sigma\left[\int_{t_I}^{t_I+T_I} |\theta(t) - \theta_c|_2^2\,dt\right]^{1/2}$

where θ_*^s denotes the Lipschitz continuous part of θ_*. ▽▽

Proof: In Appendix VI.

Such an estimator has the advantage that parameter boundedness is guaranteed without the explicit a priori knowledge of a bound on the parametric uncertainty. However, unlike an estimator using projection, it does not recover any asymptotic identification properties when the parameters become constants. An additional drawback is that the selection of σ and γ involves certain trade-offs since both σ and $1/\sigma$ appear in the various bounds. For example, $\frac{\sigma}{\gamma}|\theta_* - \theta_c|^2$ and $\frac{1}{\gamma\sigma}|\dot{\theta}_*^s|^2$ both appear as perturbations in the mean-square value of the estimation error. And although their contribution can be lessened by increasing the value of the adaptation gain, such an approach has the undesirable effect of increasing the bound of θ as well as the mean-square value of $\dot{\theta}$ and may cause instability in an adaptive control scheme employing

(6.7). On the other hand, the properties of the above theorem indicate that this estimator can be particularly effective when the (unknown) parametric uncertainty is small (i.e., $|\theta_* - \theta_c|$ is 'small').

It should be mentioned that another difference between the two techniques discussed above is that using an estimator with projection, the parameters are guaranteed to be UB with considerably weaker assumptions. In fact, if the bounds of the signals w, η depend on T, the projection still guarantees the uniform boundedness of θ, although in this case the constants in the Theorem 6.6 may depend on T. The same statement is not true for an estimator with a σ-modification, in which case, the failure of w, η to be UB results in parameter estimates whose bound may depend on T. Such a property, however, is not exploited any further and in the sequel Assumption 6.2 —or sufficient conditions for it to be satisfied— is employed whenever a parameter estimation algorithm is constructed.

6.8 Remark: It is possible to retain most of the attractive simplicity of the σ-modification and, at the same time, achieve error bounds as in Theorem 6.6 by using the a priori knowledge of M_0 to switch on the σ term only when the magnitude of the parameter estimates is large. In this case, σ is chosen as a Lipschitz function of $|\theta - \theta_c|_2$, e.g.,

$$\sigma = \sigma_0 \max\left\{0, \min\left[1, \frac{|\theta - \theta_c|_2 - M_0}{\epsilon_*}\right]\right\}$$

where σ_0 is a positive constant. This switching σ-modification has the property that, provided $|\theta_* - \theta_c|_2 < M_0$, $\sigma(\theta - \theta_c)^T(\theta - \theta_*) \geq 0$.[2] Using this property it is straightforward to verify that Theorem 6.7 is still valid and the mean-square value of ϵ_1 satisfies

$$\int_{t_I}^{t_I+T_I} \left[\epsilon_1^2(t) + \frac{2\sigma}{\gamma}(\theta(t) - \theta_c)^T(\theta(t) - \theta_*(t))\right] dt \leq C_0' + \ldots$$
$$+ \int_{t_I}^{t_I+T_I} \eta^2(t)\, dt + \frac{K_\phi'}{\gamma}\int_{t_I}^{t_I+T_I} |\dot{\theta}_*^s(t)|\, dt + \frac{K_J'}{\gamma}n_I$$

where the constants C_0', K_ϕ', K_J' have similar properties as C_0, K_ϕ, K_J of the projection algorithm. Moreover, θ is UB and $(\theta - \theta_c)$ is exponentially decaying to a set bounded (in the 2-norm) by $M_0 + O[\epsilon_*] + O[\sqrt{\frac{\gamma}{\sigma_0}}]$. Hence, the size of this set is essentially independent of the initial conditions $\theta(t_0)$ and, with an appropriate choice of ϵ_* and the ratio γ/σ_0, can be made arbitrarily close to M_0. An attractive property of the switching σ-modification is that it can be used together with a projection to accommodate a priori known time-varying

---

[2] Otherwise, the switching σ-modification acts effectively as the constant one with similar properties.

parametric uncertainty (θ_c, M_0) sets, while preserving the properties of the projection algorithms. Note that when $\mathcal{M}(t)$ changes with time, there is a possibility of θ being well outside the projection set at some time instant. In such a case, the term $-\sigma(\theta - \theta_c)$ provides means to bring θ back to $\mathcal{M}(t)$ exponentially fast. ▽▽

It is now apparent that the estimators presented in this section have the essential properties required for the successful I/O identification of a slowly TV multiplier θ_*. The quality of the identification is determined by the mean-square value of the estimation error and, as intuitively expected, improves as the speed of variations of θ_* decreases. What remains to be established is that the unknown I/O operator can be described in a form satisfying the assumptions of Theorem 6.6 and 6.7. This problem is studied in the following section.

6.3 Parametric Identification of LTV Plants

In this section we discuss the identification problem of the I/O operator of a LTV plant. Our focus is on parametric identification whereby we perform the identification of the operator by estimating the parameters of a suitable parametric model. Such parametric models for the LTV plants under consideration have been derived in Chapter 3 [e.g., see equations (3.20), (3.21)] and are in or can be transformed to the standard form (6.1). Thus, we can apply the results of the previous section to estimate the parameter vector θ_* in, say, (3.21) which in turn corresponds to the PDO coefficients of the P_L-form of the plant I/O operator. The analysis of such identification schemes and the derivation of error bounds assessing the quality of the parameter estimates as well as the quality of the identified plant I/O operator are the subjects of this section. For reasons of clarity we consider the case of smooth parameter variations first and then generalize the results to the case of non-smooth parameter variations.

6.3.1 Smooth Parameter Variations

Let us consider an LTV plant with a state-space representation given by (3.1) and satisfying Assumptions 3.1-3.3 and suppose that its order n is known. Invoking Lemma 2.32, the plant representation is topologically equivalent to the corresponding observable canonical form and, therefore, it admits an I/O description in terms of PDO's given by

$$y_p = G_p^L(s,t)[u_p] = D_p^{-1}(s,t)N_p(s,t)[u_p]$$

Using Lemma 3.10, the I/O description of the plant becomes

$$y_p = G(s)[u_p \theta_{1*}] + G(s)[y_p \theta_{2*}] \tag{6.8}$$

where $G(s) = q^T(sI - F)^{-1}$ is a filter selected by the designer and q, F are as specified in that lemma. Since the vector $\theta_* = [\theta_{1*}^T, \theta_{2*}^T]^T : \mathbf{R}_+ \mapsto \mathbf{R}^{2n}$ is directly related to the coefficients of $D_p(s,t)$ and $N_p(s,t)$ by a constant affine transformation depending on q, F, it follows that $\theta_*(t)$ is also smooth and UB.

Further, in order to allow nonzero initial conditions at $t = t_0$, in both the plant and the auxiliary filters $G(s)$, we augment equation (6.8) by an additional exponentially decaying term, as follows

$$y_p = G(s)[u_p \theta_{1*}] + G(s)[y_p \theta_{2*}] + \varepsilon \tag{6.9}$$

where $|\varepsilon(t, t_0)| \le c_0 \exp[-a(t - t_0)]$, c_0 is a constant depending on the size of the initial conditions and $-a$ is the rate of exponential stability of the auxiliary filters (e.g., see proof of Lemma 3.10).

Next, employing the notion of structured parameter variations we assume, without loss of generality, that

$$\theta_*(t) = \Pi(t)\hat{\theta}_*(t)$$

where $\Pi(t)$ is a known (not necessarily square) matrix with smooth UB entries and $\hat{\theta}_*(t)$ is a partially unknown smooth UB vector. Letting $\Pi_1(t), \Pi_2(t)$ be a partition of $\Pi(t)$ such that $\theta_{i*}(t) = \Pi_i(t)\hat{\theta}_*(t)$, equation (6.9) becomes

$$y_p = G(s)[u_p \Pi_1 \hat{\theta}_*] + G(s)[y_p \Pi_2 \hat{\theta}_*] + \varepsilon \tag{6.10}$$

We may now invoke the swapping Lemma 2.59 to express (6.10) in the general form of the parametric model (6.1)

$$\begin{aligned}
y_p &= w^T \hat{\theta}_* - \eta \tag{6.11} \\
w &= [G(s)\{u_p \Pi_1\} + G(s)\{y_p \Pi_2\}]^T \\
\eta &= w^T \hat{\theta}_* - G(s)[u_p \Pi_1 \hat{\theta}_*] - G(s)[y_p \Pi_2 \hat{\theta}_*] \\
&= G(s)\{G'(s)[u_p \Pi_1]\dot{\hat{\theta}}_*\} + G(s)\{G'(s)[y_p \Pi_2]\dot{\hat{\theta}}_*\} \\
G'(s) &= (sI - F)^{-1}
\end{aligned}$$

In order to employ Theorem 6.6 or 6.7, we need to ensure the boundedness of the regressor vector w and swapping term η in the parametric model (6.11). One way to achieve this is to normalize both sides of (6.11) by using one of the normalization signals described in Chapter 2. The general form of these

6.3. PARAMETRIC IDENTIFICATION OF LTV PLANTS

signals is $m_p^{1/p}$ where $p \in [1, \infty)$ and m_p is constructed by integrating the following differential equation

$$\dot{m}_p = -p\delta_0 m_p + |QU|^p + q_e \; ; \quad m_p(t_0) > 0 \qquad (6.12)$$

where Q is a positive definite weighting matrix, $U = [u_p, y_p]^T$, q_e is a positive constant and δ_0 is such that $(F + \delta_0 I)$ is a Hurwitz matrix. Since the g_{p,δ_0}-gains of G and G' are finite, Lemma 2.56 shows that the ratio $x/m_p^{1/p}$, x being any of the signals w, η, y_p, is UB, for as long as the truncated U_t belongs to L_∞, (implying, of course, that $U_t \in L_p(\delta)$ for $t < \infty$).

Thus, the plant I/O description after normalization becomes

$$\frac{y_p}{m_p^{1/p}} = \frac{w^T \hat{\theta}_*}{m_p^{1/p}} - \frac{\eta}{m_p^{1/p}} + \frac{\varepsilon}{m_p^{1/p}} \qquad (6.13)$$

with the normalized signals being UB and satisfying Assumption 6.2 for as long as the truncated U_t belongs to L_∞. Applying the procedure of Section 6.2 on (6.11) we define the estimation error ϵ_1 by

$$\epsilon_1 = w^T \hat{\theta} - y_p$$

and the normalized estimation error

$$\frac{\epsilon_1}{m_p^{1/p}} = \frac{w^T \hat{\theta}}{m_p^{1/p}} - \frac{y_p}{m_p^{1/p}}$$

where $\hat{\theta}(t)$ is the estimate of $\hat{\theta}_*(t)$ at time t. The last equation defines a (normalized) parametric model that satisfies Assumptions 6.1–6.3. Therefore, the results of the previous section are now applicable, providing means to design and analyze adaptive laws to estimate $\hat{\theta}_*$. Of course, once $\hat{\theta}$ becomes available, the estimate $\theta(t)$ of $\theta_*(t)$ is obtained from

$$\theta(t) = \Pi(t)\hat{\theta}(t)$$

Note that the estimation of θ via $\hat{\theta}$ may also be viewed as a Kalman filtering problem where θ is known to satisfy a differential equation $\dot{\theta} = A(t)\theta$ with unknown initial conditions. In this case the matrix $\Pi(t)$ corresponds to the completely or partially known state transition matrix associated with $A(t)$.

Before proceeding with the design and analysis of adaptive laws to estimate $\hat{\theta}_*$, let us summarize at this point the main notational conventions.

- U denotes the I/O vector $[u_p, y_p]^T$. It is assumed throughout this section that the truncated $(u_p)_t$ belongs to L_∞, for all t in a compact interval

$[t_0, t_0 + T]$; $T > 0$. For the class of plants considered it follows that $(y_p)_t$, $U_t \in L_\infty$ over the same interval; we use the notation

$$U \in L_{\infty, [t_0, t_0+T]}$$

to signify this assumption. Notice that this assumption does not require u_p or y_p to be in L_∞^e or the plant to be stable in any sense.

- $\phi, \hat{\phi}$ denote the parameter errors $\theta - \theta_*$ and $\hat{\theta} - \hat{\theta}_*$ respectively, related by $\phi(t) = \Pi(t)\hat{\phi}(t)$. Whenever necessary, the matrix Π may be partitioned as Π_i such that $\theta_{i*} = \Pi_i \hat{\theta}_*$. We also denote by $\hat{\theta}_c$ a bias term, possibly zero, being the center of the ellipsoid containing $\hat{\theta}_*$.

- $G(s) = q^T(sI - F)^{-1}$, $G'(s) = (sI - F)^{-1}$ are frequently used. Also, to simplify the various expressions, we use the shorthand notation G_π, G'_π, G_{θ_*}, G_{θ_c} to signify the operators defined by their I/O pairs:

$$\begin{aligned} G_\pi &: QU \mapsto w \triangleq G(s)[u_p \Pi_1] + G(s)[y_p \Pi_2] \\ G'_\pi &: QU \mapsto G'(s)[u_p \Pi_1] + G'(s)[y_p \Pi_2] \\ G_{\theta_*} &: QU \mapsto G(s)[u_p \theta_{1*}] + G(s)[y_p \theta_{2*}] = y_p \\ G_{\theta_c} &: QU \mapsto G(s)[u_p \Pi_1 \hat{\theta}_c] + G(s)[y_p \Pi_2 \hat{\theta}_c] \end{aligned}$$

where Q is a positive definite matrix.

- η and $\hat{\eta}$ are reserved for the 'swapping' terms

$$\begin{aligned} \eta &= w^T \hat{\theta}_* - G(s)[u_p \Pi_1 \hat{\theta}_*] - G(s)[y_p \Pi_2 \hat{\theta}_*] \\ \hat{\eta} &= w^T \hat{\theta} - G(s)[u_p \Pi_1 \hat{\theta}] - G(s)[y_p \Pi_2 \hat{\theta}] \end{aligned}$$

which, for absolutely continuous $\theta_*, \hat{\theta}$ and zero initial conditions become

$$\begin{aligned} \eta &= G(s)\{G'(s)[u_p \Pi_1]\dot{\hat{\theta}}_*\} + G(s)\{G'(s)[y_p \Pi_2]\dot{\hat{\theta}}_*\} \\ \hat{\eta} &= G(s)\{G'(s)[u_p \Pi_1]\dot{\hat{\theta}}\} + G(s)\{G'(s)[y_p \Pi_2]\dot{\hat{\theta}}\} \end{aligned}$$

- e_1 is used for the 'identification' error defined by

$$e_1 = G(s)[u_p \theta_1] + G(s)[y_p \theta_2] - y_p$$

and ϵ_1 for the 'estimation' error

$$\epsilon_1 = w^T \hat{\theta} - y_p$$

while ε denotes an exponentially decaying term describing the effect of initial conditions.

6.3. PARAMETRIC IDENTIFICATION OF LTV PLANTS

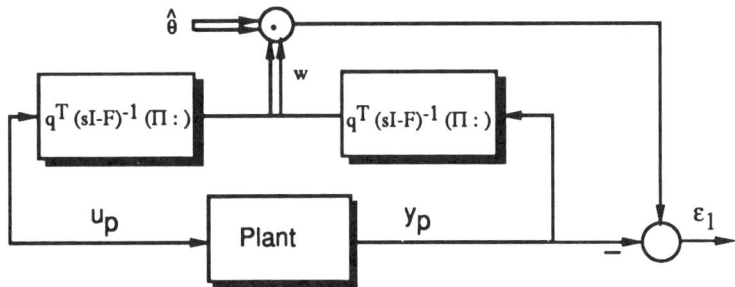

Figure 6.2: Block diagram of the plant identifier.

With this notation, the block diagram of the identifier structure is shown in Fig. 6.2. It is straightforward to verify that the identification error satisfies

$$e_1 = G(s)[u_p \phi_1] + G(s)[y_p \phi_2] + \varepsilon$$

and hence it is unbiased, in the sense that for $\hat{\theta} = \hat{\theta}_*$, e_1 is decaying to zero exponentially fast. The identification error is therefore appropriate to assess the quality of the parameter estimates interpreted as a part of the plant I/O identification algorithm, i.e., whenever $G(s)[u_p \theta_1] + G(s)[y_p \theta_2]$ is used as an estimate of the plant I/O operator. On the other hand, the estimation error, related to the parameter error via

$$\epsilon_1 = w^\top \hat{\phi} + \eta$$

has the desired for estimation purposes inner product form between the parameter error and a regressor vector (similarly for the normalized estimation error). Notice, however that, unless $\hat{\theta}_*(t)$ is constant, $\hat{\theta} = \hat{\theta}_*$ does not necessarily imply that ϵ_1 is exponentially decaying or even converging to zero.

6.9 Corollary: *Suppose that for an LTV plant satisfying Assumptions 3.1–3.3, $U \in L_{\infty,[t_0,t_0+T]}$ and $\hat{\theta}_*$ satisfies Assumption 6.3. Then, the estimator*

$$\dot{\hat{\theta}} = \mathcal{P}\left(-\gamma \frac{\epsilon_1 w}{m_p^{2/p}}\right) \quad ; \quad \hat{\theta}(t_0) \in \mathcal{M}$$

guarantees that $\hat{\theta}, \dot{\hat{\theta}}$ are UB on $[t_0, t_0 + T]$ and there exist constants C_0, C_0', C_0'', K_ϕ, independent of T, t_0, such that for any interval $I = [t_I, t_I + T_I] \subseteq [t_0, t_0 + T]$

1. $\displaystyle\int_{t_I}^{t_I+T_I} \left(\frac{\epsilon_1(t)}{m_p^{1/p}(t)}\right)^2 dt \leq C_0 + \frac{K_\phi}{\gamma} \int_{t_I}^{t_I+T_I} |\dot{\hat{\theta}}_*(t)|_i \, dt \ldots$

$$+ \left[g_{p,\delta}[G] g_{p,\delta_0}[G'_\pi] \left\{ \int_{t_I}^{t_I+T_I} \left(\int_{t_I}^{t} e^{-p(\delta-\delta_0)(t-\tau)} |\dot{\hat{\theta}}_*(\tau)|^p \, d\tau \right)^{\frac{2}{p}} dt \right\}^{\frac{1}{2}} \right.$$
$$\left. + C'_0 \right]^2;$$

2. $\displaystyle\int_{t_I}^{t_I+T_I} |\dot{\hat{\theta}}(t)|^2 \, dt \leq \left[\gamma g_{p,\delta_0}[G_\pi] \left(\int_{t_I}^{t_I+T_I} \left(\frac{\epsilon_1(t)}{m_p^{1/p}(t)} \right)^2 dt \right)^{1/2} \cdots \right.$
$$\left. + C''_0 \right]^2;$$

where $\delta > \delta_0$ is such that $F + \delta I$ is a Hurwitz matrix, $K_\phi \leq 4M_0 + 2\epsilon_*$ and the $g_{p,\delta}$-gains of the various operators are evaluated with respect to the underlying vector space norm. ▽▽

Proof: In Appendix VI.

The above corollary gives a quite general description of the mean-squared normalized estimation error in terms of the speed of the unstructured part of the plant parameter variations. Despite its generality, however, the resulting expressions are quite complicated and, perhaps, unnecessarily so. Therefore, we choose at this point to work with the normalization signal m_2 only, yielding more convenient and compact expressions.

Also, in an effort to further simplify the various bounds we introduce a parameter μ as a measure of the 'average' speed of the unstructured parameter variations. That is, we assume that

6.10 Assumption: *There exist constants c, μ such that*

(a.) $\displaystyle\int_{t_0}^{t_0+T} |\dot{\hat{\theta}}_*(t)|_i \, dt \leq c + \mu T$

(b.) $\displaystyle\int_{t_0}^{t_0+T} |\dot{\hat{\theta}}_*(t)|^2 \, dt \leq c + \mu T$

for all $t_0, T \geq 0$.[3] ■

We are now in the position to give a simplified and more intuitive, albeit more conservative, version of Corollary 6.9, stated as follows.

6.11 Corollary: *Under the assumptions of Corollary 6.9 and Assumption 6.10, the estimator*

$$\dot{\hat{\theta}} = \mathcal{P}\left(-\gamma \frac{\epsilon_1 w}{m_2} \right) \; ; \; \hat{\theta}(t_0) \in \mathcal{M}$$

---

[3] For simplicity, we use the same constant μ to characterize both the mean-absolute and mean-square speed of the unstructured parameter variations.

6.3. PARAMETRIC IDENTIFICATION OF LTV PLANTS

guarantees that $\hat{\theta}, \dot{\hat{\theta}}$ are UB on $[t_0, t_0 + T]$ ($|\hat{\theta} - \hat{\theta}_c| \le M_0 + \epsilon_*$); furthermore, there exists a constant C_0 depending on the initial conditions but independent of T, t_0, such that for any interval $I = [t_I, t_I + T_I] \subseteq [t_0, t_0 + T]$

1. $\displaystyle\int_{t_I}^{t_I+T_I} \frac{\epsilon_1^2(t)}{m_2(t)} dt \le C_0 + \left[\Gamma_1^2 + \frac{K_\phi}{\gamma}\right]\mu T_I;$

2. $|\dot{\hat{\theta}}(t)| \le \gamma \Gamma_2 \dfrac{|\epsilon_1|}{\sqrt{m_2}} + \varepsilon(t, t_0);$

where $K_\phi, \Gamma_1, \Gamma_2$ are constants depending only on operator gains and the radius of the set $\mathcal{M}$ (their expressions are given in the proof). ▽▽

Proof: For the proof of the corollary we follow the same steps as in Corollary 6.9 except that we also use the Cauchy inequality

$$(x+y)^2 \le (1+\epsilon_c)x^2 + (1+\frac{1}{\epsilon_c})y^2$$

where $\epsilon_c > 0$ is an arbitrary constant ('Cauchy constant') to group the more interesting integral terms together. From Corollary 6.9, we can easily obtain the following estimates for the various bounds

$$K_\phi = 4M_0 + 2\epsilon_* \quad ; \quad \Gamma_1 = (1+\epsilon_c)\frac{g_{2,\delta}[G]g_{2,\delta_0}[G'_\pi]}{\sqrt{2(\delta-\delta_0)}} \quad ; \quad \Gamma_2 = g_{2,\delta_0}[G_\pi]$$

where $\epsilon_c > 0$ is an arbitrary Cauchy constant and $\delta > \delta_0$ as defined in Corollary 6.9.[4]

The constant terms, depending on the initial conditions, are then combined to a single constant C_0 which contains terms $O(1/\epsilon_c)$. (Since ϵ_c is arbitrary, a single Cauchy constant is used without loss of generality.) □□

In other words, the use of a basic estimator with projection and normalization guarantees that the mean-square value of the estimation error is $O(\mu, \mu/\gamma)$ with μ representing the average speed of variation of the unstructured part of the plant parameters. It is interesting to observe that higher adaptation gains (γ) tend to decrease but not eliminate the mean-square normalized estimation error while increasing the speed of the parameter estimates, which is $O(\gamma\mu)$ in the mean-square sense. However, what we actually need to make small is the mean-square value of the identification error which can be expressed as

$$e_1 = \epsilon_1 - \hat{\eta}$$

---
[4]Notice that, as mentioned in the proof of Corollary 6.9, simpler and perhaps tighter expressions can be used when the parameter derivatives are small uniformly in time.

yielding, after some straightforward calculations,

$$\left[\int_{t_I}^{t_I+T_I} \frac{e_1^2(t)}{m_2(t)} dt\right]^{1/2} \leq C_0 + \left[\int_{t_I}^{t_I+T_I} \frac{\epsilon_1^2(t)}{m_2(t)} dt\right]^{1/2} \cdots$$
$$+ (1+\epsilon_c)\Gamma_1 \left[\int_{t_I}^{t_I+T_I} |\dot{\hat{\theta}}(t)|^2 dt\right]^{1/2}$$

with our usual notation. It is now apparent that, for 'successful' identification (in an I/O, mean-square, normalized sense) both $\epsilon_1/\sqrt{m_2}$ and $\dot{\hat{\theta}}$ should be made small in the mean-square sense, something that involves a trade-off in the selection of the adaptation gain. Of course, as $\mu \to 0$ one recovers the standard estimation results for LTI systems (e.g., see [N.A.89, S.B.89]) whereby the value of γ is irrelevant and the normalized identification and estimation errors as well as $\dot{\hat{\theta}}$ are square-integrable.

Also note that the auxiliary filters, the parameter δ_0 and the weighting matrix Q play an essential role in the success of the identification scheme. Qualitative guidelines for their selection can be derived from the above corollary so as to minimize, e.g., the identification error. However, due to the nonlinear interaction between the estimation and control laws, the effect of such parameters on the closed-loop stability/boundedness is quite complicated and cannot be analyzed until the final stage of the analysis.

6.12 Remark: In order for the parameter estimates to converge to their true values in the uniform norm, i.e., $\hat{\theta} \to \hat{\theta}_*$ if $\hat{\theta}_*$ is constant, or $|\hat{\theta}(t) - \hat{\theta}_*(t)|$ to converge to an $O(\mu)$-small residual set for a slowly varying $\hat{\theta}_*(t)$, we need to assume that the regressor vector $w/\sqrt{m_2}$ is persistently exciting ([And.77]). In the special case of slowly TV plants, criteria for persistence of excitation have been derived in [M.G.87]. In general, however, the appearance of the possibly fast TV matrix Π in the regressor vector makes the translation of this condition into a condition on the input signal rather complicated. Although persistence of excitation could be beneficial in the performance of an adaptive controller, it is not necessary for closed-loop boundedness and is not discussed any further in the present study. ▽▽

Analogous statements can be made for estimators employing the σ or switching σ-modification. While for the latter the derivations and the resulting expressions are similar to those of Corollary 6.11 and are omitted, the case of the σ-modification is somewhat more interesting. This is partly due to the fact that the a priori knowledge of the set $\mathcal{M}$ is not required and partly due to the involved trade-off's in the selection of both σ and γ. Moreover, from Theorem 6.7 it follows that the various bounds can be expressed in terms of

6.3. PARAMETRIC IDENTIFICATION OF LTV PLANTS

the mean-square value of the derivative of $\hat{\theta}_*$ and therefore only part (b) of Assumption 6.10 is needed.

6.13 Corollary: *Suppose that for an LTV plant satisfying Assumptions 3.1-3.3, $\hat{\theta}_*$ is UB and satisfies Assumption 6.10-b and $U \in L_{\infty,[t_0,t_0+T]}$. Then, the estimator*

$$\dot{\hat{\theta}} = -\gamma \frac{\epsilon_1 w}{m_2} - \sigma(\hat{\theta} - \hat{\theta}_c)$$

guarantees that $\hat{\theta}, \dot{\hat{\theta}}$ are UB on $[t_0, t_0+T]$ and $\hat{\theta}$ converges exponentially fast to the residual set

$$\left\{ \hat{\theta} : |\hat{\theta}(t) - \hat{\theta}_c|_2 \leq \sqrt{\frac{\gamma}{4\sigma}} \Gamma_3 \right\}.$$

Furthermore, there exists a constant C_0 depending on the initial conditions but independent of T, t_0, such that for any interval $I = [t_I, t_I+T_I] \subseteq [t_0, t_0+T]$

1. $\displaystyle\int_{t_I}^{t_I+T_I} \frac{\epsilon_1^2(t)}{m_2(t)} dt \leq C_0 + \left[\Gamma_1^2 \mu + \frac{\mu}{\gamma\sigma} + K_{\hat{\theta}_*}^2 \frac{\sigma}{\gamma} \right] T_I;$

2. $|\dot{\hat{\theta}}(t)|_2 \leq \gamma\Gamma_2 \dfrac{|\epsilon_1|}{\sqrt{m_2}} + \sqrt{\dfrac{\gamma\sigma}{2}} \Gamma_3 + \varepsilon(t, t_0);$

where $K_{\hat{\theta}_*}, \Gamma_1, \Gamma_2, \Gamma_3$ are constants depending only on operator gains and the radius of the set $\mathcal{M}$ (their expressions are given in the proof). ▽▽

Proof: The proof follows as a straightforward application of Theorem 6.7 using similar techniques as in Corollary 6.9. Expressions for the estimates of the various bounds can be obtained as

$$K_{\hat{\theta}_*} = \|(|\hat{\theta}_* - \hat{\theta}_c|_2)\|_\infty \quad ; \quad \Gamma_3 = g_{2,\delta_0}[G_{\theta_*} - G_{\theta_c}]$$

and Γ_1, Γ_2 are as in Corollaries 6.9 and 6.11. □□

One implication of the last corollary is that the value of σ must be carefully selected for the estimation to be successful. For example, a large ratio γ/σ tends to decrease the mean-square value of $\epsilon_1/\sqrt{m_2}$ while increasing that of $|\dot{\hat{\theta}}|_2$. On the other hand, letting $\sigma = \gamma\sigma'$, the mean-square values of $\epsilon_1/\sqrt{m_2}$ and $|\dot{\hat{\theta}}|_2$ become

$$O\left(\mu, \frac{\mu}{\gamma^2\sigma'}, \sigma'\right) \quad \text{and} \quad O\left(\gamma^2\mu, \frac{\mu}{\sigma'}, \gamma^2\sigma'\right)$$

respectively. Taking $\gamma = 1$ for simplicity, the two can be made small, say less than $\epsilon < 1$, by choosing σ' to be $O(\epsilon)$ and assuming that $\mu = O(\epsilon^2)$. With such a choice of the adaptation parameters σ and γ, however, the parameter estimates can become as large as $O(1/\epsilon)$. This, in turn, gives rise to some technical problems when the results of the corollary are used to establish the boundedness of signals in adaptive control systems. In order to avoid these problems we need to ensure that the identification error can be made small without increasing the bounds of the parameter estimates and the parameter error. One way to achieve this is to restrict the class of parameter variations to those which are slow uniformly in time, as shown by the following corollary.

6.14 Corollary: *Suppose that for an LTV plant satisfying Assumptions 3.1–3.3, θ_* is UB, $\|\dot{\theta}_*\|_\infty \leq \mu$ and $U \in L_{\infty,[t_0,t_0+T]}$. Then, the estimator*

$$\dot{\hat{\theta}} = -\gamma \frac{\epsilon_1 w}{m^2} - \sigma(\hat{\theta} - \hat{\theta}_c)$$

guarantees that $\hat{\theta}, \dot{\hat{\theta}}$ are UB on $[t_0, t_0 + T]$ and $\hat{\phi}$ converges exponentially fast to the residual set

$$\left\{ \hat{\phi} : |\hat{\phi}(t)|_2 \leq K_{\theta_*} + \Gamma_1' \sqrt{\frac{\gamma}{2\sigma}\mu + \frac{\mu}{\sigma}} \right\}$$

Furthermore, there exists a constant C_0 depending on the initial conditions but independent of T, t_0, such that for any interval $I = [t_I, t_I + T_I] \subseteq [t_0, t_0 + T]$

1. $\displaystyle\int_{t_I}^{t_I+T_I} \frac{\epsilon_1^2(t)}{m^2(t)} dt \leq C_0 + \left[\Gamma_1'^2 \mu^2 + \frac{K_{\theta_*}^2 \sigma}{\gamma} + \frac{\mu^2}{\gamma\sigma} \right] T_I;$

2. $|\dot{\hat{\theta}}(t)| \leq \gamma \Gamma_2 \dfrac{|\epsilon_1|}{\sqrt{m_2}} + 2K_{\theta_*}\sigma + \mu + \Gamma_1' \sqrt{\dfrac{\gamma\sigma}{2}\mu} + \varepsilon(t, t_0);$

where K_{θ_*}, Γ_2 are as in Corollary 6.13 and Γ_1' is a constant depending on operator gains (its expression is given in the proof). ▽▽

Proof: In Appendix VI.

From a different point of view, Corollaries 6.11, 6.13 and 6.14 describe the properties of an identification procedure which identifies a left-fractional representation of an LTV I/O operator within two stable-factor perturbations (see Fig. 6.3). These perturbations have the property that their combined output is 'small' in a mean-square, normalized sense. This can be seen by expressing y_p in terms of the identification error as

$$y_p = G(s)[u_p\theta_1] + G(s)[y_p\theta_2] - e_1 + \varepsilon$$
$$e_1 = G(s)[u_p\phi_1] + G(s)[y_p\phi_2] = \epsilon_1 - \hat{\eta}.$$

Rewriting the above equation in terms of operators we have

$$y_p = D^{-1}(s)\hat{N}_p(s,t)[u_p] + D^{-1}(s)\{D(s) - \hat{D}_p(s,t)\}[y_p]$$
$$- D^{-1}(s)\tilde{N}_p(s,t)[u_p] - D^{-1}(s)\tilde{D}_p(s,t)[y_p]$$

6.3. PARAMETRIC IDENTIFICATION OF LTV PLANTS

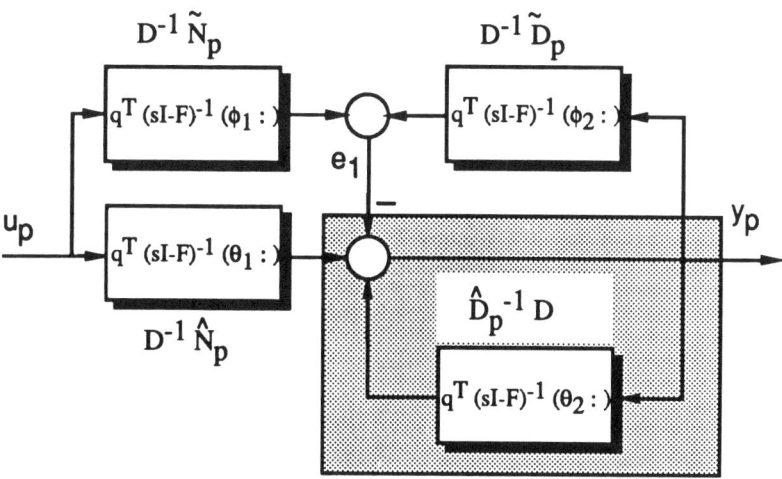

Figure 6.3: Alternative representation of the LTV plant.

where the coefficients of $\hat{D}_p(s,t), \hat{N}_p(s,t)$ and $\tilde{D}_p(s,t), \tilde{N}_p(s,t)$ are easily obtained from θ, ϕ and $D(s) = \det(sI - F)$.[5] In other words the unknown plant $D_p^{-1}(s,t)N_p(s,t)$ is effectively replaced by its estimate with the known I/O operator $\hat{D}_p^{-1}(s,t)\hat{N}_p(s,t)$ and the stable factor perturbations $D^{-1}(s)\tilde{N}_p(s,t)$ and $D^{-1}(s)\tilde{D}_p(s,t)$. The benefits are quite apparent. Since θ is known, a controller can now be designed for the estimated I/O operator of the plant $\hat{D}_p^{-1}(s,t)\hat{N}_p(s,t)$. Consequently, the closed-loop stability properties are determined by the robustness properties of the controller with respect to the stable factor perturbations e_1 which are not necessarily of small gain but are small in a mean-squared, normalized sense. Such a design is studied in more detail in a forthcoming chapter. At this point, however, we should emphasize the importance of the auxiliary filters used in the estimator. Not only do they determine the properties of the perturbation e_1 but they also affect the sensitivity operators $e_1 \mapsto y_p$ and $e_1 \mapsto u_p$ which are the key operators involved in the analysis of the properties of adaptive controllers.

6.15 Example: To demonstrate the advantages of modeling the plant as in Lemmas 3.10 or 3.11 and using the structured-parameter-variations approach, let us consider the case where the plant I/O description is given by

$$s^2[y_p] + s[a_1 y_p] + a_2 y_p = s[u_p] + b_1 u_p \qquad (6.14)$$

where

$$a_1 = 20 + 12\sin 2t \;;\; a_2 = 6\cos 2t \;;\; b_1 = -1 \qquad (6.15)$$

---

[5]Note that $\tilde{D}_p(s,t)$ is a PDO of degree $n-1$.

Let us also suppose that we know a priori that the plant parameters are of the form

$$a_1 = c_1 + c_2 \sin 2t \; ; \; a_2 = c_3 + c_4 \cos 2t \; ; \; b_1 = c_5 \qquad (6.16)$$

where $N_p(s,t)$ is known to be a monic PDO of degree one and $c_1 - c_5$ are some unknown constants whose range, say $c_{i,min} \leq c_i \leq c_{i,max}$, is known a priori.

Following the guidelines of the presented analysis let us consider the second order auxiliary filter:

$$G(s) = q^T(sI - F)^{-1}, \; q^T = [1,0] \; ; \; F = \begin{pmatrix} -5 & 1 \\ -6 & 0 \end{pmatrix} \qquad (6.17)$$

from which $D(s) = s^2 + 5s + 6$. Applying Lemma 3.10, we obtain the following parametrization for the plant

$$y_p = G(s)[u_p \theta_{1*}] + G(s)[y_p \theta_{2*}]$$

where

$$\theta_{1*} = \begin{bmatrix} 1 \\ \hat{\theta}_{1*} \end{bmatrix} \; ; \; \theta_{2*} = \begin{bmatrix} \hat{\theta}_{2*} + \hat{\theta}_{4*} \sin 2t \\ \hat{\theta}_{3*} + \hat{\theta}_{5*} \cos 2t \end{bmatrix}$$

and $\hat{\theta}_*^T = [-1, -15, 6, -12, -6]$ is an unknown constant vector, to be estimated. Next, employing the notion of structured parameter variations and since the leading coefficient of $N_p(s,t)$ is known, we may express y_p as

$$y_p = sD^{-1}(s)[u_p] + G(s)[u_p \Pi_1]\hat{\theta}_* + G(s)[y_p \Pi_2]\hat{\theta}_*$$

where,

$$\Pi_1(t) = \begin{pmatrix} 0 & 0 & 0 & 0 & 0 \\ 1 & 0 & 0 & 0 & 0 \end{pmatrix} \; ; \; \Pi_2(t) = \begin{pmatrix} 0 & 1 & 0 & \sin 2t & 0 \\ 0 & 0 & 1 & 0 & \cos 2t \end{pmatrix}$$

Thus, the estimation error is constructed as

$$\epsilon_1 = w^T \hat{\theta} - y_p + sD^{-1}(s)[u_p]$$

where $w^T = [(0,1)w_1, w_2^T, (1,0)w_3, (0,1)w_4]$,

$$\dot{w}_1 = F^T w_1 + q u_p \; ; \; \dot{w}_2 = F^T w_2 + q y_p$$
$$\dot{w}_3 = F^T w_3 + q y_p \sin 2t \; ; \; \dot{w}_4 = F^T w_4 + q y_p \cos 2t$$

and $\hat{\theta}$ being the estimate of $\hat{\theta}_*$. Since the range of each element of $\hat{\theta}_*$ can be directly deduced from the range of the c_i's, we may use the following adaptive law to update $\hat{\theta}$

$$\dot{\hat{\theta}} = \mathcal{P}\left(-\gamma \frac{\epsilon_1 w}{m^2}\right) \qquad (6.18)$$

6.3. PARAMETRIC IDENTIFICATION OF LTV PLANTS

with $\mathcal{P}$ as given by (6.5) and $\delta_0 < 2$. Finally, the estimate θ_i of θ_{i*} is

$$\theta_1 = \begin{bmatrix} 1 \\ 0 \end{bmatrix} + \Pi_1 \hat{\theta} \ ; \quad \theta_2 = \Pi_2 \hat{\theta}$$

from which the estimates of the plant parameters, denoted by $(\hat{\cdot})$, are given by

$$\hat{b}_1 = (0,1)\theta_1 \ ; \ \hat{a}_1 = -(1,0)\theta_2 + 5 \ ; \ \hat{a}_2 = -(0,1)\theta_2 + 6 \tag{6.19}$$

Thus, according to Corollary 6.11, the estimator (6.18) guarantees that $\epsilon_1/\sqrt{m_2}$ and $\dot{\hat{\theta}}$ is square integrable, for as long as $U \in L_{\infty,[t_0,t_0+T]}$. Furthermore, if $u_p/\sqrt{m_2}$ is UB, it also follows that $\epsilon_1/\sqrt{m_2}$ is uniformly continuous and therefore converge to zero as $t \to \infty$, provided of course that $u_p \in L_\infty^e$. However, whether this condition is met or not depends also on the way u_p is generated (for further discussion, see Chapter 8).

It should be noted that in the case there is some ambiguity in the frequency of the $\sin w_0 t$ terms, e.g., the true frequency is $w_0 + \delta$, we can decompose the sine terms as

$$\sin(w_0 + \delta)t = \sin w_0 t \cos \delta t + \cos w_0 t \sin \delta t$$

The last equation suggests that an additional parameter should be used to estimate the $\cos w_0 t$ terms with the corresponding elements of $\hat{\theta}_*$s being slowly TV for small δ (similarly for the cosine terms). In this case, Lemma 3.10 can still be used to establish the smallness in the mean-square of the normalized estimation error even if w_0 is large. ▽▽

6.3.2 Non-Smooth Parameter Variations

For our next topic of study we consider the identification problem for plants with non-smooth parameter variations. In this case, the plant may not be described with a convenient PDO/PIO factorization (even with non-smooth coefficients) which would allow the straightforward application of our previous results. Instead, we rely on Lemma 3.11 to parametrize an LTV plant of known order, with a state-space representation satisfying Assumptions 3.4–3.6. This lemma together with Corollary 2.60 imply that the output of the LTV plant at time t can be expressed as

$$y_p = w^\top \hat{\theta}_* + \eta_s + \tilde{\eta} + \eta_J + q^\top \Phi_F(t,t_0) x_F(t_0) \tag{6.20}$$

$$\eta_s = -G(s)\{G'_\pi[QU]\dot{\hat{\theta}}_*^s\}$$

$$\tilde{\eta} = G(s)[\tilde{A}_F x_F + \tilde{b}_F u_p]\} + \tilde{c}_F^\top x_F$$

$$\eta_J = \sum_{t_j \le t} q^\top \Phi_F(t,t_j) \left\{ [\bar{P}(t_j) - I] x_F(t_j^-) + G'_\pi[QU](t_j) \hat{\theta}_{*j}^J \right\}$$

where $\hat{\theta}_*^s$ is the Lipschitz continuous part and $\hat{\theta}_*^J = \sum_{t_j \leq t} \hat{\theta}_{*j}^J$ is the jump part of $\hat{\theta}_*$ and w is the usual regressor vector as in the smooth parameter case; the rest of the notation is as in Lemma 3.11 Although such a parametrization is quite more complicated than before, it nevertheless retains the same basic properties required for an application of Theorems 6.6 and 6.7. The only difference here is that the 'noise' signal η is composed of three terms:

1. the term η_s, which depends on the speed of variation of the Lipschitz part of $\hat{\theta}_*$ and is characterized by the parameter μ;

2. the term $\tilde{\eta}$ which incorporates the effects of smooth approximations of the plant parameters and possible short excursions of the plant parameters to regions where strong controllability or observability fail; this term is characterized by the parameter μ' which is typically expected to be very small ($\ll 1$);

3. the term η_J describing the cumulative effects of the jump part of the plant parameters and characterized by the usually very small parameter ν.

Thus, non-smooth-parameter analogs of Corollaries 6.11 and 6.13, characterizing the I/O quality of the identification by means of the three parameters μ, μ', ν, can be derived in a straightforward manner from the general theorems of Section 6.2. In the following statements, we consider an LTV plant satisfying Assumptions 3.4–3.6 and the corresponding parameter vector θ_*, as defined in Lemma 3.11. Without loss of generality, we assume that

$$\theta_* = \Pi \hat{\theta}_*$$

where Π is a piecewise smooth, UB matrix and $\hat{\theta}_*$ is a piecewise smooth vector with possible discontinuities at $t = t_j$, $j = 1, 2, \ldots$.

6.16 Corollary: Under these conditions, suppose that $\hat{\theta}_*$ and $\hat{\theta}_*^s$ satisfy Assumptions 6.3 and 6.10, respectively. Further suppose that $U \in L_{\infty,[t_0,t_0+T]}$ and consider the estimator

$$\dot{\hat{\theta}} = \mathcal{P}\left(-\gamma \frac{\epsilon_1 w}{m^2}\right) \; ; \; \hat{\theta}(t_0) \in \mathcal{M}$$

Then, there exist $\nu_0, \mu_0' > 0$, depending on the value of δ_0 used in m^2, such that for any $\nu \in [0, \nu_0)$, $\mu' \in [0, \mu_0')$, $\hat{\theta}, \dot{\hat{\theta}}$ are UB on $[t_0, t_0 + T]$ ($|\hat{\theta} - \hat{\theta}_c| \leq M_0 + \epsilon_*$); furthermore, there exist constants $C_0, \tilde{K}, K_J$ such that for any interval $I = [t_I, t_I + T_I] \subseteq [t_0, t_0 + T]$

6.3. PARAMETRIC IDENTIFICATION OF LTV PLANTS

1. $\int_{t_I}^{t_I+T_I} \frac{\epsilon_1^2(t)}{m_2(t)} dt \leq C_0 + \left[\Gamma_1^2 + \frac{K_\phi}{\gamma}\right]\mu T_I + \tilde{K}\mu' T_I + K_J^2\left(1+\frac{1}{\gamma}\right)\nu T_I;$

2. $|\dot{\hat{\theta}}(t)| \leq \gamma\Gamma_2 \frac{|\epsilon_1|}{\sqrt{m_2}} + \varepsilon(t,t_0);$

where $K_\phi, \Gamma_1, \Gamma_2$ are as in Corollary 6.11. The constants $\tilde{K}, K_J$ depend on the perturbation part of the plant and the size of jumps in the nominal plant parameters and their derivatives, respectively but they are independent of initial conditions whose effect is incorporated in C_0; all three constants are independent of T, t_0. ▽▽

Proof: In Appendix VI.

6.17 Corollary: Under the conditions stated above, suppose that $\hat{\theta}_*$ is UB and $\hat{\theta}_*^s$ satisfies Assumption 6.10-b. Further suppose that $U \in L_{\infty,[t_0,t_0+T]}$ and consider the estimator

$$\dot{\hat{\theta}} = -\gamma\frac{\epsilon_1 w}{m_2} - \sigma(\hat{\theta} - \hat{\theta}_c)$$

Then, there exist $\nu_0, \mu'_0 > 0$, depending on the value of δ_0 used in m_2, such that for any $\nu \in [0,\nu_0), \mu' \in [0,\mu'_0), \hat{\theta}, \dot{\hat{\theta}}$ are UB on $[t_0, t_0+T]$ and $\hat{\theta}$ converges exponentially fast to the residual set

$$\left\{\hat{\theta} : |\hat{\theta}(t) - \hat{\theta}_c|_2 \leq \sqrt{\frac{\gamma}{4\sigma}}\Gamma'_3\right\}$$

where Γ'_3 is a constant, approaching $\Gamma_3 = g_{2,\delta_0}[G_{\theta_*} - G_{\theta_c}]$ as the perturbation part of the plant and the jumps in the plant parameters and their derivatives vanish. Furthermore, there exist constants $C_0, \tilde{K}, K_J$ such that for any interval $I = [t_I, t_I + T_I] \subseteq [t_0, t_0 + T]$

1. $\int_{t_I}^{t_I+T_I} \frac{\epsilon_1^2(t)}{m_2(t)} dt \leq C_0 + \left[\Gamma_1^2\mu + \frac{\mu}{\gamma\sigma} + K_{\theta_*}^2\frac{\sigma}{\gamma}\right]T_I + \tilde{K}\mu' T_I$

$\qquad + K_J^2\left(1+\frac{1}{\gamma}\right)\nu T_I;$

2. $|\dot{\hat{\theta}}(t)|_2 \leq \gamma\Gamma_2\frac{|\epsilon_1|}{\sqrt{m_2}} + \sqrt{\frac{\gamma\sigma}{2}}\Gamma'_3 + \varepsilon(t,t_0);$

where $K_{\theta_*}, \Gamma_1, \Gamma_2, \Gamma_3$ are as in Corollaries 6.11 and 6.13. The constants $\tilde{K}, K_J$ depend on the perturbation part of the plant and the size of jumps in the nominal plant parameters and their derivatives, respectively but they are independent of initial conditions whose effect is incorporated in C_0; all three constants are independent of T, t_0. ▽▽

Proof: As in Corollary 6.16, but using the expressions of Corollary 6.13 rather than 6.11. □□

Thus, Corollaries 6.16 and 6.17 describe the properties of our basic estimation algorithms when used to identify an LTV plant with non-smooth parameter variations. As in the smooth parameter case, the quality of identification —in an I/O, normalized sense— depends on the average speed of variations of the smooth part of $\hat{\theta}_*$ while the effect of the discontinuities and perturbations in the state-space description of the LTV plant introduces two additional terms $O(\nu)$ and $O(\mu')$ in the various expressions.

In this case, however, μ' and ν should be sufficiently small. This requirement arises from the need to guarantee the uniform boundedness of the normalized signals, including the plant state vector, in order to obtain expressions which are uniform with respect to the interval of integration and the initial time. Notice that large μ' or ν may result in the overall plant not being uniformly observable[6] and a consequent failure of $x_F/\sqrt{m_2}$ to be UB. If, on the other hand, we assume that, in addition to Assumptions 3.4–3.6, the LTV plant (nominal and perturbation part) is uniformly observable, then no restriction on μ' and ν is necessary.

Finally, as mentioned in the previous subsection, when an estimator employing the σ-modification is used for adaptive control purposes, we need to impose some additional uniformity conditions on the size of the perturbation terms. More precisely, we assume that

$$\|\dot{\hat{\theta}}^s_*\|_\infty \leq \mu \; ; \; |\tilde{A}|, |\tilde{b}|, |\tilde{c}| \leq \mu' \; ; \; t_{j+1} - t_j \geq \frac{1}{\nu}$$

uniformly in t and j, where as usual the superscript 's' denotes the smooth part of the parameters, '~' refers to the state-space perturbations due to smooth approximations and t_j are the discontinuity points.

6.18 Corollary: *Under the Assumptions of Corollary 6.17 as restricted by the above conditions, for any $\sigma > 0$ there exist $\nu_0, \mu'_0 > 0$ such that for any $\nu \in [0, \nu_0)$ and $\mu' \in [0, \mu'_0)$, $\hat{\theta}, \dot{\hat{\theta}}$ are UB on $[t_0, t_0 + T]$ and $\hat{\phi}$ converges exponentially fast to a residual set satisfying*

$$|\hat{\phi}(t)|_2 \leq O\left(K_{\theta_*}, \frac{\mu}{\sigma}, \mu\sqrt{\frac{\gamma}{\sigma}}, \mu'\sqrt{\frac{\gamma}{\sigma}}\right) + K_J O\left(1, \sqrt{\frac{\gamma}{\sigma}}\right) e^{-\beta(t-t_j)} \; ; \; t \geq t_j$$

where K_{θ_*} is a constant as in Corollary 6.13, K_J is a constant depending on the size of the jumps in the nominal plant parameters and their derivatives and for any $\sigma > 0$, $\beta(\sigma) > 0$ is a constant.

---
[6] This is because Assumptions 3.5 and 3.6 are on the nominal part of the plant only.

Furthermore, there exists a constant C_0 depending on the initial conditions but independent of T, t_0, such that for any interval $I = [t_I, t_I + T_I] \subseteq [t_0, t_0 + T]$

1. $\displaystyle\int_{t_I}^{t_I+T_I} \frac{\epsilon_1^2(t)}{m_2(t)}\, dt \leq C_0 + O\left(\mu^2, \mu'^2, K_J^2 \nu, \frac{K_{\theta*}^2 \sigma}{\gamma}, \frac{\mu^2}{\gamma\sigma}, \frac{K_J^2 \nu}{\gamma}\right) T_I;$

2. $|\dot{\hat{\theta}}(t)| \leq O(\gamma)\dfrac{|\epsilon_1|}{\sqrt{m_2}} + O\left(K_{\theta*}\sigma, \mu, \mu\sqrt{\gamma\sigma}, \mu'\sqrt{\gamma\sigma}\right)\cdots$

$\qquad\qquad + K_J O\left(\sigma, \sqrt{\gamma\sigma}\right) e^{-\beta(t-t_j)} + \varepsilon(t, t_0) \, ; \quad t \geq t_j;$

$\nabla\nabla$

Proof: In Appendix VI.

Note that due to the complexity and increased conservatism of the expressions, we state Corollary 6.18 using the simplified notation $O(x_1, x_2, \ldots) \leq K_1 x_1 + K_2 x_2 + \cdots$ where K_i are constants. All such constants are independent of the parameter estimates. We do, however, keep track of the important adaptation parameters γ and σ and the radius of the parametric uncertainty set $K_{\theta*}$ for later discussion.

6.19 Remark: Although, for estimation purposes, it is possible to allow discontinuities in $\Pi(t)$ at instants other than at t_j, this increases the total number of discontinuity points that should be accounted for, when the estimator is used in an adaptive control context. For this reason, and in order to maintain some uniformity in the notation, we assume that the entries of $\Pi(t)$ are smooth inside each interval (t_j, t_{j+1}).

$\nabla\nabla$

APPENDIX VI

Proof of Theorem 6.6:

With the properties of $\mathcal{P}$, y, w and η, the vectorfield in (6.4) satisfies the Caratheodory assumptions, ensuring the local existence of an absolutely continuous function θ satisfying (6.4) almost everywhere. Furthermore, this solution is unique (see [C.L.55], pp. 48–51) and can be extended in the whole interval $[t_0, t_0 + T]$. Also, from the definition of $\mathcal{P}$ and with $\theta(t_0) \in \mathcal{M}$, the distance of $\theta(t)$ from the set $\mathcal{M}$ is at most ϵ_* (see Examples 6.4 and 6.5 for the definition of $\mathcal{P}$ and ϵ_* for sets $\mathcal{M}$ defined in an L_∞ or L_2 sense). Hence, the first part of the theorem follows.

For the second part, let t_i, $i = 0, 1, \ldots, n_T$, $t_i \leq t_{i+1} \leq t_0 + T$, denote the points of discontinuity of θ_* in the interval $[t_0, t_0 + T]$. Further, define the parameter error $\phi = \theta - \theta_*$. It follows that ϕ satisfies the differential equation

$$\dot{\phi} = \mathcal{P}(-\gamma\epsilon_1 w) - \dot{\theta}_*^s \ ; \quad t \in [t_i, t_{i+1})$$

with initial conditions

$$\phi(t_0) = \theta(t_0) - \theta_*(t_0)$$

and boundary conditions at each discontinuity point

$$\phi(t_i) = \phi(t_i^-) - [\theta_*(t_i^+) - \theta_*(t_i^-)]$$

where t_i^-, t_i^+ are used to denote left and right limits respectively.

Next, consider the positive definite function $V = \frac{1}{2\gamma}\phi^T\phi$. The derivative of V along the trajectories of (6.4) is given by

$$\dot{V} = -\phi^T \mathcal{P}(\epsilon_1 w) - \frac{1}{\gamma}\phi^T \dot{\theta}_*^s \ ; \quad t \in [t_i, t_{i+1})$$

From the definitions of $\mathcal{P}$ and $\mathcal{M}$ and after some simple geometry, it follows that when the projection is active ($\neq 1$), V decreases faster than the V corresponding to a vectorfield without projection. Hence,

$$\dot{V} \leq -\epsilon_1 \phi^T w - \frac{1}{\gamma}\phi^T \dot{\theta}_*^s \ ; \quad t \in [t_i, t_{i+1})$$

from which, using (6.3) and taking norms of the sign-indefinite terms we obtain

$$\dot{V} \leq -\epsilon_1^2 + |\epsilon_1\eta| + \frac{1}{\gamma}\|\phi\|_\infty |\dot{\theta}_*^s|_i \ ; \quad t \in [t_i, t_{i+1})$$

where the subscript i denotes the functional norm induced by the norm of ϕ (typically the norm used to define $\mathcal{M}$). Moreover, V being absolutely continuous in $[t_i, t_{i+1})$, a simple completion of squares and integration yields

$$\int_{t_i}^{t} \epsilon_1^2(\tau)\,d\tau \leq 2[V(t_i^+) - V(t)] + \int_{t_i}^{t} \eta^2(\tau)\,d\tau + \frac{K_\phi}{\gamma}\int_{t_i}^{t} |\dot{\theta}_*^s(\tau)|_i\,d\tau \ ; \quad t \in [t_i, t_{i+1})$$

from which, V being UB, property (1) follows. Notice that expressing $\dot{V}$ in terms of $w^T\phi$ rather than ϵ_1, the same inequality holds for $(w^T\phi)^2$. Finally, property (2) follows immediately from (6.4) and Assumption 6.2. $\square\square$

Proof of Theorem 6.7:

While the existence and uniqueness of an absolutely continuous θ satisfying (6.7) follow as in Theorem 6.6, boundedness can be shown by considering the positive definite function $V_1 = \frac{1}{2}(\theta - \theta_c)^T(\theta - \theta_c)$. Then, along the trajectories of (6.7)

$$\dot{V}_1 \leq \frac{\gamma}{4}(y - w^T\theta_c)^2 - 2\sigma V_1$$

APPENDIX VI. 175

from which, θ is uniformly ultimately bounded with an upper bound as given in the theorem. Further, a bound on $\dot{\theta}$ is simply derived by taking norms on both sides of (6.7). It is important to notice that, modulo exponentially decaying terms, these bounds are independent of the initial conditions $\theta(t_0)$.

For the rest of the properties, let $V = \frac{1}{2\gamma}\phi^\mathsf{T}\phi$ where $\phi = \theta - \theta_*$. Then, as in Theorem 6.6

$$\dot{V} \leq -\epsilon_1^2 + |\epsilon_1\eta| - \frac{\sigma}{\gamma}|\phi|^2 + \frac{\sigma}{\gamma}|\phi||\theta_* - \theta_c + \frac{1}{\sigma}\dot{\theta}_*^s| \; ; \quad t \in [t_i, t_{i+1})$$

while, from the boundedness of θ and θ_*, V is UB. The inequalities of the theorem are now easily obtained after completing the squares of the above inequality and integrating both sides. □□

Proof of Corollary 6.9:

The boundedness of $\hat{\theta}$ and its derivative are obtained as a straightforward application of Theorem 6.6 with the affine model (6.13), which also implies that

$$\int_{t_I}^{t_I+T_I} \left(\frac{\epsilon_1(t)}{m_p^{1/p}(t)}\right)^2 dt \leq C_0 + K_\phi \int_{t_I}^{t_I+T_I} |\dot{\hat{\theta}}_*(t)|_i \, dt + \int_{t_I}^{t_I+T_I} \left(\frac{\eta(t) + \varepsilon(t,t_0)}{m_p^{1/p}(t)}\right)^2 dt$$

where C_0 is a constant $O(M_0^2/\gamma)$, M_0 being the radius of the set $\mathcal{M}$ and K_ϕ is a constant due to pulling $|\phi(t)|$ outside the integral. When the projection $\mathcal{P}$ is given by (6.5), $\|\theta - \theta_c\|_\infty \leq M_0 + \epsilon_*$ which implies that $K_\phi \leq 4M_0 + 2\epsilon_*$, while the induced norm of $\dot{\hat{\theta}}_*(t)$ is the l_1 vector norm. Similar relations are also obtained if the set $\mathcal{M}$ and the projection $\mathcal{P}$ are defined through a Euclidean norm.

Further, for the first property of the corollary, we notice that $\eta/m_p^{1/p}$ is UB and

$$m_p(t) \geq \exp[-p\delta_0(t-\tau)]m_p(\tau) \, , \quad \forall t_0 \leq \tau \leq t \, . \tag{6.21}$$

Hence, we can express $\eta(t)$, $t \in [t_I, t_I + T_I]$ as

$$\eta(t) = (G(s)\{G'_\pi[QU]\dot{\hat{\theta}}_*\})(t) + \beta(t_I)e^{-\delta(t-t_I)}$$

where the functions $U, \dot{\hat{\theta}}_*$ are taken as zero for $t < t_I$. This, of course, introduces an effective initial condition term $\beta(t_I)$ for which we have that $\beta(t_I)/m_p^{1/p}(t_I)$ is UB. We may now invoke Lemma 2.56, with operator gains induced by the norms of the $L_p(\delta)$ as well as the underlying $\mathbf{R}^n$ spaces, to write

$$\frac{|\eta(t)|}{m_p^{1/p}(t)} \leq g_{p,\delta}[G]\frac{\mathcal{E}_{-\delta}(t)\|(G'_\pi[QU]\dot{\theta}_*)_t\|_{p,\delta}}{m_p^{1/p}(t)} + \frac{|\beta(t_I)|}{m_p^{1/p}(t_I)}e^{-(\delta-\delta_0)(t-t_I)}$$

$$\leq g_{p,\delta}[G]\left[\int_{t_I}^t e^{-p(\delta-\delta_0)(t-\tau)}\frac{|G'_\pi[QU]|^p(\tau)}{m_p(\tau)}|\dot{\theta}_*(\tau)|^p\,d\tau\right]^{1/p} \cdots$$

$$+\frac{|\beta(t_I)|}{m_p^{1/p}(t_I)}e^{-(\delta-\delta_0)(t-t_I)}$$

$$\leq g_{p,\delta}[G]g_{p,\delta_0}[G'_\pi]\left[\int_{t_I}^t e^{-p(\delta-\delta_0)(t-\tau)}|\dot{\theta}_*(\tau)|^p\,d\tau\right]^{1/p} \cdots$$

$$+\frac{|\beta(t_I)|}{m_p^{1/p}(t_I)}e^{-(\delta-\delta_0)(t-t_I)}$$

from which, property (1) follows as an immediate application of the Schwarz inequality.

Note that, alternatively, one may also derive a bound for $|\eta|$ in terms of the $g_{1,\delta}$-gain of G, the γ_{2,δ_0}-gain of G_π and the conjugate-index $L_q(\delta)$-norm of $\dot{\theta}_*$. Omitting for simplicity the exponentially decaying terms, this procedure is outlined below.

$$|\eta(t)| \leq g_{1,\delta}[G]\mathcal{E}_{-\delta}\|(G'_\pi[QU]\dot{\theta}_*)_t\|_{1,\delta}$$

$$\leq g_{1,\delta}[G]\mathcal{E}_{-\delta}\|(G'_\pi[QU])_t\|_{p,\delta_0}\|(\dot{\theta}_*)_t\|_{q,\delta}\,;\quad \begin{cases}\delta > \delta_0 \\ p^{-1}+q^{-1}=1\end{cases}$$

$$\leq g_{1,\delta}[G]\gamma_{p,\delta_0}[G'_\pi]m_p^{1/p}(t)\mathcal{E}_{-\delta+\delta_0}\|(\dot{\theta}_*)_t\|_{q,\delta-\delta_0}$$

More efficient bounds can be derived in the special case of TV parameters whose derivatives are small, uniformly in time. Assuming, for example, that $|\dot{\theta}_*| \leq \mu$ and following the same procedure we get

$$\frac{|\eta(t)|}{m_p^{1/p}(t)} \leq g_{p,\delta_0}[G]\frac{\mathcal{E}_{-\delta_0}(t)\|(G'_\pi[QU]\dot{\theta}_*)_t\|_{p,\delta_0}}{m_p^{1/p}(t)} + \frac{|\beta(t_I)|}{m_p^{1/p}(t_I)}e^{-\delta_0(t-t_I)}$$

$$\leq g_{p,\delta_0}[G]\gamma_{p,\delta_0}[G'_\pi]\mu + \frac{|\beta(t_I)|}{m_p^{1/p}(t_I)}e^{-\delta_0(t-t_I)}$$

Next, property (2) follows from Theorem 6.6 and using Lemma 2.56 to write $|w|/m_p^{1/p} \leq g_{p,\delta_0}[G_\pi]+\beta'\exp[-\delta(t-t_0)]$. The constant C''_0 is then used to absorb the integral of the exponentially decaying term (notice that $\epsilon_1/m_p^{1/p}$ is UB since $\hat{\theta}$ is UB).

APPENDIX VI. 177

Finally, it is straightforward to extend the proof to admit non-zero initial conditions in the auxiliary filters. In this case

$$\epsilon_1 = w^\top \hat{\phi} + \eta + \varepsilon + \varepsilon' \hat{\theta}$$

where the last term is due to the initial conditions in the auxiliary filters and the rest of the quantities are as before. The same arguments can now be used to show the boundedness of $\hat{\theta}$ from which the contribution of $\varepsilon'\hat{\theta}$ can be included in the C_0 constants. □□

Proof of Corollary 6.14:

While the existence and uniqueness of an absolutely continuous $\hat{\theta}$ satisfying the estimator ODE follow as in Theorem 6.6, boundedness can be shown by considering the positive definite function $V = \frac{1}{2}\hat{\phi}^\top \hat{\phi}$. Then,

$$\dot{V} = -\gamma \frac{\epsilon_1^2 - \epsilon_1 \eta}{m_2} - \sigma \|\hat{\phi}\|^2 + (\sigma \hat{\phi}_c - \dot{\hat{\theta}}_*)^\top \hat{\phi} \qquad (6.22)$$

where $\hat{\phi}_c = \hat{\theta}_c - \hat{\theta}_*$. Completing the squares in (6.22) we obtain

$$\dot{V} \leq -\sigma V + \frac{\gamma}{4} \frac{|\eta|^2}{m_2} + \frac{1}{2\sigma}(\sigma K_{\theta*} + \mu)^2 \qquad (6.23)$$

where $K_{\theta*}$ is as in Corollary 6.13. Further, as in Corollary 6.9, $\eta^2/m_2 \leq \Gamma_1'^2 \mu^2 + \varepsilon$ where $\Gamma_1' = (1 + \epsilon_c)g_{2,\delta_0}[G]\gamma_{2,\delta_0}[G_\pi']$, ϵ_c is a Cauchy constant and ε is an exponentially decaying term due to initial conditions. Substituting this expression in (6.23), the first part of the corollary follows.

For the rest of the properties, we use again (6.22) and complete the squares to obtain

$$2\dot{V} \leq -\gamma \frac{\epsilon_1^2 - \eta^2}{m_2} + \frac{1}{2\sigma}(\sigma K_{\theta*} + \mu)^2$$

Integrating both sides of the last inequality we obtain property (1) while property (2) follows by taking the norms of both sides of the estimator ODE. □□

Proof of Corollary 6.16:

From (6.20), y_p has the same general form as in the smooth parameter case and therefore, in order to apply the results of Theorem 6.6 and Corollary 6.11, it suffices to show that η_s, $\tilde{\eta}$ and η_J, normalized by $\sqrt{m_2}$, are UB and have similar average properties as η. Indeed, $\eta_s/\sqrt{m_2}$ can be handled the same way as $\eta/\sqrt{m_2}$ in the smooth parameter case, with θ_*^s replacing θ_* in the various expressions. On the other hand, for $\tilde\eta$ we have that

$$|G(s)[\tilde A_F x_F](t)| \leq g_{2,\delta}[G]\left\{\int_{t_0}^t e^{-2\delta(t-\tau)}|\tilde A_F(\tau)|^2 |x_F(\tau)|^2 \, d\tau\right\}^{1/2}$$

Since for μ', ν sufficiently small, x_F^2/m_2 is UB for as long as $U \in L_{\infty,[t_0,t_0+T]}$ (see Lemma 3.11) and using the fact

$$m_2(t) \geq e^{-2\delta_0(t-\tau)} m_2(\tau); \quad t \geq \tau$$

we have that for all $\mu' \in [0, \mu'_0)$, $\nu \in [0, \nu_0)$

$$\frac{|G(s)[\tilde A_F x_F](t)|^2}{m_2} \leq g_{2,\delta}^2[G] K_F \int_{t_0}^t e^{-2(\delta-\delta_0)(t-\tau)} |\tilde A_F(\tau)|^2 \, d\tau + \varepsilon(t,t_0)$$

for some constant K_F, independent of the initial conditions which are included in the exponentially decaying term $\varepsilon(t,t_0)$. Next,

$$|G(s)[\tilde b_F u_p](t)| \leq g_{1,\delta}[G] \int_{t_0}^t e^{-\delta(t-\tau)} |\tilde b_F(\tau)| |u_p(\tau)| \, d\tau$$

and therefore, using the Schwarz inequality, we obtain

$$\frac{|G(s)[\tilde b_F u_p](t)|^2}{m_2(t)} \leq g_{1,\delta}^2[G] \int_{t_0}^t e^{-2(\delta-\delta_0)(t-\tau)} |\tilde b_F(\tau)|^2 \, d\tau \ldots$$
$$\frac{\int_{t_0}^t e^{-2\delta_0(t-\tau)} |u_p(\tau)|^2 \, d\tau}{m_2}$$

the last factor being UB. Finally,

$$|\tilde c_F^T(t) x_F(t)|^2 / m_2 \leq K_F |\tilde c_F(t)|^2 + \varepsilon(t,t_0).$$

Accumulating all the terms it follows that $\tilde\eta^2(t)/m_2(t)$ is UB and, using Assumption 3.4, we get that in an interval $I \subseteq [t_0, t_0 + T]$

$$\int_{t_I}^{t_I+T_I} \frac{\tilde\eta^2(t)}{m_2(t)} \, dt \leq C_0 + \tilde K' \mu' T_I$$

for some constant $\tilde K'$ independent of the initial conditions.

Further, a similar property can be established for the term η_J under Assumption 3.6. For this, we make use of the following intermediate result

- Let t_j, $j = 1, 2, \ldots$ be an increasing sequence and for $t, T \geq 0$ define k, l such that $t_k \leq t < t + T \leq t_m$. Suppose that there exist constants

APPENDIX VI.

$C, T > 0$ such that $l - k \leq C \ \forall t \geq 0$. Then, for any $\alpha > 0$, there exists a constant C' such that

$$\sum_{j=1}^{N} e^{-\alpha(t_N - t_j)} \leq C' \ ; \ \forall N > 0$$

Proof: Consider the intervals $[t_N - (\lambda+1)T, t_N - \lambda T]$, $\lambda = 0, 1, 2 \ldots, \Lambda$; $\Lambda : t_N - \Lambda T \leq 0$. Then each one of these intervals contains at most C terms of the sequence and therefore

$$\sum_{j=1}^{N} e^{-\alpha(t_N - t_j)} \leq C \sum_{\lambda=0}^{\Lambda} e^{-\alpha \lambda T} \leq C' = C \frac{1}{1 - e^{-\alpha T}} \ ; \ \forall N > 0$$

□□

In view of the above statement, we may now use the properties of the normalizing signal and the fact that $x_F/\sqrt{m_2}$ is UB, to write

$$\frac{|\eta_J(t)|}{\sqrt{m_2(t)}} \leq e^{-(\delta - \delta_0)(t - t_N)} \sum_{j=1}^{N} e^{-(\delta - \delta_0)(t_N - t_j)} [K + \varepsilon(t, t_0)] \ ; \ t_N \leq t$$

where N is the total number of discontinuities in $[t_0, t]$, and ε is a term due to the initial conditions in x_F. Hence, under Assumption 3.6, we have that $\eta_J/\sqrt{m_2}$ is UB and integrating inside the intervals (t_j, t_{j+1}), $j = 1, 2, \ldots, N$ we get

$$\int_{t_I}^{t_I + T_I} \frac{\eta_J^2(t)}{m_2(t)} dt \leq C_0 + K_J'^2 \nu T_I$$

where K_J' is a constant depending on the size of the jumps in the plant parameters and their derivatives but independent of the initial conditions. It is now quite straightforward to verify the inequalities of the corollary, working as in the previous cases. Notice that although it is not difficult to keep the various Cauchy constants in the equations, resulting in sharper bounds, the inequalities tend to become rather messy. Moreover, the bounds related to the perturbation part of the plant and the parameter discontinuities are quite conservative, as they involve norms of —often sparse— perturbation matrices. And since the parameters μ', ν are typically expected to be very small, there is little intuition to be gained from such an approach. On the other hand, we must keep track of the adaptation parameters which eventually affects the upper bounds of μ, ν and μ' for the BIBO stability of the adaptive closed-loop system. And although the overall stability problem becomes very complicated at the final stage of the analysis, these bounds reveal some design guidelines which may be useful in various special cases. □□

Proof of Corollary 6.18:

Working as in Theorem 6.6 we first establish the existence and uniqueness of an absolutely continuous $\hat{\theta}$ satisfying the estimator ODE, at least in a subinterval of $[t_0, t_0 + T]$. Next, consider the positive definite function $V = \frac{1}{2\gamma}\tilde{\phi}^T\tilde{\phi}$. Then, for $t \in (t_j, t_{j+1})$ and inside the subinterval where the solution exists

$$\dot{V} \leq -\sigma V - \frac{\epsilon_1^2 - \bar{\eta}^2}{2m_2} + \frac{1}{2\sigma\gamma}(\sigma K_{\theta*} + \mu)^2 \tag{6.24}$$

where $\bar{\eta} = \eta_s + \tilde{\eta} + \eta_J + \varepsilon$ Further, from the proof of Corollary 6.16 we have that for μ', ν sufficiently small so that $\|x_F\|^2/m_2$ is UB,

$$\frac{\bar{\eta}^2(t)}{m_2(t)} \leq O\left(\mu^2, \mu'^2\right) + K_j^2 e^{-2\delta_0(t-t_j)} + \varepsilon(t, t_0)$$

inside the interval (t_j, t_{j+1}). Integrating both sides of (6.24) inside (t_j, t_{j+1}) with the initial condition $V(t_j^+)$ we obtain

$$\begin{aligned} V(t) &\leq V(t_j^+) e^{-\sigma(t-t_j)} + O\left(\frac{K_j^2}{\sigma}\right) e^{-\beta(\sigma)(t-t_j)} + \ldots \\ &\quad + O\left(\frac{K_{\theta*}^2}{\gamma}, \frac{\mu^2}{\gamma\sigma^2}, \frac{\mu^2}{\sigma}, \frac{\mu'^2}{\sigma}\right) + \varepsilon(t, t_0) \end{aligned}$$

where $\beta < \min[\sigma, 2\delta_0]$ is a positive constant. Observing that $|V(t_j^+) - V(t_j^-)| \leq O\left(K_j^2/\gamma\right)$, it now follows that for any constant A there exists $\nu_1 > 0$ sufficiently small such that $\forall \nu \in [0, \nu_1]$

$$O\left(\frac{K_j^2}{\sigma}\right) e^{-\beta(\sigma)/\nu}, \quad O\left(\frac{K_j^2}{\gamma}\right) e^{-\sigma/\nu} \leq A$$

Hence, for ν sufficiently small, $V(t_j)$ is a bounded sequence from which the first part of the corollary follows by extending the solutions to the whole interval $[t_0, t_0 + T]$.

Finally, properties (1) and (2) are obtained by following the same procedure as in the proof of Corollary 6.16. □

Chapter 7

Model Reference Adaptive Control

7.1 Introduction

A crucial assumption in designing the MRC schemes of Chapter 4 is that the TV plant parameters are known functions of time. Such an assumption can be quite restrictive in applications especially since the need to use LTV plant models is often accompanied by a poor knowledge of the LTV plant parameters. Of course, the MRC's of Chapter 4 possess some robustness properties with respect to modeling error and parametric uncertainty, but they may fail to even ensure closed-loop stability if the uncertainty becomes large enough. This problem becomes more pronounced in cases where the plant parameters may vary in an unknown fashion over a wide range. In this chapter, our objective is to develop and analyze MRC schemes that can tolerate large parametric uncertainty. Our approach evolves around the so called *Certainty Equivalence Principle* whereby the partially unknown controller parameters are replaced by their estimates, generated by a suitable adaptive law. The resulting controller is referred to as an *adaptive* controller and in particular, when the underlying control law is an MRC one, as a Model Reference Adaptive Controller (MRAC).

Depending on the way the estimates of the controller parameters are generated, we distinguish two types of MRAC. In the first one, the adaptive law generates estimates of the plant parameters which are then used to calculate the corresponding MRC parameters according to the techniques discussed in Chapter 4. This procedure can also be used with other control laws, e.g., PPC, and the resulting controller is referred to as an *indirect* adaptive controller. In the second type of adaptive controllers, the adaptive law generates the estimates of the controller parameters on line, without any intermediate calculations. This approach has been particularly successful in generating adaptive controllers with underlying control laws of the MRC kind, referred to as a *direct* MRAC.

In this chapter we focus our attention on the direct MRAC, starting with Section 7.2 where we design adaptive laws that generate on-line estimates of the MRC parameters. In Section 7.3 we combine these adaptive laws with the MRC structures of Chapter 4 to form direct MRAC schemes and study the stability properties of the resulting closed-loop plant. We conclude with Section 7.4 where we present examples and simulations illustrating the design and properties of direct MRAC's.

7.2 Parameter Estimation in Direct MRAC

The distinguishing property of a direct MRAC is that the controller parameters are estimated from I/O data without requiring the prior identification of the LTV plant. This is achieved by considering the closed-loop description of an MRAC, that is, the closed-loop plant obtained when the LTV plant is controlled by MRC whose parameters are determined on-line by a suitable adaptive law. Manipulating the I/O expressions of the various signals, we can then arrive at an affine, in terms of the controller parameters, model. Having obtained such an affine model, the estimator part of a MRAC has similar properties with the estimators analyzed in Chapter 6. There are, however, a few rather important differences. One is arising due to the need to identify the multiplier c_{0*}, related to the high-frequency gain of the plant. Traditionally, this parameter is estimated separately, requiring the a priori knowledge of its sign (or, equivalently, the sign of the high-frequency gain of the plant). Although several studies have produced adaptive laws removing this requirement [Nus.83, M.M.85] we will not pursue this direction at present. Another difference is exactly due to the fact that the unknown parameters are those of the controller. As a consequence, the assumptions and conditions under which our estimation results have been derived, must be imposed on the controller rather than the plant parameters (e.g., bounding ellipsoids, structure of time variations). Since the relation between the two is highly nonlinear, except some simple cases, the translation of the plant parameter properties to properties of the controller parameters is very complicated. However, due to the general character of our assumptions, this problem only affects the conservatism of our approach but not its validity.

7.2. PARAMETER ESTIMATION IN DIRECT MRAC

Again, we begin our presentation with the analytically simpler case of smooth parameters and then generalize the results to the non-smooth parameter case.

7.2.1 Smooth Parameter Variations

In Chapter 4 we studied the design and properties of a TV MRC, given complete knowledge of the plant parameters. We now employ these results in order to design an adaptive algorithm to estimate the controller parameters in the presence of parametric uncertainty. Throughout this development we assume that:

7.1 Assumption: *(Plant and Reference Model) The LTV plant (3.1) satisfies Assumptions 3.1-3.3, 4.1-4.2 and the reference model is selected to satisfy Assumptions 4.3 and 4.4.* ∎

Under these assumptions, Theorem 4.10 and Corollary 4.9 guarantee the existence of smooth, UB controller parameters $c_{0*}(t)$ and $\theta_*(t)$ for which the TV MRC objective is satisfied and the closed-loop system is ES. For the controller parameters $c_{0*}(t)$ and $\theta_*(t)$ we assume that:

7.2 Assumption: *(Controller Parameters) There exist a vector $\hat{\theta}_*(t)$ and a scalar $\hat{c}_{0*}(t)$ such that $\theta_*(t) = \Pi(t)\hat{\theta}_*(t)$ and $c_{0*}(t) = \pi_0(t)\hat{c}_{0*}(t)$ where $\pi_0(t)$ and the entries of $\Pi(t)$ are known, smooth UB functions of time. The vector $\hat{c}_{0*}\hat{\theta}_*(t)$ satisfies Assumption 6.3 with an ellipsoid center denoted by $\hat{\theta}_c$, while a similar assumption holds for $\hat{c}_{0*}$; that is*

$$\hat{c}_{0*}(t) \in \mathcal{C} \triangleq [\hat{c}_{0min}, \hat{c}_{0max}], \quad \forall t \geq 0; \quad \hat{c}_{0min} > 0$$

and the interval $\mathcal{C}$ is known a priori. ∎

Furthermore, we use the parameter μ as a measure of the average speed of the unstructured part of the MRC parameters, i.e.,

7.3 Assumption: *(Speed of Variations) There exist constants c, μ such that, for all $t_0, T \geq 0$,*

$$\int_{t_0}^{t_0+T} |\dot{\hat{c}}_{0*}(t)|\, dt \leq c + \mu T \quad ; \quad \int_{t_0}^{t_0+T} |\dot{\hat{c}}_{0*}(t)|^2\, dt \leq c + \mu T$$

$$\int_{t_0}^{t_0+T} |\frac{d}{dt}(\hat{c}_{0*}\hat{\theta}_*)(t)|_i\, dt \leq c + \mu T \quad ; \quad \int_{t_0}^{t_0+T} |\dot{\hat{\theta}}_*(t)|_i\, dt \leq c + \mu T$$

$$\int_{t_0}^{t_0+T} |\frac{d}{dt}(\hat{c}_{0*}\hat{\theta}_*)(t)|^2\, dt \leq c + \mu T \quad ; \quad \int_{t_0}^{t_0+T} |\dot{\hat{\theta}}_*(t)|^2\, dt \leq c + \mu T$$

where, as usual, $|\cdot|_i$ is the linear functional norm induced by the norm employed in the definition of the set $\mathcal{M}$. ∎

Notice that, for technical reasons, it is convenient to pose the speed of variation assumptions on both $\hat{c}_{0*}\hat{\theta}_*(t)$ and $\hat{\theta}_*(t)$ rather than $\hat{\theta}_*(t)$ alone.

It should be emphasized that these Assumptions require the a priori knowledge of the plant relative degree n^* and the sign of its high-frequency gain $k_p(t)$ as well as its upper and lower bounds; they do not require, however, knowledge of the rate of exponential stability of its zero dynamics.

7.4 Remark: The structure of the time-variations of the control parameter vector $\theta_*(t)$ can be deduced from the structure of the plant parameters by using the corresponding MRC design equations from Chapter 4. For plants in the P_L-form whose parameter variations are 'fully structured,' the simpler form of the design equations allows the description of the controller parameters in a 'fully structured' form. For example, consider an nth order TV plant

$$y_p = G_p^L(s,t)[u_p] = D_p^{-1}(s,t)N_p(s,t)k_p(t)[u_p]$$

with $deg[N_p(s,t)] = m$ and $k_p(t) = 1$[1] whose parameters vary as sinusoids with a known frequency w_0 and unknown amplitude. From Lemma 4.5 and Corollary 4.9, $\theta_*(t)$ can be expressed as

$$\theta^*(t) = \hat{\theta}_{0*} + \sum_{1}^{n-m} \hat{\theta}_{1i*}\sin(iw_0 t) + \sum_{1}^{n-m} \hat{\theta}_{2i*}\cos(iw_0 t)$$

where $\hat{\theta}_{0*}$, $\hat{\theta}_{ji*}$ are constant vectors.

On the other hand, for plants in the P_R-form, the nonlinear dependence of $\theta_*(t)$ on the plant parameters may force the use of a TV $\hat{\theta}_*$, reducing thus the effectiveness of the approach. ▽▽

Preserving as many similarities as possible with the LTV plant identification case of Chapter 6, we use the notation:

- The I/O vector $U = [u_p, y_p]^\mathsf{T}$.

- The auxiliary controller filters $G(s) \triangleq q^\mathsf{T}(sI-F)^{-1}$, $G'(s) \triangleq (sI-F)^{-1}$.

- The operators

$$\begin{aligned}
G_\pi &: QU \mapsto G(s)[u_p \Pi_1] + G(s)[y_p \Pi_2] + y_p \Pi_3 \\
G'_\pi &: QU \mapsto G'(s)[u_p \Pi_1] + G'(s)[y_p \Pi_2] \\
G_{\theta*} &: QU \mapsto G(s)[u_p \theta_{1*}] + G(s)[y_p \theta_{2*}] + y_p \theta_{3*}
\end{aligned}$$

---
[1] Taking $k_p(t)$ to be TV would only result in more complicated but similar expressions.

7.2. PARAMETER ESTIMATION IN DIRECT MRAC

where the terms with subscripts '3' appear in the case of a non-strictly proper TV MRC only.[2]

- The exponentially decaying term ε describing the effect of initial conditions.

Further, in order to design the adaptive law, let us consider the TV MRC structure, as given in Chapter 4:

$$u_p = c_{0*} u_1 \; ; \; u_1 = G_{\theta*}[QU] + r$$

Although it is possible to perform the identification of $\hat{c}_{0*}$ and $\hat{\theta}_*$ by simply replacing the desired parameters with their estimates in this control law (e.g., see [T.I.89]), it is more convenient to express the control input so that the unstructured part of the unknown parameters appears in an affine form:

$$u_p = \pi_0 \left\{ \hat{c}_{0*} r + G_\pi[QU](\hat{c}_{0*}\hat{\theta}_*) \right\} - \pi_0 \hat{c}_{0*} \eta_1$$

$$\eta_1 = G_\pi[QU]\hat{\theta}_* - G_{\theta*}[QU] = G(s)\{G'_\pi[QU]\dot{\hat{\theta}}_*\}$$

We may now design the MRAC law by replacing the unknown parameters by their estimates while treating the term η_1 as a perturbation.

$$u_p = \pi_0 u_1 \; ; \; u_1 = \hat{c}_0 r + G_\pi[QU]\hat{\theta} \tag{7.1}$$

where $\hat{c}_0$ is the estimate of $\hat{c}_{0*}$ and $\hat{\theta}$ is the estimate of $\hat{c}_{0*}\hat{\theta}_*$.

It should be pointed out that the control law (7.1) is similar to the realization of the PW MRC and in fact reduces to it in the case of completely unstructured parameter variations ($\Pi = I$, $\pi_0 = 1$). In the case of partially or completely structured variations, however, it does contain the known part of the parameter variations at the correct location. This is necessary in order to avoid the swapping of any of the possibly fast time-varying components of the controller parameters and take full advantage of the available a priori knowledge of their structure. We also note that, compared with a control law of the form

$$u_p = c_0 u_1 \; ; \; u_1 = G_\theta[QU] + r \tag{7.2}$$

where $c_0 = \pi_0 \hat{c}_0$, $\theta = \Pi \hat{\theta}$, the control (7.1) suffers the drawback of introducing the seemingly unnecessary perturbation η_1. In fact, using (7.2), a perturbation due to the swapping of $\hat{\theta}$ must be introduced later in the analysis (e.g., see [T.I.89]). Since we require both perturbations to be small in a mean-square normalized sense, the use of either one of the control laws produces similar

---

[2] See Section 4.3, Example 4.7.

CHAPTER 7. MODEL REFERENCE ADAPTIVE CONTROL

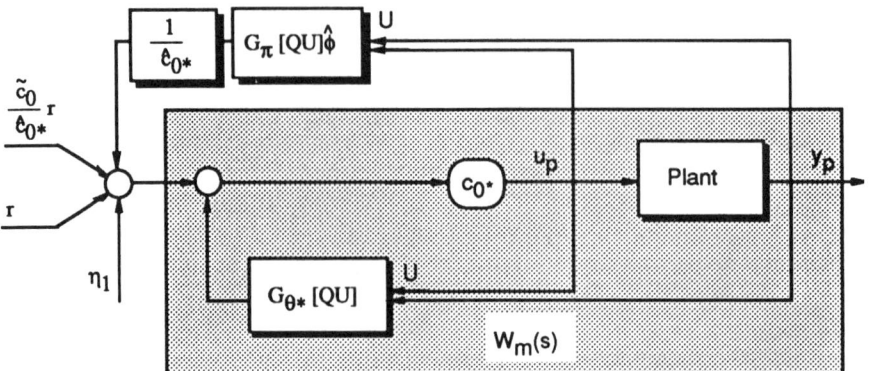

Figure 7.1: The MRAC closed-loop system.

results. It turns out, however, that the analysis for the control law (7.1) is simpler and possibly less conservative.

Thus, we arrive at a description of the closed-loop TV MRAC system, shown in Fig. 7.1, in terms of the nominal closed-loop plant and the perturbation, entering at the same node as the reference input,

$$\frac{1}{\hat{c}_{0*}} \left(G_\pi[QU]\hat{\phi} + \tilde{c}_0 r \right) + \eta_1$$

due to the swapping term η_1 and the parametric uncertainty, where

$$\hat{\phi} \triangleq \hat{\theta} - \hat{c}_{0*}\hat{\theta}_* \quad ; \quad \tilde{c}_0(t) \triangleq \hat{c}_0 - \hat{c}_{0*}$$

In this setup, the goal of the adaptation is simply to update the controller parameters as to minimize (in some sense) the contribution of this perturbation. For this purpose, we define the estimation error ϵ_1 by

$$\epsilon_1 = \hat{c}_0 y_p + \zeta^\mathsf{T} \hat{\theta} - W_m(s)[u_1] \tag{7.3}$$

where

$$\zeta^\mathsf{T} = W_m G_\pi[QU] = W_m(s)\{G(s)[u_p \Pi_1] + G(s)[y_p \Pi_2] + y_p \Pi_3\} \tag{7.4}$$

and the terms with subscript '3' appear in the cases of a non-strictly proper MRC law.

With this definition, the estimation error ϵ_1 can be written as an affine model in the parameter error vector for which our previous development of estimation algorithms is applicable. More precisely,

7.2. PARAMETER ESTIMATION IN DIRECT MRAC

7.5 Lemma: Under Assumption 7.1 the estimation error ϵ_1, defined by (7.3) satisfies

$$\epsilon_1 = \bar{\phi}^T \bar{\zeta} + \hat{c}_{0*}\eta + \varepsilon$$

where

$$\bar{\phi} = \begin{pmatrix} \tilde{c}_0 \\ \hat{\phi} \end{pmatrix} \quad ; \quad \bar{\zeta} = \begin{pmatrix} y_p \\ \zeta \end{pmatrix}$$

and η is a 'swapping' term:

$$\eta = W_m(s)[\eta_1] + \zeta^T \hat{\theta}_* - W_m(s)\left[G_\pi[QU]\hat{\theta}_*\right] \ldots$$
$$+ W_m(s)\left[\frac{1}{\hat{c}_{0*}}u_1\right] - \frac{1}{\hat{c}_{0*}}W_m(s)[u_1]$$

$\triangledown\triangledown$

Proof: In view of Theorem 4.10, Corollary 4.9 and Example 4.7 the nominal closed-loop system (shaded part of Fig. 7.1) is ES and has an I/O operator equal to $W_m(s)$. Thus,

$$y_p = W_m(s)\left[\frac{\tilde{c}_0}{\hat{c}_{0*}}r + r + \frac{1}{\hat{c}_{0*}}G_\pi[QU]\hat{\phi} + \eta_1\right] + \varepsilon$$
$$= W_m(s)\left[\frac{1}{c_{0*}}u_p\right] - W_m(s)G_{\theta*}[QU] + \varepsilon$$

Hence,

$$\epsilon_1 = \bar{\phi}^T \bar{\zeta} + \hat{c}_{0*}W_m(s)\left[\frac{1}{\hat{c}_{0*}}u_1\right] - W_m(s)[u_1] \ldots$$
$$+ \hat{c}_{0*}\left(\zeta^T \hat{\theta}_* - W_m(s)G_{\theta*}[QU]\right) + \varepsilon$$

from which the proof follows immediately by performing the swapping of $\hat{\theta}_*$ in the last two terms. Note that the term ε, describing the effect of initial conditions, is exponentially decaying since the nominal closed-loop system is ES. □

It is straightforward to see that a normalized version of the estimation error equation (7.3) can be put in the general form of the affine model studied earlier. For example, we may choose the normalizing signal m_2 defined by

$$\dot{m}_2 = -2\delta_0 m_2 + |QU|^2 + q_r r^2 + q_e; m_2(t_0) > 0 \tag{7.5}$$

where Q is a positive definite weighting matrix, q_r, q_e are positive constants[3] and δ_0 is such that the poles of $G(s - \delta_0)$ and $W_m(s - \delta_0)$ are in the open left

---
[3] The term $q_r r^2$ is only needed in one case, to ensure that the various constants are independent of the bound of r. Also, certain bounding procedures may require the use of such a term for the same purpose. In our study, however, and in order to preserve the similarities with the plant identification case, we avoid such derivations and this term is unnecessary except in Corollary 7.8.

half-plane.

The properties of the estimator part of an MRAC are described by the following corollary.

7.6 Corollary: Suppose that Assumptions 7.1, 7.2 and 7.3 hold and consider an MRAC closed-loop plant where the parameters of the control law (7.1) are updated by

$$\dot{\hat{c}}_0 = \mathcal{P}_c \left(-\gamma \frac{\epsilon_1 y_p}{m^2} \right) \; ; \; \hat{c}_0(t_0) \in \mathcal{C}$$

$$\dot{\hat{\theta}} = \mathcal{P}_\theta \left(-\gamma \frac{\epsilon_1 \zeta}{m^2} \right) \; ; \; \hat{\theta}(t_0) \in \mathcal{M}$$

and the projections $\mathcal{P}_c$ and $\mathcal{P}_\theta$ correspond to the sets $\mathcal{C}$ and $\mathcal{M}$ respectively. Further, suppose that $U \in L_{\infty,[t_0,t_0+T]}$. Then, $\hat{c}_0, \hat{\theta}, \dot{\hat{c}}_0, \dot{\hat{\theta}}$ are UB on $[t_0, t_0+T]$ ($\hat{c}_0, \hat{\theta}$ are within distance ϵ_* from $\mathcal{C}, \mathcal{M}$); furthermore, there exists a constant C_0 depending on the initial conditions but independent of T, t_0, such that for any interval $I = [t_I, t_I + T_I] \subseteq [t_0, t_0 + T]$

1. $\int_{t_I}^{t_I+T_I} \frac{\epsilon_1^2(t)}{m_2(t)} dt \leq C_0 + \left[\Gamma_1^2 + \frac{K_\phi}{\gamma} \right] \mu T_I;$

   (also valid for $(\bar{\phi}^\top \bar{\zeta})^2/m_2$.)

2. $|\dot{\hat{\theta}}(t)| \leq \gamma \Gamma_2 \frac{|\epsilon_1|}{\sqrt{m_2}} + \varepsilon(t, t_0);$

3. $|\dot{\hat{c}}_0(t)| \leq \gamma \Gamma_3 \frac{|\epsilon_1|}{\sqrt{m_2}} + \varepsilon(t, t_0);$

where $K_\phi, \Gamma_1, \Gamma_2, \Gamma_3$ are constants depending only on operator gains and the radii of the sets $\mathcal{C}, \mathcal{M}$ (their expressions are given in the proof). ▽▽

Proof: In Appendix VI.

The performance of an estimator employing the σ-modification to ensure parameter boundedness can be characterized in a similar fashion. In this case, however, and for technical reasons associated with the forthcoming study of the BIBO stability of MRAC schemes, we assume that the derivatives of $\hat{c}_{0*}$ and $\hat{\theta}_*$ are μ-small, uniformly in time. If we keep the projection in the estimation of $\hat{c}_{0*}$, the same estimation error can be used and the following result is obtained.

7.7 Corollary: Suppose that Assumptions 7.1, 7.2 hold[4] and

$$\|\dot{\hat{\theta}}_*\|_\infty \leq \mu \; ; \; \|\dot{\hat{c}}_{0*}\|_\infty \leq \mu$$

---
[4] In this case, knowledge of the radius of the set $\mathcal{M}$ is not required.

7.2. PARAMETER ESTIMATION IN DIRECT MRAC

Further, consider an MRAC closed-loop system where the parameters of the control law (7.1) are updated by

$$\dot{\hat{c}}_0 = \mathcal{P}_c\left(-\gamma \frac{\epsilon_1 y_p}{m^2}\right) ; \quad \hat{c}_0(t_0) \in \mathcal{C}$$

$$\dot{\hat{\theta}} = -\gamma \frac{\epsilon_1 \zeta}{m^2} - \sigma(\hat{\theta} - \hat{\theta}_c)$$

and the projection $\mathcal{P}_c$ corresponds to the set $\mathcal{C}$. Also suppose that $U \in L_{\infty,[t_0,t_0+T]}$. Then, $\hat{c}_0, \hat{\theta}, \dot{\hat{c}}_0, \dot{\hat{\theta}}$ are UB on $[t_0, t_0 + T]$, $\hat{c}_0$ is within distance ϵ_* from $\mathcal{C}$ and $\hat{\phi}$ converges exponentially fast to a residual set[5]

$$\left\{\hat{\phi} : |\hat{\phi}(t)|_2 \leq O\left(K_\phi, \frac{\mu}{\sigma}, \mu\sqrt{\frac{\gamma}{\sigma}}\right)\right\}$$

where K_ϕ is a constant depending on the radius of the parametric uncertainty set. Furthermore, there exists a constant C_0 depending on the initial conditions but independent of T, t_0, such that for any interval $I = [t_I, t_I + T_I] \subseteq [t_0, t_0 + T]$

1. $\displaystyle\int_{t_I}^{t_I+T_I} \frac{\epsilon_1^2(t)}{m^2(t)} dt \leq C_0 + O\left(\mu^2, \frac{K_\phi^2 \sigma}{\gamma}, \frac{\mu^2}{\gamma\sigma}\right) T_I;$

(also valid for $(\bar{\phi}^\top \bar{\zeta})^2 / m^2$.)

2. $|\dot{\hat{\theta}}(t)| \leq O(\gamma) \dfrac{|\epsilon_1|}{\sqrt{m^2}} + O(K_\phi \sigma, \mu, \mu\sqrt{\gamma\sigma}) + \varepsilon(t, t_0);$

3. $|\dot{\hat{c}}_0(t)| \leq O(\gamma) \dfrac{|\epsilon_1|}{\sqrt{m^2}} + \varepsilon(t, t_0);$

▽▽

Proof: In Appendix VI.

Similar results can also be derived without requiring the explicit knowledge of the set $\mathcal{C}$ by using an additional parameter, updated as an estimate of $1/\hat{c}_{0*}$. This technique, discussed in [N.V.78] for the LTI case and [T.I.89] for the LTV case, has the advantage that it requires less a priori knowledge about the plant, an advantage that can be exploited by adaptation algorithms employing the σ-modification. It does, however, have a disadvantage associated with the

---

[5]Due to the complexity and increased conservatism of the expressions, we used the simplified notation $O(x_1, x_2, \ldots) \leq K_1 x_1 + K_2 x_2 + \cdots$ where K_i are constants. All such constants depend only on operator gains. We do, however, keep track of the important adaptation parameters γ and σ and the radius of the parametric uncertainty set K_ϕ for later discussion.

overparametrization of the controller that prohibits parameter convergence [S.B.89] and possibly increases its susceptibility to noise.

In this approach, the estimation error is constructed as

$$\epsilon_1 = y_p - y_m + \psi\xi$$

where ψ is a scalar parameter used as an estimate of $\psi_* = 1/\hat{c}_{0*}$ and ξ is the signal

$$\xi = \zeta^T \hat{\theta} + \hat{c}_0 y_m - W_m(s)[u_1]$$

and, as usual, y_m is the output of the reference model $y_m = W_m(s)[r]$. It follows that ϵ_1 can be expressed as

$$\epsilon_1 = \psi_* \left(\zeta^T \hat{\phi} + \tilde{c}_0 y_m \right) + \tilde{\psi}\xi + \eta + \varepsilon$$

where $\tilde{\psi} = \psi - \psi_*$ and η is a swapping term as in Lemma 7.5.

The so-constructed estimation error is in the familiar affine-model form for which an estimator can be designed along the lines of our previous discussions, employing either a projection or a σ-modification to ensure parameter boundedness. Here, we only consider the latter, since the value of this approach is primarily in cases where the a priori knowledge of the sets $\mathcal{C}, \mathcal{M}$ is very poor. Thus, the parameter updates are performed by

$$\dot{\hat{\theta}} = -\gamma \frac{\epsilon_1 \zeta}{m^2} - \sigma(\hat{\theta} - \hat{\theta}_c)$$

$$\dot{\hat{c}}_0 = -\gamma \frac{\epsilon_1 y_m}{m^2} - \sigma(\hat{c}_0 - \hat{c}_{0c})$$

$$\dot{\psi} = -\gamma \frac{\epsilon_1 \xi}{m^2} - \sigma(\psi - \psi_c)$$

where, as usual, the subscripts 'c' denote the center of the ellipsoid containing the unknown parameters. Of course, if such information is unavailable, these terms are zero. Further, assuming that the usual MRAC assumptions are satisfied as in Corollary 7.7 and that the derivatives of the unknown parameters are uniformly small, i.e.,

$$\|\dot{\hat{\theta}}_*\|_\infty \leq \mu \ ; \ \|\dot{\hat{c}}_{0*}\|_\infty \leq \mu \ ; \ \|\dot{\psi}_*\|_\infty \leq \mu$$

we obtain the following result.

7.8 Corollary: Under these conditions, suppose that $U \in L_{\infty,[t_0,t_0+T]}$. Then, there exists $\mu_0 > 0$ such that $\forall \mu \in [0, \mu_0]$, $\hat{c}_0, \hat{\theta}, \psi$ and their derivatives are UB on $[t_0, t_0 + T]$. The parameter estimates converge exponentially fast to a residual set

$$\left\{ \hat{\phi}, \tilde{c}_0, \tilde{\psi} : |\hat{\phi}(t)|_2, |\tilde{c}_0(t)|, |\tilde{\psi}(t)| \leq O\left(K_{\theta*}, \frac{\mu}{\sigma}, \mu\sqrt{\frac{\gamma}{\sigma}} \right) \right\}$$

7.2. PARAMETER ESTIMATION IN DIRECT MRAC

where the constants in $O(\cdot)$ are well defined for μ_0 sufficiently small and
$K_{\theta*} = \|\hat{c}_{0*}\hat{\theta}_* - \hat{\theta}_c\|_\infty + \|\hat{c}_{0*} - \hat{c}_{0c}\|_\infty + \|\psi_* - \psi_c\|_\infty$.

Furthermore, for any interval $I = [t_I, t_I + T_I] \subseteq [t_0, t_0 + T]$

1. $\displaystyle\int_{t_I}^{t_I+T_I} \frac{\epsilon_1^2(t)}{m_2(t)} dt \leq C_0 + O\left(\mu^2, \frac{K_{\theta*}^2 \sigma}{\gamma}, \frac{\mu^2}{\gamma\sigma}\right) T_I;$

2. $|\dot{\hat{\theta}}(t)| \leq O(\gamma) \dfrac{|\epsilon_1|}{\sqrt{m_2}} + O(K_{\theta*}\sigma, \mu, \mu\sqrt{\gamma\sigma}) + \varepsilon(t, t_0);$

3. $|\dot{\hat{c}}_0(t)| \leq O(\gamma) \dfrac{|\epsilon_1|}{\sqrt{m_2}} + O(K_{\theta*}\sigma, \mu, \mu\sqrt{\gamma\sigma}) + \varepsilon(t, t_0);$

4. $|\dot{\psi}(t)| \leq O\left(\gamma, \dfrac{\gamma\mu}{\sigma}, \mu\gamma\sqrt{\dfrac{\gamma}{\sigma}}\right) \dfrac{|\epsilon_1|}{\sqrt{m_2}} + O(K_{\theta*}\sigma, \mu, \mu\sqrt{\gamma\sigma}) + \varepsilon(t, t_0);$

where C_0 is a constant depending on the initial conditions but independent of T, t_0. ▽▽

Proof: In Appendix VI.

The above results characterize the average performance of parameter estimation schemes operating in a direct MRAC setting, and bear many similarities with the respective parameter estimators for the identification of LTV plants. We note, however, that the separate estimation of the parameter $\hat{c}_{0*}$ and the more involved construction of the estimation error, tend to increase the complexity of the various expressions.

Further, the properties of certain appropriately defined forms of 'identification' errors are of particular interest. For example, such an identification error is simply the tracking error

$$e_1 = y_p - y_m = y_p - W_m(s)[r]$$

which is the difference between the plant output and the output of the reference model. As discussed in more detail in the next section, the tracking error can be expressed in terms of the estimation error and a 'swapping' term $\hat{\eta}$. The latter has the same form as η of Lemma 7.5 except that the estimates $\hat{\theta}, \hat{c}_0$ are also involved in the swapping. As a consequence, in order to make $e_1/\sqrt{m_2}$ small in the mean-square sense, we must ensure that $\epsilon_1/\sqrt{m_2}, \dot{\hat{\theta}}$ and $\dot{\hat{c}}_0$ are all small in the mean-square sense. For algorithms using projection, this introduces a trade-off between the adaptation gain and the speed of the unstructured parameter variations, very similar to the indirect case. We note, however, that for any choice of γ the mean-square tracking error can be made arbitrarily small for sufficiently small μ. This, in turn, implies that, irrespective of the choice of the adaptation parameters, a projection algorithm can always guarantee 'good' performance of the estimation process for some non-trivial class of time-variations.

On the other hand, for algorithms using the σ-modification a much more careful selection of the adaptation parameters is necessary. For example, Corollary 7.7 indicates that the mean-square value of $e_1/\sqrt{m_2}$ is of the form

$$f(\mu) + O\left(\gamma\sigma, \sigma/\gamma, \sigma^2\right)$$

where $f(\mu) \to 0$ as $\mu \to 0$. Hence, in order to guarantee that the mean-square identification error is small, we must not only require μ to be sufficiently small but select σ to be small enough as well. In other words, estimators using the σ modification may fail to guarantee small estimation/identification errors —even for constant plant parameters— if the adaptation parameters are arbitrarily selected.

Another error signal that should be made small, at least in a mean-square sense, is the 'input error'

$$\begin{aligned}
W_m^{-1}(s)[e_1] &= \frac{\tilde{c}_0}{\hat{c}_{0*}} r + \frac{1}{\hat{c}_{0*}} G_\pi[QU]\hat{\phi} + \eta_1 \\
&= \frac{1}{c_{0*}} u_p - (G_{\theta*}[QU] + r)
\end{aligned}$$

which is exactly the difference between the actual signal produced by the MRAC at the reference input node and the TV MRC input if the controller parameters were known. In fact, the input error is the effective loop perturbation introduced by the parametric uncertainty in the controller parameters (see Fig. 7.1). Consequently, its properties are of paramount importance in determining the boundedness and overall behavior of the closed-loop system. Notice that, in general, we cannot expect that the parameter errors $\hat{\phi}, \tilde{c}_0$ become small after some time; this requires persistent excitation conditions which may not be satisfied for arbitrary reference inputs. Hence, we may not rely on simple small-gain arguments as, in general, the perturbation operators $\hat{\phi}$ and $\tilde{c}_0$ may have large gains. On the other hand, our previous results ensure that the tracking error is small in the mean-square sense for sufficiently slow variations of the unstructured part of the controller parameters. Unfortunately, to take advantage of this, we need to invert the strictly proper operator $W_m(s)$, a procedure which has been a traditional difficulty in the analysis of direct MRAC schemes. An elegant approach to solve this problem can be given through the Operator Inversion Lemma which is an important part of our study in the next section.

7.2.2 Non-Smooth Parameter Variations

The previous results can be extended, in a rather straightforward manner, to the case of LTV plants with non-smooth parameter variations. In this case, however, the appearance of the complete plant state as a perturbation at every point of discontinuity introduces a requirement for the a priori knowledge of the rate of exponential stability of the plant zero dynamics so that an effective normalization signal can be realized. Although in applications this requirement may not be too severe, it constitutes a major qualitative difference from the case of smooth parameter variations. Thus, for the design and analysis of an MRAC for LTV plants with non-smooth parameter variations our assumptions are summarized as follows:

7.9 Assumption: *(i) The LTV plant (3.8) satisfies Assumptions 3.4–3.6 and 4.1–4.4 inside every interval, uniformly in j; the reference model is selected to satisfy Assumptions 4.3 and 4.4 hold.*

(ii) An upper bound on the rate of exponential stability of the nominal plant zero dynamics, $-\alpha$, (see Assumption 4.2) is known a priori and the normalizing signal m_2, given by (7.5), is designed so that

$$\delta_0 < \alpha$$

(iii) Inside every interval (t_j, t_{j+1}), the controller parameters satisfy Assumption 7.2, uniformly in j. ∎

Under these assumptions, there exist piecewise smooth controller parameters θ_* and c_{0*} for which the closed-loop is ES (Theorem 4.18). Furthermore, by Lemma 3.11, the complete state vector of the plant is bounded by $\sqrt{m_2}$, for ν, μ' sufficiently small.[6] This property is instrumental in the proof of all the results associated with the MRAC in the case of LTV plants with non-smooth parameters.

7.10 Corollary: *Suppose that Assumption 7.9 holds, the smooth part of the controller parameters satisfies Assumption 7.3 and consider an MRAC closed-loop system where the parameters of the control law (7.1) are updated by*

$$\dot{\hat{c}}_0 = \mathcal{P}_c \left(-\gamma \frac{\epsilon_1 y_p}{m_2} \right) \; ; \; \hat{c}_0(t_0) \in \mathcal{C}$$

---

[6] As in the previous cases, ν, μ' are used to denote the average frequency of parameter discontinuities and the mean-square value of the perturbation resulting from smooth approximation, respectively.

$$\dot{\hat{\theta}} = \mathcal{P}_\theta\left(-\gamma\frac{\epsilon_1\zeta}{m_2}\right) \; ; \; \hat{\theta}(t_0) \in \mathcal{M}$$

and the projections $\mathcal{P}_c$ and $\mathcal{P}_\theta$ correspond to the sets $\mathcal{C}$ and $\mathcal{M}$ respectively. Further, suppose that $U \in L_{\infty,[t_0,t_0+T]}$. Then, there exist ν_0, μ_0' such that for any $\nu \in [0, \nu_0)$ and $\mu' \in [0, \mu_0')$, $\hat{c}_0, \hat{\theta}, \dot{\hat{c}}_0, \dot{\hat{\theta}}$ are UB on $[t_0, t_0 + T]$ ($\hat{c}_0, \hat{\theta}$ are within distance ϵ_* from $\mathcal{C}, \mathcal{M}$). Furthermore, there exists a constant C_0 depending on the initial conditions but independent of T, t_0, such that for any interval $I = [t_I, t_I + T_I] \subseteq [t_0, t_0 + T]$

1. $\int_{t_I}^{t_I+T_I} \frac{\epsilon_1^2(t)}{m_2(t)} dt \leq C_0 + \left[\Gamma_1^2 + \frac{K_\phi}{\gamma}\right]\mu T_I + K_J^2\left(1 + \frac{1}{\gamma}\right)\nu T_I + \tilde{K}\mu' T_I;$
(also valid for $(\bar{\phi}^T\bar{\zeta})^2/m_2$.)

2. $|\dot{\hat{\theta}}(t)| \leq \gamma\Gamma_2 \frac{|\epsilon_1|}{\sqrt{m_2}} + \varepsilon(t, t_0);$

3. $|\dot{\hat{c}}_0(t)| \leq \gamma\Gamma_3' \frac{|\epsilon_1|}{\sqrt{m_2}} + \varepsilon(t, t_0);$

where $K_\phi, \Gamma_1, \Gamma_2$ are as in Corollary 7.6 and $\Gamma_3', K_J, \tilde{K}$ are constants depending only on operator gains, the radii of the sets $\mathcal{C}, \mathcal{M}$ the magnitude of parameter discontinuities and the magnitude of parameter perturbations due to smooth approximation but independent of T, t_0. ▽▽

Proof: In Appendix VI.

Analogous, though more complicated, results are obtained when the σ-modification is used to ensure parameter boundedness. Again, we restrict ourselves to the study of this case under some uniformity conditions on the size of the perturbation terms, namely,

$$\|\dot{\hat{\theta}}_*^s\|_\infty \leq \mu \; ; \; \|\dot{\hat{c}}_{0*}^s\|_\infty \leq \mu$$
$$|\tilde{A}|, |\tilde{b}|, |\tilde{c}| \leq \mu'$$
$$t_{j+1} - t_j \geq \frac{1}{\nu}$$

uniformly in t and j, where as usual the superscript 's' denotes the smooth part of the parameters, '~' refers to the state-space perturbations due to smooth approximations and t_j are the discontinuity points.

7.11 Corollary: Suppose that Assumption 7.9 holds and the smooth part of the controller parameters satisfies Assumption 7.3 as restricted by the conditions above. Further, consider an MRAC closed-loop system where the parameters of the control law (7.1) are updated by

$$\dot{\hat{c}}_0 = \mathcal{P}_c\left(-\gamma\frac{\epsilon_1 y_p}{m_2}\right) \; ; \; \hat{c}_0(t_0) \in \mathcal{C}$$

7.3. MRAC: DESIGN AND STABILITY ANALYSIS

$$\dot{\hat{\theta}} = -\gamma \frac{\epsilon_1 \zeta}{m_2} - \sigma(\hat{\theta} - \hat{\theta}_c)$$

and the projection $\mathcal{P}_c$ corresponds to the set $\mathcal{C}$. Further, suppose that $U \in L_{\infty,[t_0,t_0+T]}$. Then, there exist ν_0, μ'_0 such that for any $\nu \in [0, \nu_0)$ and $\mu' \in [0, \mu'_0)$, $\hat{c}_0, \hat{\theta}, \dot{\hat{c}}_0, \dot{\hat{\theta}}$ are UB on $[t_0, t_0 + T]$, $\hat{c}_0$ is within distance ϵ_* from $\mathcal{C}$ and $\hat{\phi}$ converges exponentially fast to a residual set satisfying

$$|\hat{\phi}(t)|_2 \leq O\left(K_\phi, \frac{\mu}{\sigma}, \mu\sqrt{\frac{\gamma}{\sigma}}, \mu'\sqrt{\frac{\gamma}{\sigma}}\right) + O\left(K_J, \sqrt{\frac{\gamma}{\sigma}}\right) e^{-\beta(t-t_j)} \ ; \ t \geq t_j$$

where K_ϕ is a constant depending on the radius of the parametric uncertainty set (see Corollary 7.6), K_J is a constant depending on the size of the parameter jumps and for any $\sigma > 0$, $\beta(\sigma) > 0$ is a constant.

Furthermore, there exists a constant C_0 depending on the initial conditions but independent of T, t_0, such that for any interval $I = [t_I, t_I + T_I] \subseteq [t_0, t_0 + T]$

1. $\displaystyle\int_{t_I}^{t_I+T_I} \frac{\epsilon_1^2(t)}{m_2(t)} dt \leq C_0 + O\left(\mu^2, \mu'^2, \nu, \frac{K_\phi^2 \sigma}{\gamma}, \frac{\mu^2}{\gamma\sigma}, \frac{K_J^2 \nu}{\gamma}\right) T_I;$

   (also valid for $(\bar{\phi}^T \bar{\zeta})^2/m_2$.)

2. $|\dot{\hat{\theta}}(t)| \leq O(\gamma) \dfrac{|\epsilon_1|}{\sqrt{m_2}} + O\left(K_\phi \sigma, \mu, \mu\sqrt{\gamma\sigma}, \mu'\sqrt{\gamma\sigma}\right) \ldots$

   $\qquad\qquad + O\left(\sigma K_J, \sqrt{\gamma\sigma}\right) e^{-\beta(t-t_j)} + \varepsilon(t, t_0) \ ; \ t \geq t_j;$

3. $|\dot{\hat{c}}_0(t)| \leq O(\gamma) \dfrac{|\epsilon_1|}{\sqrt{m_2}} + \varepsilon(t, t_0);$

$\triangledown\triangledown$

Proof: In Appendix VI.

The essence of all the results in this section is the description of the closed-loop plant in terms of known I/O operators and a perturbation related to an estimation/identification error. This is quite obvious in the MRC case and it is also true in the indirect control case. Consequently, in the remainder of our study, we need to discuss the robustness properties of such systems subject to persistent but small in the mean-square sense perturbations.

7.3 MRAC: Design and Stability Analysis

In this section we study the global properties of direct MRAC. Our primary objective is to establish, under some general conditions, the boundedness

(BIBO stability) of the closed-loop system. In addition, we would like to describe the closed-loop performance, in some sense, characterizing thus the effectiveness of the adaptive control scheme. The latter is an issue largely unresolved even for LTI plants without using persistent excitation and/or local analysis concepts [A++.86, Gaw.87]. Moreover, standard remedies in the form of dead-zone modifications are not directly applicable in our case due to the continuous variation of the plant parameters over possibly large sets of parametric uncertainty. Guided by the results of the previous section, our expectation is to show that the mean-square or, equivalently, the Root-Mean-Square (RMS) value of a normalized identification error is kept small, for slow variations of the unstructured part of the unknown parameters. Of course, this is a rather weak measure of performance and should be carefully interpreted.

As mentioned in the Introduction, the design of an MRAC scheme is performed in two stages. In the first one, the control input is defined by

$$u_p = \pi_0 u_1 \ ; \ u_1 = \hat{c}_0 r + w^\top \hat{\theta}$$

where

$$\begin{aligned} w^\top &= G_\pi[QU] \\ G_\pi &: QU \mapsto G(s)[u_p \Pi_1] + G(s)[y_p \Pi_2] + y_p \Pi_3 \\ G(s) &= q^\top (sI - F)^{-1} \end{aligned}$$

and $\hat{c}_0, \hat{\theta}$ are the MRAC adjustable parameters. In the second stage, an adaptive law, of the type studied is Section 7.2, is selected to update these parameters on-line. In our analysis, we consider two types of estimators, one using projection and one using the σ-modification to ensure the boundedness of the parameter estimates. Briefly described, for the projection modification, the update law equations are (see Section 7.2 and Chapter 5 for details)

$$\dot{\hat{c}}_0 = \mathcal{P}_c \left(-\gamma \frac{\epsilon_1 y_p}{m_2} \right) \ ; \ \hat{c}_0(t_0) \in \mathcal{C}$$

$$\dot{\hat{\theta}} = \mathcal{P}_\theta \left(-\gamma \frac{\epsilon_1 \zeta}{m_2} \right) \ ; \ \hat{\theta}(t_0) \in \mathcal{M}$$

where $\gamma > 0$ is the adaptation gain, $\mathcal{P}_c$ and $\mathcal{P}_\theta$ denote projection operators and

$$\begin{aligned} \epsilon_1 &= \hat{c}_0 y_p + \zeta^\top \hat{\theta} - W_m(s)[u_1] \\ \zeta &= W_m(s)[w] \\ \dot{m}_2 &= -2\delta_0 m_2 + |QU|^2 + q_r r^2 + q_e; m_2(t_0) > 0 \end{aligned}$$

7.3. MRAC: DESIGN AND STABILITY ANALYSIS

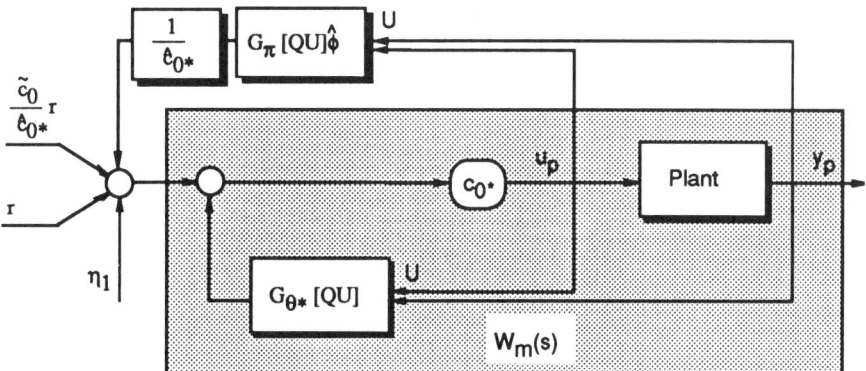

Figure 7.2: The MRAC closed-loop system.

For the σ-modification, the update equation for $\hat{\theta}$ is defined by

$$\dot{\hat{\theta}} = -\gamma \frac{\epsilon_1 \zeta}{m_2} - \sigma(\hat{\theta} - \hat{\theta}_c)$$

where $\sigma > 0$ is a constant and the rest of the quantities are as before.

Our approach to analyze the stability properties of the MRAC closed-loop plant relies on the closed-loop I/O description, shown in Fig. 7.2, where the parameter error appears as a perturbation of an ES nominal closed-loop system. In this framework, boundedness can be established provided that the —appropriately normalized— external perturbation is small in the mean-square sense. This key idea is essentially expressed by Lemma 2.45 although a slightly more refined version is used here.

The description of Fig. 7.2 is obtained by a straightforward decomposition of the control input into the nominal part $G_{\theta_*}[QU]$, corresponding to the TV MRC law if the plant parameters were precisely known, and the 'input error' part. The latter is a perturbation due to the parameter errors $\tilde{c}_0$ and $\hat{\phi}$ and the swapping term η_1 which, for the adaptive control systems considered, can be decomposed into two parts: the first is the output of an operator having small gain and input the I/O pair; the second is a signal which, by virtue of the properties of the estimation algorithm, is small in an average, normalized sense.[7]

Motivated by such a decomposition, we study the boundedness properties of the closed-loop system by combining small-gain and Bellman-Gronwall arguments to obtain the following intermediate result.

---

[7] A similar description is obtained when the control input is defined by $u_p = c_0 u_1$, $u_1 = G_\theta[QU] + r$ where c_0, θ are the adjustable parameters [T.I.89]. However, the analysis for this control law is slightly more involved and is not discussed here.

7.12 Lemma: Let U, v, E, R, r be, possibly vector valued, signals in $L_{\infty,[t_0,t_1]}$ with $R, r, 1/r$ being UB.[8] Also, let m_f be a signal defined by

$$m_f = \mathcal{E}_{-2\delta}\left(\|(U)_t\|_{2,\delta}^2 + \|(r)_t\|_{2,\delta}^2 + qe^{2\delta t_0}\right)$$

where $q > 0$ is a constant and suppose that

$$\|(U)_t\|_{2,\delta} \leq \|(R)_t\|_{2,\delta} + a\|(U)_t\|_{2,\delta} + \|(v)_t\|_{2,\delta} + \|(E)_t\|_{2,\delta}; \; \forall t \in [t_0, t_1]$$

$$\int_t^{t+T} \frac{|v(\tau)|^2}{m_f(\tau)} d\tau \leq C + \mu^2 T; \; \forall t \in [t_0, t_1], \; \forall 0 \leq T \leq (t_1 - t)$$

$$\frac{|E(t)|^2}{m_f(t)} \leq C'e^{-b(t-t_0)}; \; \forall t \in [t_0, t_1]$$

$$\frac{|v|^2}{m_f}, \; \frac{|E|^2}{m_f} \; \text{are UB on } [t_0, t_1]$$

where a, b, μ, C, C' are positive constants and $(\cdot)_t$ denotes truncation at t.
Then, a sufficient condition for m_f to be UB on $[t_0, t_1]$ is

$$a + \frac{\mu}{\sqrt{2\delta}} < 1 \tag{7.6}$$

▽▽

Proof: In Appendix VII.

It is interesting to observe that this result can be interpreted as a weak version of the small-gain theorem, applicable to ES systems perturbed by an operator of small gain and an operator whose output has a small normalized RMS value. That is, a can be interpreted as the $L_2(\delta)$ gain of a perturbation operator in cascade with the corresponding system sensitivity operator and μ as the RMS value of $v/\sqrt{m_f}$, v being the output of another, not necessarily small perturbation operator. Notice that the ES property of the unperturbed system is, at this point, indirectly required through the term R which is the nominal output of the system and the assumption that a is a finite constant. On the other hand, δ can be interpreted as the exponential stability margin of the nominal system or, alternatively, as its minimum rate of energy dissipation. Thus, the first term of (7.6) imposes a small-gain condition on the first perturbation operator while the second ascertains the worst-case maximum average energy that can be dissipated by the system.

Note that Lemma 7.12 does not guarantee by itself that m_f is UB for all t nor that the closed-loop signals are UB. Such conclusions, however, can be drawn by making some technical arguments regarding the ODE's that describe the evolution of the closed-loop system. As usual, we begin our discussion with the analytically simpler case of smooth parameter variations.

---
[8] Again, we use the term UB to signify that the L_∞ bounds are independent of t_0, t_1.

7.3.1 Smooth Parameter Variations

In the case of smooth parameter variations and invoking Lemma 4.6, we express the signal $U = [u_p, y_p]^T$ as

$$QU = QS_r[\tilde{u}] + \varepsilon$$

where

$$\tilde{u} = \frac{\hat{c}_0}{\hat{c}_{0*}}r + \frac{1}{\hat{c}_{0*}}w^T\hat{\phi} + \eta_1$$

$$\eta_1 = w^T\hat{\theta}_* - G_{\theta*}[QU] = G(s)\{G'_\pi[QU]\dot{\hat{\theta}}_*\}$$

and $S_r = [S_{ru}, S_{ry}]^T$ is an $L_p(\delta)$-stable operator. Further, since $S_{ry} = W_m(s)$, we have that

$$W_m(s)[\tilde{u}] = y_p$$

while from equation (7.3)

$$y_p = \frac{1}{\hat{c}_0}\left(\epsilon_1 - W_{m1}\{W_{m2}[w^T]\dot{\hat{\theta}}\} + W_m[\hat{c}_0 r]\right) \quad (7.7)$$

Thus, combining the expressions above and employing Lemma 2.61, we arrive at the following description of the closed-loop system

$$\begin{aligned}
QU &= QS_r(1-\Lambda)[\tilde{u}] + (1-\rho)QS_r\Lambda W_m^{-1}[y_p] + \rho QS_r\Lambda[\tilde{u}] + \varepsilon \\
&= R + QS_r\Lambda_1\left[\frac{d}{dt}\left(\frac{1}{\hat{c}_{0*}}w^T\hat{\phi}\right)\right] + QS_r(1-\Lambda)[\eta_1] \\
&\quad + (1-\rho)QS_r\Lambda W_m^{-1}\frac{1}{\hat{c}_0}[\epsilon_1] \\
&\quad + (1-\rho)QS_r\Lambda W_m^{-1}\frac{1}{\hat{c}_0}W_{m1}\left\{W_{m2}[w^T](-\dot{\hat{\theta}})\right\} \\
&\quad + \rho QS_r\Lambda[\eta_1] + \rho QS_r\Lambda\left[\frac{1}{\hat{c}_{0*}}w^T\hat{\phi}\right]
\end{aligned} \quad (7.8)$$

where R includes the contribution of all terms depending on initial conditions or the UB reference input, $\rho \in [0,1]$ is a free parameter[9] and Λ, Λ_1 are operators as in Lemma 2.61 such that the γ_2-gain of Λ_1 is $O(1/a)$ with 'a' being another free parameter to be selected.

Equation (7.8) establishes a connection between the estimation process and the nominal closed-loop perturbation due to the TV parametric uncertainty

---

[9] The usefulness of ρ is in recovering a small-gain condition when the parametric uncertainty is small, e.g., see [Tsa.89]. In our case, however, the general bounding procedure invoked for the perturbation term η_1 makes such an approach rather inconvenient. Instead, the analysis can be performed in two stages. One for large parametric uncertainty sets ($\rho = 0$) and another for small parametric uncertainty sets ($\rho = 1$).

$\hat{\phi}$. Clearly, a stability argument can be made using the small-gain theorem alone. Such an argument, however, would involve the worst-case parametric uncertainty which, in general, is of the order of the radius of the parametric uncertainty set and thus limiting the usefulness of the result. On the other hand, the estimator adjusts on-line the controller parameters and therefore $\hat{\phi}$ in an attempt to minimize the estimation error ϵ_1. Translating the properties of the estimation error, established in the previous chapter, to properties of the effective perturbation $w^\top \hat{\phi}$ involves the non-proper inverse of the reference model. This operation can be interpreted as a decomposition of $w^\top \hat{\phi}$ into a low and a high frequency component, the former being small in a mean-square normalized sense. The decomposition is performed by the fictitious filter Λ whose cut-off frequency 'a' determines, in an indirect way, the worst-case frequency content of the large-in-magnitude components of $w^\top \hat{\phi}$. The rest of the analysis is a tedious but rather straightforward procedure where a constraint on the free parameter 'a' and an upper bound on μ are determined via Lemma 7.12. The final result describes the boundedness and performance, in a mean-square sense, properties of the adaptive closed-loop system and can be stated in a simple form as follows.

7.13 Theorem: *Let K_ϕ be a constant denoting the size of the parametric uncertainty set as in Corollary 7.6 and suppose that the conditions of Corollary 7.6 are satisfied with the projection $\mathcal{P}_c$ designed so that $\hat{c}_{0min} - \epsilon_* > 0$. Then, for any finite K_ϕ, there exists $\mu_0 > 0$ such that for any $\mu \in [0, \mu_0)$ all signals in the closed-loop are UB for all $t \geq t_0$.*

Furthermore, there exists a constant $C_0 > 0$ such that the tracking error $e_1 = y_p - y_m$ satisfies

$$\int_t^{t+T} \frac{e_1^2(\tau)}{m_2(\tau)} d\tau \leq C_0 + O\left(1, \frac{K_\phi}{\gamma}, \gamma^2, K_\phi \gamma\right) \mu T$$

for all $t \geq t_0$ and all $T \geq 0$. ▽▽

Proof: In Appendix VII.

Clearly, this result confirms our intuitive expectation that an adaptive controller should be able to guarantee boundedness and good performance on the average even if the plant description contains unknown but slowly varying elements. Next, to gain some further insight on the implications of Theorem 7.13, we make several interesting observations following either directly from the Theorem itself or its proof.[10]

---

[10] Theorem 7.13 can also be stated in a form reminiscent of the small-gain theorem, e.g., see [Tsa.89] from which all the observations follow as corollaries. Such a form, however, is notationally complicated and perhaps unnecessary. In an attempt to keep the presentation simple we refer the interested reader to the proof for further details.

7.3. MRAC: DESIGN AND STABILITY ANALYSIS

7.14 Remark: As a first observation we note that the critical parameter for the boundedness of the closed-loop signals is the speed of variation of only the unstructured part of the controller parameters. This in turn implies that an adaptive controller can be used with plants that are not necessarily slowly TV, as long as the structure of the fast TV elements has been correctly incorporated in the control law. And although the structure of the fast TV elements (Π, π_0) affects the value of μ_0 in an indirect way, through the gains of the various operators, boundedness can always be guaranteed for some non-trivial class of TV plants. On the other hand, μ_0 is very sensitive to the size of parametric uncertainty and especially M_0. A large radius of the parametric uncertainty set tends to decrease μ_0 quite rapidly and even more so if the relative degree of the plant and the reference model is high. This can be attributed to the increasing loss of high-frequency information, during the estimation process, as the relative degree of the plant increases. ▽▽

7.15 Remark: In the limit as $\mu \to 0$, boundedness is guaranteed for arbitrarily large parametric uncertainty sets ($K_\phi \to \infty$) and the mean-square value of the tracking error also approaches zero without excluding, however, the possibility of burst phenomena. In addition, when $\mu = 0$, (fully structured variations) the adaptive controller recovers the performance of the ideal case (known parameters), in the sense that $e_1 \to 0$ as $t \to \infty$ irrespective of the size of the parametric uncertainty. This result is an immediate consequence of the square integrability and Lipschitz continuity of the tracking error.[11] Again, the result is valid even if the overall plant is TV. ▽▽

7.16 Remark: In the case of completely unstructured parameter variations, $\Pi = I$, $\pi_0 = 1$, the MRAC design is identical to the standard one used for LTI plants. This observation establishes the robustness of a MRAC with respect to slow or slow-in-the-mean variations of the plant parameters and provides a theoretically concrete justification for the use of adaptive controllers in a TV environment. ▽▽

7.17 Remark: When the parametric uncertainty becomes sufficiently small, boundedness can be guaranteed independent of the speed of variation of the unknown parameters. This result can be seen via a straightforward small-gain argument (for further details, see the proof of the theorem). ▽▽

7.18 Remark: It is both theoretically and practically interesting to note the role of the adaptation gain γ in the behavior of the adaptive closed-loop systems. Since quantities depending on both γ or $1/\gamma$ appear in the condition

---

[11]Note that the convergence of the tracking error to zero is not necessarily exponential.

for stability as well as the performance characterization, the adaptation gain should not be selected as too small or too large. This observation agrees well with intuition as estimators with too small adaptation gains are unable to track fast parameter variations. On the other hand, large adaptation gains introduce large perturbations since the derivative of the parameter estimate is not necessarily an estimate of the derivative of the parameter. An 'optimum' selection of the adaptation gain is not clear at this point, especially in view of the fact that the stability condition in Theorem 7.13 is only sufficient and quite conservative. It seems reasonable, however, that an initial selection of the adaptation parameters should be approached from a performance point of view. That is, design the adaptive controller as to minimize the RMS value of the tracking error, for which relatively tight bounds can be derived, assuming that the plant parameter variations are sufficiently slow to ensure the closed-loop BIBO stability. ▽▽

7.19 Remark: It cannot be overemphasized that the results described in this and the next chapter are sufficient conditions for the boundedness of signals in the adaptive closed-loop plant. And although these conditions are intuitively appealing, it is by no means implied that their violation leads to an unbounded closed-loop plant. In fact, as shown in [A.N.89], there are special cases where the signals in the adaptive closed-loop plant remain bounded despite fast unstructured variations over a wide range. These results, however, are not directly applicable to the general case and since general instability theorems are extremely hard to establish, the derivation of necessary and sufficient or even 'tight' conditions for boundedness is still an unresolved problem.
▽▽

7.20 Remark: Following similar analytical steps, one may also establish the robustness of an MRAC with respect to different types of perturbations (e.g. unmodeled dynamics, unstructured non-parametric uncertainty [Tsa.89]), although for such a result one should rederive the estimator properties in order to include the effects of the additional perturbations. In this case, the properties of the various loop sensitivity operators are of critical importance and loop shaping procedures should be used to enhance both robustness and performance of the adaptive system. However, a systematic design methodology is yet unavailable except in special cases, e.g., local analysis and/or persistent excitation [A++.86]. ▽▽

When the parameters of an MRAC are updated by an estimator using the σ-modification, the analysis becomes more complicated due to the additional $O(\sigma)$ perturbation introduced in the parameter estimation algorithm. The

7.3. MRAC: DESIGN AND STABILITY ANALYSIS

net result is that, in order to guarantee boundedness, we need to impose conditions on the magnitude of both σ and μ.

7.21 Theorem: Let K_ϕ be a constant denoting the size of the parametric uncertainty set and suppose that the conditions of Corollary 7.7 are satisfied with the projection $\mathcal{P}_c$ designed so that $\hat{c}_{0min} - \epsilon_* > 0$. Then, for any finite K_ϕ, there exist $\sigma_0 > 0$ and $\mu_0(\sigma)$ [12] such that for any $\sigma \in (0, \sigma_0)$, $\mu_0 > 0$ and for any $\mu \in [0, \mu_0)$ all signals in the closed-loop are UB for all $t \geq t_0$.

Furthermore, there exists a constant $C_0 > 0$ such that the tracking error $e_1 = y_p - y_m$ satisfies

$$\int_t^{t+T} \frac{e_1^2(\tau)}{m_2(\tau)} d\tau \leq C_0 + O\left(\frac{K_\phi^2}{\gamma}, K_\phi^2 \sigma, K_\phi^2 \gamma\right) \sigma T \ldots$$
$$+ O\left(1, \gamma^2, \frac{\gamma}{\sigma}, \frac{1}{\gamma\sigma}, \gamma\sigma\right) \mu^2 T$$

for all $t \geq t_0$ and all $T \geq 0$. ▽▽

Proof: In Appendix VII.

7.22 Remark: As it was intuitively expected, the boundedness of the closed loop signals can be guaranteed provided that σ is sufficiently small. This condition appears even in the LTI case, [I.S.88], as well as in the case of completely structured parameter variations. One interpretation is that the parameter σ adjusts the size of the parameter search set in an indirect way. Small values of σ allow the parameters to move further away from the 'nominal' point θ_c. In general, this set must be large enough (its diameter should be $O(K_\phi)$) so that θ_* is close to it, implying in return that the set contains the parameters of a stabilizing compensator. Indeed, even though our analysis is quite conservative, σ_0 decreases rapidly as K_ϕ increases, indicating that the size of parametric uncertainty must be taken into account in the selection of σ. On the other hand, the previous theorem can be stated in a somewhat different form to indicate the usefulness of the algorithm in the case of poor a priori knowledge of the size of parametric uncertainty. Namely,

"for any $\sigma > 0$ there exist $\mu_0 > 0$ and $K_0 > 0$ such that for any $\mu \in [0, \mu_0)$ and $K_\phi \in [0, K_0)$ all signals in the loop are UB. Furthermore, as $\sigma \to 0$, $\mu_0 \to 0$ and $K_0 \to \infty$."

The proof of this statement follows directly from the proof of Theorem 7.21 by considering K_ϕ as a variable instead of σ. ▽▽

---
[12] μ_0, as derived in the proof of the theorem, is a continuous function of σ.

7.23 Remark: When the parametric uncertainty K_ϕ becomes sufficiently small, it is shown in the proof of the theorem that one way to guarantee boundedness is to increase the value of σ so that $\sigma \gg \mu, \mu^2$. That is, boundedness can be guaranteed for any μ, provided that σ is sufficiently large. Indeed, this result has a quite intuitive interpretation. Small parametric uncertainty implies that our initial guess for the controller parameters $\hat{\theta}_c$ is sufficiently close to $\hat{c}_{0*}\hat{\theta}_*$ for all t and therefore it is a stabilizing one. Thus, all we need to do is to make sure that the parameter estimates remain close to the initial point in the presence of an $O(\mu)$ perturbation. This can be ensured by choosing σ to be sufficiently large compared to μ. Notice that such a restriction is unnecessary in the case of estimators using projection since the latter maintains a 'hard-bound' on the parameter estimates. On the other hand, what is interesting about the σ-modification is that the critical parameter is the *actual* value of K_ϕ, i.e., the actual worst-case difference between the desired controller parameters $\hat{c}_{0*}\hat{\theta}_*$ and their initial guess $\hat{\theta}_c$. When a projection algorithm is used, this difference is (conservatively) estimated *a priori* and remains fixed throughout the operation of the controller. In other words, if it happens that, for the particular plant, $\hat{\theta}_c$ is sufficiently close to the desired parameters $\hat{c}_{0*}\hat{\theta}_*$, then an algorithm using a σ-modification may be superior to one using projection. But the reverse should be expected when one considers the worst-case scenario where the parameters are allowed to vary inside a certain bounded set. ▽▽

In an analogous fashion, closed-loop boundedness can be established for MRAC algorithms designed as in Corollary 7.8 while robustness with respect to other types of perturbations can be shown using the same analytical tools. The details, however, extend beyond the scope of this discussion and are omitted.

7.3.2 Non-Smooth Parameter Variations

The approach used in the previous subsection to analyze the MRAC properties is also applicable when the variations of the plant parameters include discontinuities, according to the models and assumptions discussed in Chapter 3. Again, the main idea is to establish the boundedness of the closed-loop signals by invoking Lemma 7.12 and using the properties of the estimation error and estimated parameters, derived in the corresponding corollaries of Chapter 5. Indeed, one can easily verify that the RMS values of both $\epsilon_1/\sqrt{m_2}$ and $\dot{\hat{\theta}}$ are simultaneously small, provided that

7.3. MRAC: DESIGN AND STABILITY ANALYSIS

- the speed of variation of the smooth part of $\hat{\theta}_*, \hat{c}_{0*}$, parametrized by μ, is small;

- the frequency of discontinuities, parametrized by ν, is small;

- any perturbation terms caused by smooth approximations, whose size is parametrized by μ', are small;

- when the σ-modification is used in the estimator, σ is small.

Consequently, Theorems 7.13 and 7.21 can be generalized to admit plants with discontinuous or non-smooth parameters. In view of the results obtained so far, this generalization is actually quite straightforward despite the fact that the necessary 'bookkeeping' is rather tedious. The final outcome, showing the effect of such perturbations, is stated below.

7.24 Theorem: *Let K_ϕ be a constant denoting the size of the parametric uncertainty set and suppose that the conditions of Corollary 7.10 are satisfied with the projection $\mathcal{P}_c$ designed so that $\hat{c}_{0min} - \epsilon_* > 0$. Then, for any finite K_ϕ, there exist $\mu_0, \nu_0, \mu_0' > 0$ such that for any $\mu \in [0, \mu_0)$, $\nu \in [0, \nu_0)$, $\mu' \in [0, \mu_0')$ all signals in the closed-loop are UB for all $t \geq t_0$.*

Furthermore, there exists a constant $C_0 > 0$ such that the tracking error $e_1 = y_p - y_m$ satisfies

$$\int_t^{t+T} \frac{e_1^2(\tau)}{m_2(\tau)} d\tau \leq C_0 + O\left(1, \frac{K_\phi}{\gamma}, \gamma^2, K_\phi \gamma\right) \mu T \ldots$$
$$+ O\left(\nu, \mu', \frac{\nu}{\gamma}, \gamma^2 \nu, \gamma \nu, \gamma^2 \mu'\right) T$$

for all $t \geq t_0$ and all $T \geq 0$. ▽▽

Proof: In Appendix VII.

7.25 Theorem: *Let K_ϕ be a constant denoting the size of the parametric uncertainty set and suppose that the conditions of Corollary 7.11 are satisfied with the projection $\mathcal{P}_c$ designed so that $\hat{c}_{0min} - \epsilon_* > 0$. Then, for any finite K_ϕ, there exist $\sigma_0 > 0$ and $\mu_0(\sigma), \nu_0(\sigma), \mu_0'(\sigma)$ such that for any $\sigma \in (0, \sigma_0)$, $\mu_0 > 0$, $\nu_0 > 0$, $\mu_0' > 0$ and for any $\mu \in [0, \mu_0)$, $\nu \in [0, \nu_0)$, $\mu' \in [0, \mu_0')$ all signals in the closed-loop are UB for all $t \geq t_0$.*

Furthermore, there exists a constant $C_0 > 0$ such that the tracking error $e_1 = y_p - y_m$ satisfies

$$\int_t^{t+T} \frac{e_1^2(\tau)}{m_2(\tau)} d\tau \leq C_0 + O\left(\frac{K_\phi^2}{\gamma}, K_\phi^2 \sigma, K_\phi^2 \gamma\right) \sigma T$$

$$+ O\left(1, \gamma^2, \frac{\gamma}{\sigma}, \frac{1}{\gamma\sigma}, \gamma\sigma\right)\mu^2 T$$

$$+ O\left(1, \gamma^2, \gamma, \gamma K_\phi^2, \frac{K_\phi^2}{\gamma}, \sigma K_\phi^2\right)\nu T$$

$$+ O\left(1, \gamma^2, \gamma\sigma\right)\mu'^2 T$$

for all $t \geq t_0$ and all $T \geq 0$. ▽▽

Proof: In Appendix VII.

The above results are quite similar to those derived for the case of smooth parameter variations. They can all be classified under the same intuitive comment, that is, under some technical conditions ensuring that the adaptive closed-loop is well-behaved, perturbations which are small in an RMS sense cause an RMS-small deviation from the desired behavior. In this framework, the interesting property of adaptive schemes is that they translate parametric uncertainty, in a certain class of LTV parametric models, into a perturbation of the order of the parameter derivatives in an RMS sense. However, it should be emphasized at this point that, although this result can be established for a nontrivial and practically interesting class of plants, the performance characterization in an RMS sense may not prove to be what the designer had originally in mind. Infrequent but large burst phenomena are a common cause of concern, an issue for which only partial remedies are currently available.

7.4 Examples

We conclude this study of MRAC for LTV plants with a few examples illustrating its design procedure and some of its properties encountered during closed-loop operation. In the following discussion we consider the same plant as in Examples 4.21 and 4.22, i.e.,

$$[s^2 + a_1 s + a_2] y_p = u_p \qquad (7.9)$$

where a_1, a_2 are TV parameters. More specifically, we take

$$a_1 = \hat{a}_1 \ ; \quad a_2 = \hat{a}_2 \sin(\mu t)$$

where $\hat{a}_1, \hat{a}_2$ are constants with nominal unknown values -6 and 2 respectively. Thus, the speed of variations of the plant parameters can be specified by the parameter μ. The control objective is to make the plant output y_p track the output of the LTI reference model

$$[s^2 + 3s + 2] y_m = r \qquad (7.10)$$

7.4. EXAMPLES

for any bounded, piecewise continuous reference input signal r.

7.26 Example: *(TV MRAC, 'fast', structured parameter variations)*
Let us first consider the case where $\mu = 1$ and it is known a priori. Such parameter variations can be characterized as 'fast' since, from the examples of Chapter 4, a PW MRC is unable to maintain closed-loop stability. From Example 4.22, a TV MRC satisfying the above control objective, irrespective of the speed of the parameter variations, is designed as

$$\dot{\omega}_1 = -\omega_1 + \theta_{*1} u_p \; ; \; \dot{\omega}_2 = -\omega_2 + \theta_{*2} y_p$$
$$u_p = \omega_1 + \omega_2 + \theta_{*3} y_p + r$$

with the desired control parameters given by

$$\theta_{*1} = \hat{a}_1 - 3$$
$$\theta_{*2} = \{(\hat{a}_1 - 4)\hat{a}_1 + 3\} + \{(3 - \hat{a}_1)\hat{a}_2\}\sin(\mu t)$$
$$\theta_{*3} = \{(4 - \hat{a}_1)\hat{a}_1 - 5\} + \hat{a}_2 \sin(\mu t)$$

Following our analysis, we may use the alternative construction of the control input

$$\dot{w}_1 = -w_1 + q_1 u_p$$
$$\dot{w}_2 = -w_2 + q_2 y_p$$
$$\dot{w}_3 = -w_3 + q_3 y_p \sin(\mu t)$$
$$w_4 = q_4 y_p$$
$$w_5 = q_5 y_p \sin(\mu t)$$
$$u_p = r + \hat{\theta}_*^T w$$

where $w = [w_1, \ldots w_5]^T$, q_i denotes fixed, constant weights and $\hat{\theta}_*$ is the (constant) parameter vector with nominal value

$$\hat{\theta}_* = \begin{pmatrix} -9/q_1 \\ 63/q_2 \\ 18/q_3 \\ -65/q_4 \\ 2/q_5 \end{pmatrix}$$

Further, let us consider the case where the plant parameters are not precisely known, but there is some parametric uncertainty associated with the values of $\hat{a}_i$. As shown in Chapter 4, the TV MRC guarantees that the nominal closed-loop system is ES and, consequently, it remains so for sufficiently

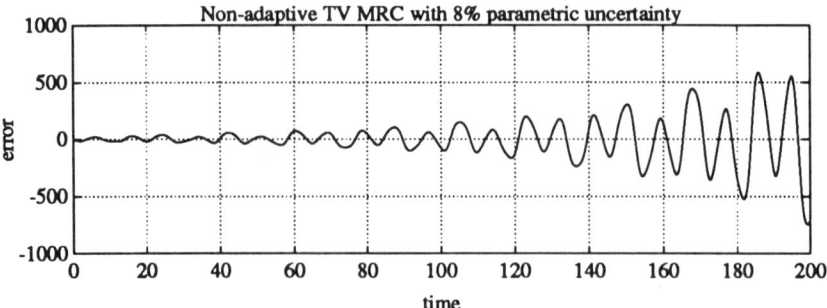

Figure 7.3: Closed-loop response with a non-adaptive TV MRC in the presence of parametric uncertainty.

small values of the parametric uncertainty. However, in our example, a mere 8% parametric uncertainty is sufficient to destabilize the closed-loop system. That is, taking $\hat{a}_1 = -5.52$ and $\hat{a}_2 = 1.84$ (i.e., an 8% difference from the actual) the TV MRC parameters are calculated as

$$\hat{\theta}'_* = \left[\frac{-8.52}{q_1}, \frac{55.5504}{q_2}, \frac{15.6768}{q_3}, \frac{-57.5504}{q_4}, \frac{1.84}{q_5}\right]^\mathsf{T}$$

Using these values of the control parameter vector to control the original plant ($\hat{a}_1 = -6, \hat{a}_2 = 2$), a simulation of its response to a sinusoid $r = 10\sin t$ demonstrates that the closed-loop is unstable (see Fig. 7.3). The destabilization of the closed-loop system due to large parametric uncertainty is, of course, not surprising and appears in the case of LTI plants as well. On the other hand, selecting the auxiliary filters and the reference model more carefully may increase the amount of parametric uncertainty that can be tolerated. Nevertheless, such optimization is still an open problem, even in the LTI case, while the resulting bounds of parametric uncertainty are always finite leaving the main points of this discussion qualitatively unchanged.

Here we follow a different approach to handle the parametric uncertainty problem. According to the procedure discussed in Chapter 6 and assuming knowledge of μ, we design an adaptive law to estimate the unknown controller parameters $\hat{\theta}_*$. Letting $\zeta_i = (s^2 + 3s + 2)^{-1}[w_i]$, we update the controller parameter estimates by

$$\dot{\hat{\theta}} = \mathcal{P}_\theta\left(-\gamma\frac{\epsilon_1\zeta}{m_2}\right)$$

where

$$\epsilon_1 = y_p + \hat{\theta}^\mathsf{T}\zeta - (s^2 + 3s + 2)^{-1}u_p$$
$$\dot{m}_2 = -1.4m_2 + |u_p|^2 + 5|y_p|^2 + 0.1|r|^2 + 1 \ ; \ m_2(0) = 1$$

7.4. EXAMPLES

and the projection $\mathcal{P}_\theta$ is designed to keep the parameter estimates in a bounded set reflecting a considerably larger, 20% parametric uncertainty in $\hat{a}_i$. Notice that the constant weights q_i used in the definition of the signals w_i, can now be given a loose interpretation as normalizers of the uncertainty in each component of θ_i or, alternatively, as normalizers of the adaptation gain of each component. For this example we use

$$q_1 = q_5 = 1 \ ; \ q_2 = q_4 = 5 \ ; \ q_3 = 3$$

for which the projection intervals for each component of $\hat{\theta}$ are easily found to be

$$\hat{\theta} \in \begin{Bmatrix} [-10.2 & , & -7.8] \\ [9.048 & , & 16.728] \\ [4.16 & , & 8.16] \\ [-17.128 & , & -9.448] \\ [1.6 & , & 2.4] \end{Bmatrix}$$

Finally the control input of the TV MRAC is generated by (notice that $c_{0*} = 1$)

$$u_p = r + \hat{\theta}^\top w \tag{7.11}$$

with the vector w defined above and correctly incorporating the available a priori knowledge of the structure of the variations of $\hat{\theta}_*$. Then, according to Theorem 7.13 and this being a case of fully structured variations, the TV MRAC guarantees boundedness of all signals in the loop and convergence of the tracking error to zero, asymptotically with time. This is demonstrated in Fig. 7.4 where $r = 10\sin t$, $\gamma = 10$ and all initial conditions are taken as zero, except

$$\begin{aligned} y(0) &= 1 \\ \hat{\theta}(0) &= [-8.52, 11.11008, 5.2256, -11.51008, 1.84]^\top \end{aligned}$$

It is worthwhile to observe that although the initial convergence of the tracking error to the corresponding residual set is quite rapid, its convergence to zero inside the residual set set is not exponential and can be very slow. This behavior is, of course, similar to the LTI plant case. And although the underlying system dynamics and properties of the nominal sensitivity operators and excitation issues become a lot more complicated, the net result is that the estimation problem is reduced to the estimation of some unknown but constant parameters. In other words, one may consider the adaptive control of LTI plants as a special case of the adaptive control of LTV plants with fully structured parameter variations.

It should be mentioned that if there were some ambiguity in the frequency μ, e.g., a case of partially structured variations, two additional parameters

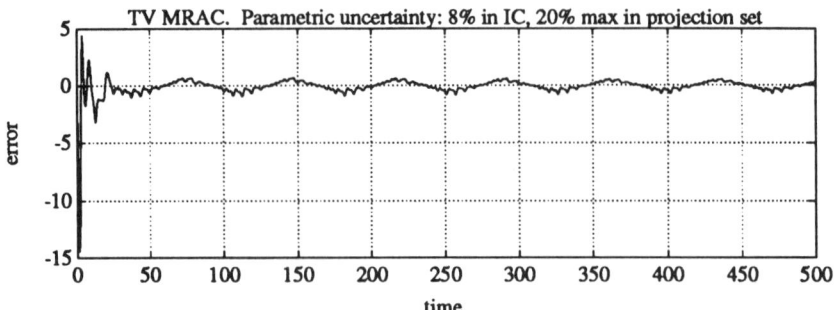

Figure 7.4: Closed-loop response with a TV MRAC. Fast, fully structured parameter variations, $\mu = 1$.

would be needed to estimate the amplitudes of the $\cos \mu t$ terms (see Example 6.15). In this case, the uncertain TV components introduce a perturbation whose size is related to the derivative of the unstructured part of the estimated parameters. This situation is considered in the next example. ▽▽

7.27 Example: *(Slow, unstructured parameter variations)* In this example we consider the more interesting case where the estimated parameters are actually TV. This situation arises when there is partial or no a priori knowledge of the structure of the parameter variations. Such problems have been a traditional motivation for the use of adaptive controllers. Here, and in order to simplify the presentation, we only discuss the case of unstructured parameter variations. Of course, partially structured parameter variations present no additional conceptual difficulty, except perhaps in that the underlying nominal closed-loop structured is an LTV system.

Thus, let us consider again the MRC problem for the plant given in the beginning of this section, where now μ is an unknown, small constant. Following the analysis of Example 4.22, a TV MRC is given by

$$\dot{\omega}_1 = -\omega_1 + \theta_{*1} u_p \; ; \; \dot{\omega}_2 = -\omega_2 + \theta_{*2} y_p$$
$$u_p = \omega_1 + \omega_2 + \theta_{*3} y_p + r$$

with the desired control parameters

$$\theta_{*1} = a_1 - 3$$
$$\theta_{*2} = a_1^2 - 4a_1 - a_1 a_2 + 3a_2 + 3 - 5\dot{a}_1 + 2a_1 \dot{a}_1 - \ddot{a}_1$$
$$\theta_{*3} = 4a_1 + a_2 - a_1^2 - 5 + \dot{a}_1$$

Since no information is available on the structure of parameter variations it is reasonable to assume that they are sufficiently slow so that the plant

7.4. EXAMPLES

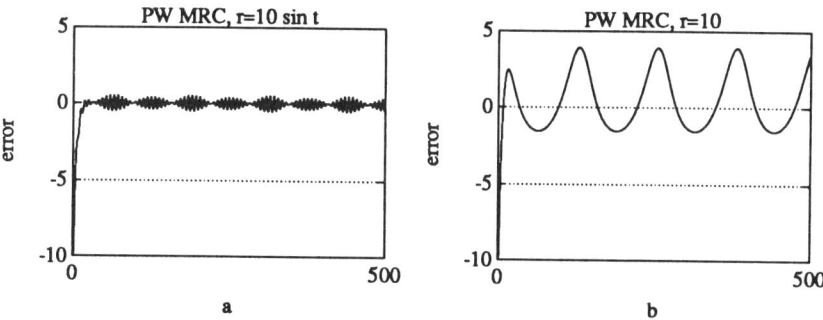

Figure 7.5: Closed-loop response with a PW MRC. Slow, unstructured parameter variations: a. Reference input $r = 10 \sin t$; b. Reference input $r = 10$.

as well as the controller parameters can be treated pointwise as constants. Consequently, we can write the control law as

$$u_p = r + \theta^\top w \qquad (7.12)$$

where

$$\dot{w}_1 = -w_1 + q_1 u_p \ ; \ \dot{w}_2 = -w_2 + q_2 y_p \ ; \ w_3 = q_3 y_p$$

and, as in the previous example, q_i denote fixed weights. Note that since the unknown parameters are TV with no a priori available model of variations, we do not expect to recover the asymptotic tracking property of either the LTI or fully structured TV case. Moreover, as discussed in Chapter 6, the more convenient definition of the TV MRAC control input given by (7.12) is identical to the PW MRC structure in the case of completely unstructured parameter variations (although different in the partially and fully structured cases). With this in mind, it is more appropriate to use the PW MRC response as an indicator of how well we expect the adaptive controller to perform. For example, the tracking error to a sinusoid and a step reference input ($r = 10 \sin t$ and $r = 10$) is shown in Fig. 7.5, where $\mu = 0.05$.

Next, let us assume a 20% uncertainty in the parameters $\hat{a}_i$, as in the previous example. It is straightforward albeit tedious to calculate the ranges for the controller parameters. Selecting the weighting constants as

$$q_1 = 1 \ ; \ q_2 = 10 \ ; \ q_3 = 5$$

we find that the projection intervals for each component of the control parameter vector is

$$\theta \in \left\{ \begin{array}{lll} [-10.2 & , & -7.8] \\ [2.964 & , & 10.404] \\ [-17.568 & , & -9.048] \end{array} \right\}$$

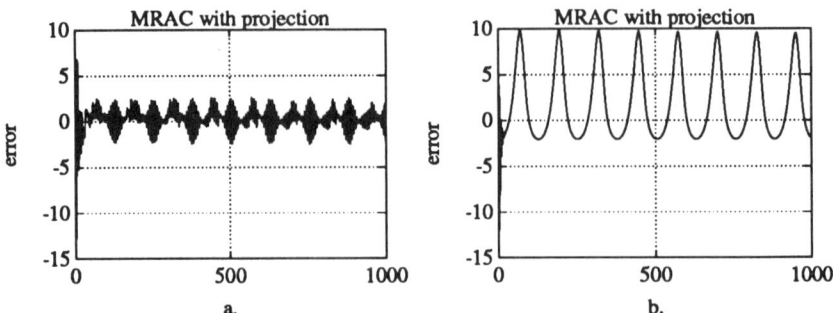

Figure 7.6: Closed-loop response of an MRAC with projection. Slow, unstructured parameter variations, $\mu = 0.0.05$.
a. Reference input $r = 10\sin t$; b. Reference input $r = 10$.

The closed-loop response of an MRAC with projection is shown in Fig. 7.6, for $r = 10\sin t$ and $r = 10$. For these simulations, we used $\gamma = 10$ and initial conditions

$$y(0) = 1$$
$$\theta(0) = [-8.52, 5.55504, -11.51008]^T$$

Notice that although the error is kept bounded, it is still large compared to the reference trajectory. This indicates that the sensitivity of the closed-loop system to parameter mismatch is quite high; to improve the performance, we may need to redesign the controller and/or the reference model so as to improve the properties of the nominal sensitivity operators (e.g., see Chapter 4, Section 4.3). On the other hand, these simulations also indicate that for the particular example, the speed of the parameter variations (governed by μ) is large enough to cause a considerable deterioration of the tracking performance.

The design of an MRAC with σ-modification follows a similar procedure. In this case, instead of a projection interval, we need to determine a nominal control parameter vector and the value of σ. According to the previous example, the former is chosen as

$$\theta_c = \theta(0) = [-8.52, 5.55504, -11.51008]^T$$

The latter defines, in a dynamic way, the interval of parametric uncertainty and, as seen from Theorem 7.21, its selection involves a performance versus robustness trade-off. For our simulations we select $\sigma = 0.01, \gamma = 10$. The closed-loop response of a MRAC with a σ-modification to a sinusoid and a step input is shown in Fig. 7.7.

APPENDIX VII.

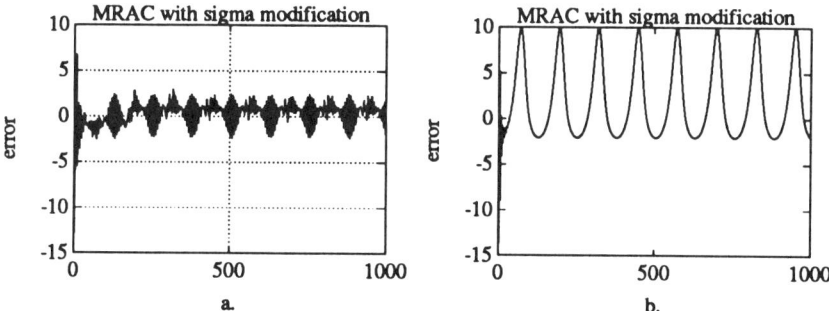

Figure 7.7: Closed-loop response of an MRAC with σ-modification. Slow, unstructured parameter variations, $\mu = 0.05$.
a. Reference input $r = 10\sin t$; b. Reference input $r = 10$.

We should mention at this point that, although our simulations show a reasonable behavior, one should expect that in the general case all properties of adaptive schemes found in the LTI-plant case, desirable and undesirable ones, may occur. Moreover, burst phenomena become even more likely to appear due to the time variations of the desired controller parameters. A possible scenario is that the desired controller parameters (θ_*) vary in such a way that they leave periodically the estimated parameters in a destabilizing region, and thus produce bursts of large amplitude. In fact, this indicates that the adaptation gain should be selected sufficiently large so that the estimation algorithm responds fast enough to prevent such behavior. On the other hand, large values of the adaptation gain are not always desirable as they induce more oscillatory behavior and destabilize the closed-loop system. These qualitative ideas are illustrated in Fig. 7.8, using the same example as before and $r = 10\sin t$.

APPENDIX VII

Proof of Corollary 7.6:

The proof of the corollary follows along the lines of Corollary 6.11, with minor modifications due to the separate estimation of $\hat{c}_{0*}$, something that is also responsible for the increased complexity of the various expressions.

First, as in Theorem 6.6, we establish the boundedness of the estimated parameters $\hat{c}_0, \hat{\theta}$ at least in a subinterval of $[t_0, t_0+T]$. By the properties of the projection, they remain within distance ϵ_* of the sets $\mathcal{C}, \mathcal{M}$ respectively and consequently the solutions can be extended in the whole interval $[t_0, t_0 + T]$.

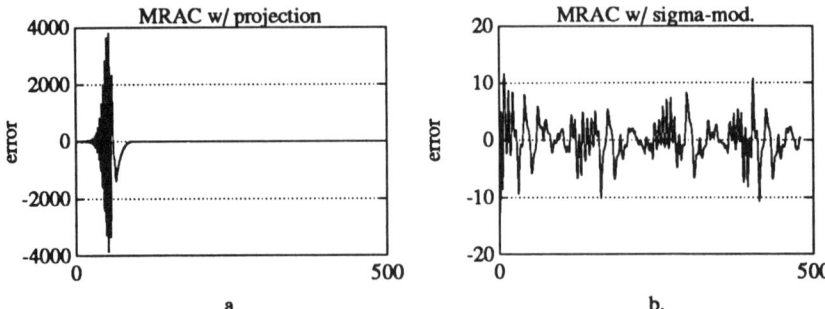

Figure 7.8: Closed-loop response of an MRAC. Slow, unstructured parameter variations: a. MRAC with projection; adaptation gain too small ($\gamma = 1$); b. MRAC with σ-modification; adaptation gain too large ($\gamma = 50$).

It then follows that the parameter derivatives are also UB since ϵ_1, ζ and y_p are all bounded by $\sqrt{m_2}$.

Letting $V = \frac{1}{2\gamma}(\tilde{c}_0^2 + \hat{\phi}^\top \hat{\phi})$ we have that

$$\begin{aligned}\dot{V} &= -\tilde{c}_0 \mathcal{P}_c \left(\frac{\epsilon_1 y_p}{m_2}\right) - \frac{1}{\gamma}\tilde{c}_0 \dot{\hat{c}}_{0*} - \hat{\phi}^\top \mathcal{P}_\theta \left(\frac{\epsilon_1 \zeta}{m_2}\right) - \frac{1}{\gamma}\hat{\phi}^\top \frac{d}{dt}\left(\hat{c}_{0*}\hat{\theta}_*\right) \\ &\leq -\frac{\hat{\phi}^\top \bar{\zeta}\epsilon_1}{m_2} + \frac{1}{\gamma}\left(|\tilde{c}_0||\dot{\hat{c}}_{0*}| + |\hat{\phi}|\left|\frac{d}{dt}\left(\hat{c}_{0*}\hat{\theta}_*\right)\right|\right)_i\end{aligned}$$

From the boundedness of $\hat{c}_0, \hat{\theta}$ it follows that V is also UB; hence, invoking Lemma 7.5 and after a completion of squares and integration, we obtain

$$\int_{t_I}^{t_I+T_I} \frac{\epsilon_1^2(t)}{m_2(t)} dt \leq C_0 + \int_{t_I}^{t_I+T_I} \frac{(\hat{c}_{0*}\eta)^2(t)}{m_2(t)} dt + \frac{K_\phi}{\gamma}\mu T_I$$

where an upper bound for the constant K_ϕ is

$$K_\phi = 4M_0 + 2(\hat{c}_{0max} - \hat{c}_{0min}) + 4\epsilon_*$$

Next, working as in Corollaries 6.9 and 6.11 to evaluate the contribution of the swapping terms, we express η as

$$\begin{aligned}\eta &= W_m(s)[\eta_1] + W_m(s)G_\pi[QU]\hat{\theta}_* - W_m(s)\left[G_\pi[QU]\hat{\theta}_*\right]\ldots \\ &\quad + \frac{1}{\hat{c}_{0*}}\left(\hat{c}_{0*}W_m(s)\left[\frac{1}{\hat{c}_{0*}}u_1\right] - W_m(s)[u_1]\right) \\ &= W_m(s)G(s)\left\{G'_\pi[QU]\dot{\hat{\theta}}_{1,2*}\right\} + W_{m1}(s)\left\{W_{m2}(s)G_\pi[QU]\dot{\hat{\theta}}_*\right\}\ldots \\ &\quad + \frac{1}{\hat{c}_{0*}}W_{m1}(s)\left\{W_{m2}\left[\frac{1}{c_{0*}}u_p\right]\dot{\hat{c}}_{0*}\right\}\end{aligned}$$

APPENDIX VII. 215

where we used Lemma 2.59, omitting the exponentially decaying terms, which can be incorporated in the constant C_0 through the usual Cauchy constants ϵ_c. We may now integrate the swapping term as in Corollary 6.9 and, using Assumption 7.3, obtain the expression given in the Corollary, with

$$\Gamma_1 = \frac{(1+\epsilon_c)\|\hat{c}_{0*}\|_\infty}{\sqrt{2(\delta-\delta_0)}} \left(g_{2,\delta}[W_m G]g_{2,\delta_0}[G'_\pi] + g_{2,\delta}[W_{m1}]g_{2,\delta_0}[W_{m2}G_\pi]\right)$$
$$+ \frac{(1+\epsilon_c)}{\sqrt{2(\delta-\delta_0)}} g_{2,\delta}[W_{m1}]g_{2,\delta_0}[W_{m2c_{0*}}]$$

where $\epsilon_c > 0$ is an arbitrary Cauchy constant and, for convenience, we used the operator definitions

$$W_{m1}(s) \triangleq c_m(sI - A_m)^{-1}$$
$$W_{m2}(s) \triangleq (sI - A_m)^{-1} b_m$$
$$W_{m2c_{0*}} : QU \mapsto W_{m2}(s)\left[\frac{1}{c_{0*}}u_p\right]$$

(A_m, b_m, c_m) being a minimal state-space representation of the reference model $W_m(s)$ and $\delta > \delta_0$ a constant such that the poles of $G(s-\delta)$ and $W_m(s-\delta)$ are in the open left-half plane.[13]

Finally, from Lemma 2.56, the bound for $\dot{\hat{\theta}}$ is quite straightforward, with

$$\Gamma_2 = g_{2,\delta_0}[W_m G_\pi]$$

while for the bound of $\dot{\hat{c}}_0$ we may express y_p as

$$y_p = W_m(s)\left[\frac{1}{c_{0*}}u_p\right] - W_m(s)G_{\theta*}[QU]$$

Letting

$$W_{mc_{0*}} : QU \mapsto W_m(s)\left[\frac{1}{c_{0*}}u_p\right]$$

we obtain that

$$\Gamma_3 = g_{2,\delta_0}[W_{mc_{0*}}] + g_{2,\delta_0}[W_m G_{\theta*}]$$

□□

Proof of Corollary 7.7:

From the adaptive law, the solutions for $\hat{c}_0, \hat{\theta}$ exist at least in a subinterval of $[t_0, t_0 + T]$ and due to the projection $\hat{c}_0(t) \in [\hat{c}_{0min} - \epsilon_*, \hat{c}_{0max} + \epsilon_*]$ (see also Theorem 6.7).

---

[13] Again, as in Corollary 6.9, simpler expressions can be obtained if the parameter derivatives are small uniformly in time.

CHAPTER 7. MODEL REFERENCE ADAPTIVE CONTROL

Next, consider the function $V = \frac{1}{2\gamma}(\tilde{c}_0^2 + \tilde{\phi}^\top \tilde{\phi})$; taking its derivative along the trajectories of the adaptive law,

$$\begin{aligned}
\dot{V} &= -\tilde{c}_0 \mathcal{P}_c \left(\frac{\epsilon_1 y_p}{m_2}\right) - \frac{1}{\gamma}\tilde{c}_0 \dot{\hat{c}}_{0*} - \frac{\hat{\phi}^\top \zeta \epsilon_1}{m_2} - \frac{\sigma}{\gamma}\hat{\phi}^\top(\hat{\theta} - \hat{\theta}_c) - \frac{1}{\gamma}\hat{\phi}^\top \frac{d}{dt}(\hat{c}_{0*}\hat{\theta}_*) \\
&\leq -\frac{\epsilon_1^2}{2m_2} + \frac{(\hat{c}_{0*}\eta)^2}{2m_2} - \sigma V \ldots \\
&\quad + \frac{\sigma}{\gamma}\left(|\hat{c}_{0*}\hat{\theta}_* - \hat{\theta}_c| + \frac{1}{\sigma}\left|\frac{d}{dt}(\hat{c}_{0*}\hat{\theta}_*)\right|\right)^2 + \frac{\sigma}{\gamma}\left(\tilde{c}_0^2 + \frac{1}{\sigma^2}\dot{\hat{c}}_{0*}^2\right)
\end{aligned}$$

Since $\dot{\hat{c}}_{0*}$ and $\dot{\hat{\theta}}_*$ are uniformly $O(\mu)$-small, the swapping operators $QU \mapsto \eta$ have $O(\mu)$-small g_{2,δ_0} gains. Hence, $|\eta|/\sqrt{m_2}$ is also $O(\mu)$, within an exponentially decaying term, from which the first part of the corollary follows.

From the boundedness of the parameter estimates it follows that the parameter derivatives are also UB since ϵ_1, ζ and y_p are all bounded by $\sqrt{m_2}$.

Further, from the above inequality and since V is bounded, we obtain property 1 of the corollary after integration. Finally, properties 2 and 3 are obtained by a straightforward bounding procedure as in Corollary 7.6.

Note that in this case the size of the parametric uncertainty is $K_\phi = \|\hat{c}_{0*}\hat{\theta}_* - \hat{\theta}_c\|_\infty + \hat{c}_{0max} - \hat{c}_{0min} + 2\epsilon_*$ which decreases when $\hat{\theta}_c$, the initial guess of the desired parameters $\hat{c}_{0*}\hat{\theta}_*$, becomes more accurate uniformly in time. This is not the case with the projection estimators where K_ϕ depends on the a priori selected projection set. However, since a projection is used in updating $\hat{c}_0$, K_ϕ is lower bounded by the size of the projection interval $\mathcal{C}$. For this reason, we simply use the same constant to characterize the parametric uncertainty in both cases, noting that under Assumption 6.3 the radius of the set $\mathcal{M}$ is an upper bound of $\|\hat{c}_{0*}\hat{\theta}_* - \hat{\theta}_c\|_\infty$. Of course, when the σ-modification is used, the knowledge of the radius of $\mathcal{M}$ is not required.

As a final comment, it should be mentioned that the reason behind the more restrictive assumptions for this case is that σ-modification performs a 'dynamic' adjustment of the radius of the estimated parameters set. It is not hard to verify that the arguments used in our general estimation results are still applicable in this case. That is, one may also show that under the more general assumptions of Corollary 7.6, the parameter estimates are also UB. However, since the perturbations may be large in short time intervals, the bound on $\dot{\phi}$ can be as large as $O\left(\frac{1}{\sigma}\right)$. Although this does not alter the fact that the estimation error is small for small μ and σ, it does induce certain problems in the stability analysis of the MRAC. Thus, for the sake of simplicity, we only study the MRAC/σ-modification case under some uniformity conditions on the smallness of the perturbation terms. □□

APPENDIX VII. 217

Proof of Corollary 7.8:

From the adaptive law, the solutions for $\hat{c}_0, \hat{\theta}, \psi$ exist at least in a subinterval of $[t_0, t_0 + T]$. Next, consider the function $V = \frac{1}{2\gamma}(\tilde{c}_0^2 + \hat{c}_{0*}\tilde{\psi}^2 + \hat{\phi}^\top \hat{\phi})$; taking its derivative along the trajectories of the adaptive law and after some calculations, we find that

$$\dot{V} \leq -\sigma V - \frac{\hat{c}_{0*}\epsilon_1^2}{2m_2} + \frac{\hat{c}_{0*}\eta_1^2}{2m_2} + \frac{1}{2\gamma}\tilde{\psi}^2 \dot{\hat{c}}_{0*} \cdots$$

$$+ \frac{\sigma}{2\gamma}\left\{\left(|\hat{c}_{0*}\hat{\theta}_* - \hat{\theta}_c| + \frac{1}{\sigma}\left|\frac{d}{dt}(\hat{c}_{0*}\hat{\theta}_*)\right|\right)^2 \cdots\right.$$

$$\left. + \hat{c}_{0*}\left(|\psi_* - \psi_c| + \frac{1}{\sigma}|\dot{\psi}_*|\right)^2 + \left(|\hat{c}_{0*} - \hat{c}_{0c}| + \frac{1}{\sigma}|\dot{\hat{c}}_{0*}|\right)^2\right\}$$

Under the assumption that $|\dot{\hat{\theta}}_*|, |\dot{\psi}_*|, |\dot{\hat{c}}_{0*}|$ are $O(\mu)$, uniformly in t, it follows that $\frac{1}{2\gamma}\tilde{\psi}^2 \dot{\hat{c}}_{0*} \leq \psi_* \mu V$. Hence, for any $\mu \in [0, \mu_0]$, with $\mu_0 < \sigma/\hat{c}_{0min}$, V and therefore the parameter estimates are bounded and converge exponentially to a residual set as given in the corollary. Further, we may now extend the solutions for the parameters to the whole interval $[t_0, t_0 + T]$. Note that η is also $O(\mu)$, modulo an exponentially decaying term due to initial conditions, since the uniformly $O(\mu)$-small derivatives of the unstructured part of the parameters imply that the swapping operator $QU \mapsto \eta$ has an $O(\mu)$-small g_{2,δ_0} gain.

From the boundedness of the parameter estimates it also follows that the derivatives of the parameter estimates are UB since ϵ_1, ζ, y_p and ξ are all bounded by $\sqrt{m_2}$.

Finally, integrating $\dot{V}$ in an interval $I \subseteq [t_0, t_0 + T]$ the properties of the adaptive law follow after a straightforward bounding procedure and using the previously derived parameter bounds (see also Corollaries 7.6 and 7.7). □□

Proof of Corollary 7.10:

The proof follows the same steps as the proof of Corollary 6.16, although with an increased degree of complexity since the closed-loop system is not minimal.

Thus, after establishing parameter boundedness as in Corollary 7.6, we have that inside an interval (t_j, t_{j+1}) where the plant parameters are smooth, the plant output can be expressed as

$$y_p(t) = [c_c(t) + \tilde{c}_c(t)]^\top \Phi_c(t, t_j) x_c(t_j)$$

$$+ [c_c(t) + \tilde{c}_c(t)] \int_{t_j}^{t} \Phi_c(t, \tau) \left[\tilde{A}_c(\tau) x_c(\tau) + \tilde{b}_c(\tau) \tilde{u}(\tau)\right] d\tau$$

$$+c_c(t)\int_{t_j}^{t}\Phi_c(t,\tau)b_c(\tau)\tilde{u}(\tau)\,d\tau+\tilde{c}_c(t)\int_{t_j}^{t}\Phi_c(t,\tau)b_c(\tau)\tilde{u}(\tau)\,d\tau$$

where the subscript 'c' applies to the nominal closed-loop state-space representation and

$$\tilde{u}=r+\frac{1}{\tilde{c}_{0*}}(G_\pi[QU]\hat{\phi}+\tilde{c}_0 r)+\eta_1=\frac{1}{c_{0*}}u_p-G_{\theta*}[QU]$$

is the 'external' input to the system (see Fig. 7.1). From Lemma 3.11 it follows that the plant state is bounded by $\sqrt{m_2}$ for any $\nu \in [0,\nu_0)$ and any $\mu' \in [0,\mu'_0)$ with ν_0, μ'_0 sufficiently small.[14] The same conclusion is true for the state of the controller filters since they are chosen to be ES and have input either u_p or y_p. Hence, for sufficiently small ν_0, μ'_0, there exists a constant K_x such that the entire state vector x_c of the closed-loop system satisfies

$$\frac{|x_c|}{\sqrt{m_2}}\le K_x+\varepsilon$$

Further, consider the system

$$\dot{x}_{mx}=A_m x_{mx}+b_m\tilde{u}\;;\;y_{mx}=c_m x_{mx}$$

where (A_m, b_m, c_m) is a minimal state-space representation of the reference model $W_m(s)$. Since inside every interval (t_j, t_{j+1}) the I/O operator of the closed-loop system is equal to $W_m(s)$ (see Theorem 4.18) and in view of Lemma 7.5 we can express the estimation error as

$$\epsilon_1=\bar{\phi}^T\bar{\zeta}+\hat{c}_{0*}\eta+\eta_J+\tilde{\eta}$$

where

$$\eta_J=[c_c(t)+\tilde{c}_c(t)]^T\Phi_c(t,t_j)x_c(t_j)-c_m^T\Phi_m(t,t_j)x_m(t_j)$$

$$\tilde{\eta}=[c_c(t)+\tilde{c}_c(t)]\int_{t_j}^{t}\Phi_c(t,\tau)\left[\tilde{A}_c(\tau)x_c(\tau)+\tilde{b}_c(\tau)\tilde{u}(\tau)\right]d\tau$$

$$+\tilde{c}_c(t)\int_{t_j}^{t}\Phi_c(t,\tau)b_c(\tau)\tilde{u}(\tau)\,d\tau$$

For the rest of the proof, the arguments used in Corollary 7.6 are applicable with η being decomposed into two terms, one of which contains the derivatives of the absolutely continuous part of the controller parameters and the other contains the contribution of the jump part (see Corollary 2.60). Notice that the effect of the parameter discontinuities also appears in the term

---

[14] Although this bound is not uniform as ν, μ' approach the limit where the plant becomes non-minimal, this can easily be avoided by taking ν_0, μ'_0 to be smaller than the respective limit.

APPENDIX VII.

$[V(t_i^+) - V(t_i^-)]$ when we combine the intervals (t_j, t_{j+1}) (see also Theorem 6.6). On the other hand, observing that $\tilde{u}$ is the sum of a term $O(u_p)$ and a term $O(\sqrt{m_2})$ the part of the proof related to the parameter jumps and their smooth approximation follows by integrating inside the intervals (t_j, t_{j+1}). Finally, in estimating the bounds for the derivatives of the parameter estimates, the only difference from Corollary 7.6 is in $\dot{\hat{c}}_0$. In this case, the bound of $|y_p|/\sqrt{m_2}$ is $O(K_x)$, obtained directly from the corresponding bound of the plant state.

Notice that a qualitative difference in this proof, compared with the indirect case, is that the perturbation due to parameter jumps need not disappear as the size of the jumps decreases. The roots of this problem are in the fact that the MRC closed-loop system is not minimal and the related arguments used in the indirect identification case do not apply here. At this point, it is not clear whether this difference can be attributed to technical reasons and how it can be resolved in the general case where fast (but structured) time-variations are allowed. □□

Proof of Corollary 7.11:

As in Corollary 7.7, the solutions for $\hat{c}_0, \hat{\theta}$ exist at least in a subinterval of $[t_0, t_0 + T]$ and due to the projection $\hat{c}_0(t) \in [\hat{c}_{0min} - \epsilon_*, \hat{c}_{0max} + \epsilon_*]$ (see also Theorem 6.7).

Again, consider the function $V = \frac{1}{2\gamma}(\tilde{c}_0^2 + \hat{\phi}^T \hat{\phi})$; taking its derivative along the trajectories of the adaptive law, for $t \in (t_j, t_{j+1})$ and inside the subinterval where the solutions exist

$$\dot{V} = -\tilde{c}_0 \mathcal{P}_c \left(\frac{\epsilon_1 y_p}{m_2} \right) - \frac{1}{\gamma} \tilde{c}_0 \dot{\hat{c}}_{0*}^s - \frac{\hat{\phi}^T \zeta \epsilon_1}{m_2} - \frac{\sigma}{\gamma} \hat{\phi}^T (\hat{\theta} - \hat{\theta}_c) - \frac{1}{\gamma} \hat{\phi}^T \frac{d}{dt}(\hat{c}_{0*}^s, \hat{\theta}_*^s)$$

$$\leq -\frac{\epsilon_1^2}{2m_2} + \frac{(\hat{c}_{0*}\eta + \tilde{\eta} + \eta_J)^2}{2m_2} - \sigma V \ldots$$

$$+ \frac{\sigma}{\gamma} \left(|\hat{c}_{0*}\hat{\theta}_* - \hat{\theta}_c| + \frac{1}{\sigma} \left| \frac{d}{dt}(\hat{c}_{0*}^s, \hat{\theta}_*^s) \right| \right)^2 + \frac{\sigma}{\gamma} \left(\tilde{c}_0^2 + \frac{1}{\sigma^2}(\dot{\hat{c}}_{0*}^s)^2 \right)$$

where as in Corollary 7.10 and with the same notation

$$\epsilon_1 = \bar{\phi}^T \bar{\zeta} + \hat{c}_{0*}\eta + \eta_J + \tilde{\eta}$$

$$\eta_J = [c_c(t) + \tilde{c}_c(t)]^T \Phi_c(t, t_j) x_c(t_j) - c_m^T \Phi_m(t, t_j) x_m(t_j)$$

$$\tilde{\eta} = [c_c(t) + \tilde{c}_c(t)]^T \int_{t_j}^{t} \Phi_c(t, \tau) \left[\tilde{A}_c(\tau) x_c(\tau) + \tilde{b}_c(\tau) \tilde{u}(\tau) \right] d\tau$$

$$+ \tilde{c}_c(t) \int_{t_j}^{t} \Phi_c(t, \tau) b_c(\tau) \tilde{u}(\tau) d\tau$$

Also, for sufficiently small ν_0, μ'_0, there exists a constant K_x such that $\frac{|x_c|}{\sqrt{m_2}} \leq K_x + \varepsilon$ and using the uniform bounds on the parameter derivatives and the perturbations due to smooth approximations we find that

$$\dot{V} \leq -\sigma V + O\left(\frac{K_\phi^2 \sigma}{\gamma}, \frac{\mu^2}{\gamma\sigma}, \mu^2, {\mu'}^2\right) + O\left(K_x^2\right) e^{-2\alpha_c(t-t_j)}$$

where $-\alpha_c$ is the rate of exponential decay of $\Phi_c(.,.)$ (which, by Corollary 3.8, is exponentially decaying for ν sufficiently small). Integration using the initial condition $V(t_j^+)$ inside (t_j, t_{j+1}) yields

$$V(t) \leq V(t_j^+)e^{-\sigma(t-t_j)} + O\left(\frac{1}{\sigma}\right)e^{-\beta(\sigma)(t-t_j)} + \ldots$$

$$+ O\left(\frac{K_\phi^2}{\gamma}, \frac{\mu^2}{\gamma\sigma^2}, \frac{\mu^2}{\sigma}, \frac{{\mu'}^2}{\sigma}\right)$$

where $\beta < \min[\sigma, 2\alpha_c]$ is a positive constant. It now follows that for any constant A there exists $\nu_1 > 0$ such that $\forall \nu \in [0, \nu_1]$

$$O\left(\frac{1}{\sigma}\right)e^{-\beta(\sigma)/\nu}, \quad O\left(\frac{K_J^2}{\gamma}\right)e^{-\sigma/\nu} \leq A$$

where K_J is a constant denoting the supremum of the size of parameter jumps and $|V(t_j^+) - V(t_j^-)| \leq O(K_J^2/\gamma)$. Hence, for sufficiently small ν, $V(t_j)$ is a bounded sequence from which the first part of the corollary follows by extending the solutions to the whole interval $[t_0, t_0 + T]$.

For the rest of the proof, and as in Corollary 7.10, property 1 is obtained by integrating inside the intervals (t_j, t_{j+1}) and properties 2 and 3 after a straightforward bounding procedure. □□

Proof of Lemma 7.12:

Assuming that $a < 1$, we have that $\forall t \in [t_0, t_1]$

$$\|(U)_t\|_{2,\delta} \leq \frac{1}{1-a}\left(\|(R)_t\|_{2,\delta} + \|(v)_t\|_{2,\delta} + \|(E)_t\|_{2,\delta}\right)$$

and squaring on both sides

$$\|(U)_t\|_{2,\delta}^2 \leq \lambda_0^2\|(R)_t\|_{2,\delta}^2 + \lambda_1^2\|(v)_t\|_{2,\delta}^2 + \lambda_2^2\|(E)_t\|_{2,\delta}^2$$

where $\lambda_i = \frac{1}{1-a}k(\epsilon_c)$ and $k(\epsilon_c)$ denotes dependence on Cauchy constants. Adding $\|(r)_t\|_{2,\delta}^2 + qe^{2\delta t_0}$ on both sides and operating by $\mathcal{E}_{-2\delta}$ we get

$$m_f \leq qe^{-2\delta(t-t_0)} + \int_{t_0}^t \left(\lambda_0^2|R|^2 + |r|^2\right)(\tau)e^{-2\delta(t-\tau)}\,d\tau$$

$$+ \int_{t_0}^t \left(\lambda_1^2|v|^2 + \lambda_2^2|E|^2\right)(\tau)e^{-2\delta(t-\tau)}\,d\tau$$

Next, since $q > 0$ and $1/r$ is UB, we have that $m_f \geq c > 0$ and therefore we can rewrite the last inequality as

$$M(t) \leq q + \frac{R'}{2\delta}e^{2\delta(t-t_0)} + \int_{t_0}^{t} \frac{\left(\lambda_1^2|v|^2 + \lambda_2^2|E|^2\right)(\tau)}{m_f(\tau)} M(\tau) d\tau$$

where $M(t) = m_f(t)e^{2\delta(t-t_0)}$ and R' is a constant depending on the L_∞ bounds of R, r (its precise expression is not needed). Applying the Bellman-Gronwall lemma, we get that

$$m_f(t) \leq qe^{-2\delta(t-t_0)} + \frac{R'}{2\delta} + \int_{t_0}^{t} V(\tau) \left[qe^{-2\delta(t-t_0)} + \frac{R'}{2\delta}e^{-2\delta(t-\tau)}\right] e^{\int_\tau^t V(s)ds} d\tau$$

where

$$V(\tau) = \frac{\left(\lambda_1^2|v|^2 + \lambda_2^2|E|^2\right)(\tau)}{m_f(\tau)}$$

is UB according to the assumptions of the lemma and

$$e^{\int_\tau^t V(s)ds} \leq C'' e^{\lambda_1^2 \mu^2 (t-\tau)}$$

where C'' is a constant depending on b, C, C' and λ_2. It now follows that, if

$$-2\delta + \lambda_1^2 \mu^2 < 0$$

then m_f is UB on $[t_0, t_1]$ with a bound that is independent of t_0, t_1. We may now choose the Cauchy constants such that $\lambda_1 = \frac{1+\epsilon_c}{1-a}$, where ϵ_c is a positive constant. The choice of ϵ_c will of course affect the actual bound of m_f through R', C'' but this is irrelevant for the boundedness result as long as $\epsilon_c > 0$. Thus, since ϵ_c can be arbitrarily small, it can be absorbed under the strict inequality sign from which the result of the lemma follows. □□

Proof of Theorem 7.13:

First, observe that under the assumptions of the theorem, the vectorfield of the ODE describing the evolution of the adaptive closed-loop system is locally Lipschitz and therefore there exists $T > 0$ such that the solution exists and is unique on the interval $[t_0, t_0 + T]$. Hence, the assumptions of Corollary 7.6 are satisfied, implying that the parameter estimates are UB, at least on $[t_0, t_0 + T]$.

Further, for $t \in [t_0, t_0 + T]$, define the fictitious signal m_f by

$$\dot{m}_f = -2\delta m_f + |QU|^2 + q_r r^2 + q_e; \quad m_f(t_0) = m_2(t_0)$$

where $\delta \in (0, \min[\delta_0, \alpha])$. Note that, by construction, $m_f \geq m_2$ on any interval U is well defined.

Next, consider the closed-loop description given by (7.8). Since the parameter estimates are UB on $[t_0, t_0 + T]$ it follows that R is also UB (independent of t_0 and T). Moreover, the second term in the right-hand side of (7.8) can be expanded as follows

$$\begin{aligned}\frac{d}{dt}\left(\frac{1}{\hat{c}_{0*}}w^T \hat{\phi}\right) &= \frac{d}{dt}\left(\frac{1}{\hat{c}_{0*}}\right) w^T \hat{\phi} + \frac{1}{\hat{c}_{0*}}\dot{w}^T \hat{\phi} \\ &\quad + \frac{1}{\hat{c}_{0*}}w^T \left(\frac{d}{dt}[\hat{\theta} - \hat{c}_{0*}\hat{\theta}_*]\right)\end{aligned} \quad (7.13)$$

Substituting the above expression in (7.8) and taking $L_2(\delta)$-norms of truncated signals on both sides, we obtain (for simplicity we set $\rho = 0$ at this stage)

$$\begin{aligned}\|(QU)_t\|_{2,\delta} &\leq \|(R)_t\|_{2,\delta} + \|\hat{\phi}\|_\infty \gamma_{2,\delta}[QS_r\Lambda_1]\gamma_{2,\delta}\left[\frac{1}{\hat{c}_{0*}}sG_\pi\right]\|(QU)_t\|_{2,\delta} \\ &\quad + \|\hat{\phi}\|_\infty \gamma_{2,\delta}[QS_r\Lambda_1]g_{2,\delta}[G_\pi]\|\frac{1}{\hat{c}_{0*}^2}\|_\infty \|(|\dot{\hat{c}}_{0*}|\sqrt{m_f})_t\|_{2,\delta} \\ &\quad + \gamma_{2,\delta}[QS_r\Lambda_1]g_{2,\delta}\left[\frac{1}{\hat{c}_{0*}}G_\pi\right]\left\|\left(\left|\frac{d}{dt}[\hat{\theta} - \hat{c}_{0*}\hat{\theta}_*]\right|\sqrt{m_f}\right)_t\right\|_{2,\delta} \\ &\quad + \gamma_{2,\delta}[QS_r\Lambda_1 sG]g_{2,\delta}[G'_\pi]\|(|\dot{\hat{\theta}}_*|\sqrt{m_f})_t\|_{2,\delta} \\ &\quad + \gamma_{2,\delta}\left[QS_r\Lambda W_m^{-1}\frac{1}{\hat{c}_0}\right]\|(\epsilon_1)_t\|_{2,\delta} \\ &\quad + \gamma_{2,\delta}\left[QS_r\Lambda W_m^{-1}\frac{1}{\hat{c}_0}W_{m1}\right]g_{2,\delta}[W_{m2}G_\pi]\|(|\dot{\hat{\theta}}|\sqrt{m_f})_t\|_{2,\delta} \\ &\quad + C\|(\varepsilon\sqrt{m_f})_t\|_{2,\delta}\end{aligned} \quad (7.14)$$

where C is some constant and ε denotes exponentially decaying terms which are collected and shown under the last term only. Notice that in view of Lemma 7.12, their precise expressions are unnecessary for our purposes. We may now combine the similar terms to get

$$\begin{aligned}\|(QU)_t\|_{2,\delta} &\leq \|(R)_t\|_{2,\delta} + H_0\|(QU)_t\|_{2,\delta} + H_1\|(\epsilon_1)_t\|_{2,\delta} \\ &\quad + H_2\|(|\dot{\hat{\theta}}|\sqrt{m_f})_t\|_{2,\delta} + H_3\|(|\dot{\hat{c}}_{0*}| + |\dot{\hat{\theta}}_*|\sqrt{m_f})_t\|_{2,\delta} \\ &\quad + C\|(\varepsilon\sqrt{m_f})_t\|_{2,\delta}\end{aligned} \quad (7.15)$$

APPENDIX VII.

where H_i are constants with order of magnitude as described below

$$H_0 \leq O\left(\frac{M_0}{a}\right) \quad ; \quad H_1 \leq O\left(a^{n^*}\right)$$

$$H_2 \leq O\left(\frac{1}{a}\right) + O\left(a^{n^*}\right) \quad ; \quad H_3 \leq O\left(\frac{M_0}{a}\right) + O\left(\frac{1}{a}\right)$$

The above relationships follow by inspection of the corresponding operator gains and the fact that, due to the projection, $\|\hat{\phi}\|_\infty \leq 2M_0 + \epsilon_*$.[15] Note that the derivations are quite straightforward in the case of strictly proper MRC laws but they become more involved when a non-strictly proper MRC is considered. In the latter case, one needs to verify that the operator gains $g_{2,\delta}[G_\pi]$ and $\gamma_{2,\delta}[sG_\pi]$ are finite. The validity of this statement becomes apparent by expressing y_p as $W_m[\tilde{u}]$ although the derivation of precise bounds is rather messy and is omitted.

Invoking Lemma 7.12 and using the expressions derived in the identification part (Corollary 7.6), our sufficient condition for m_f to be UB becomes

$$\inf_{a,\delta} \left\{ H_0 + \left(H_1 \sqrt{\Gamma_1^2 + \frac{K_\phi}{\gamma}} + H_2 \Gamma_2 \sqrt{\gamma^2 \Gamma_1^2 + \gamma K_\phi} + H_3 \right) \sqrt{\frac{\mu}{2\delta}} \right\} < 1 \tag{7.16}$$

The condition (7.16) is actually sufficient to guarantee the boundedness of all the signals in the closed-loop. Since, under the assumptions of the theorem, m_f bounds the overall state vector of the closed-loop system, the latter is UB on $[t_0, t_0 + T]$ with a bound independent of t_0, T. Thus, the solution can be extended on an interval $[t_0 + T, t_0 + T + T_\delta]$ for some $T_\delta > 0$ [Vid.78]. Since all the bounds derived above were independent of t_0, T, they are also valid on $[t_0+T, t_0+T+T_\delta]$. Consequently, the solutions can be extended on an interval $[t_0+T+T_\delta, t_0+T+2T_\delta]$ etc. from which the first part of the theorem follows. Notice that Lemma 3.11 (or Lemma 2.36) can be invoked at this point to establish the internal boundedness (BIBS stability) of the closed-loop system.

Thus, we only need to verify that (7.16) can be satisfied for any given size of the parametric uncertainty set (M_0, K_ϕ) by requiring μ to be sufficiently small. This can be seen by choosing the free parameter 'a' to be sufficiently large, e.g.,

$$a > \frac{M_0}{\epsilon}$$

where ϵ is a small constant making H_0 small, say less than $1/2$. The rest of the left-hand side terms can then be made small, less than $1/2$, provided that

---

[15] For simplicity, the small parameter ϵ_* is also chosen to be $O(M_0)$.

$\mu < \mu_0$ where μ_0 satisfies simultaneously inequalities of the form

$$O\left(a^{n^*}, a^{n^*}\sqrt{\frac{K_\phi}{\gamma}}\right)\sqrt{\mu_0} \leq \epsilon \ ; \ O\left(\frac{1}{a}, \frac{M_0}{a}\right)\sqrt{\mu_0} \leq \epsilon$$

$$O\left(a^{n^*}\sqrt{\gamma^2 + \gamma K_\phi}, \frac{\sqrt{\gamma^2 + \gamma K_\phi}}{a}\right)\sqrt{\mu_0} \leq \epsilon$$

something that is always possible for some $\mu_0 > 0$, proving the first part of the theorem.

In the special case where $\mu \to 0$, our stability condition is satisfied with $a \to \infty$ indicating that in the limit, boundedness is guaranteed independent of the size of the parametric uncertainty set (see also Remark 7.15).

On the other hand, when the parametric uncertainty becomes sufficiently small, we may invoke similar arguments starting from (7.8) with $\rho = 1$. The reason for this is that depending on the value of μ, the bound of $(1 - \Lambda)[w^\top \hat{\phi}/\hat{c}_{0*}]$ in terms of derivatives of the input may be quite conservative. Thus, in order to avoid any restrictions on μ, we can decompose the controller parameters $\hat{\theta}_*$ as $\hat{\theta}_c + \tilde{\theta}_*$, the second term being bounded by $O(M_0)$. Using this decomposition in the expression for η_1 we get

$$\|(\eta_1)_t\|_{2,\delta} \leq O(M_0)\|(QU)_t\|_{2,\delta}$$

Hence,

$$\|(QU)_t\|_{2,\delta} \leq \|(R)_t\|_{2,\delta} + O(M_0)\|(QU)_t\|_{2,\delta}$$

from which, as in the previous case, we obtain that boundedness is guaranteed independent of the value of μ, provided that M_0 is sufficiently small (see also Remark 7.17). Two observations are worthwhile at this point. One is that the proof of stability can be derived by a small-gain argument on an $L_2(\delta)$-space, with δ being arbitrarily small; boundedness is then obtained via Lemma 2.55. The second is that the stability condition now depends only on M_0, something that should have been expected since $\hat{c}_{0*}$ is not involved in any feedback signals. However, in the general case, any uncertainty in $\hat{c}_{0*}$ affects the stability condition since it introduces a perturbation in the estimation process.

Finally, the properties of the tracking error are easily derived from (7.7), yielding

$$y_p - y_m = \frac{1}{\hat{c}_0}\left(\epsilon_1 - W_{m1}\{W_{m2}[w^\top]\dot{\hat{\theta}}\} - W_{m1}\{W_{m2}[r]\dot{\hat{c}}_0\}\right)$$

APPENDIX VII.   225

Working as in Corollaries 6.9 and 6.11 we obtain for $\delta > \delta_0$

$$\left(\int_t^{t+T} \frac{e_1^2(\tau)}{m_2(\tau)} d\tau\right)^{1/2} \leq C_0 + \frac{1}{\hat{c}_{0min} - \epsilon_*} \left(\int_t^{t+T} \frac{\epsilon_1^2(\tau)}{m_2(\tau)} d\tau\right)^{1/2}$$

$$+ B_1 \left(\int_t^{t+T} |\dot{\hat{\theta}}|^2(\tau) d\tau\right)^{1/2}$$

$$+ B_2 \left(\int_t^{t+T} |\dot{\hat{c}}_0|^2(\tau) d\tau\right)^{1/2}$$

where

$$B_1 = \frac{g_{2,\delta}[W_{m1}]\, g_{2,\delta_0}[W_{m2}G_\pi]}{(\hat{c}_{0min} - \epsilon_*)\sqrt{2(\delta - \delta_0)}} \quad ; \quad B_2 = \frac{g_{2,\delta}[W_{m1}]\, g_{2,\delta_0}[W_{m2}]}{q_r(\hat{c}_{0min} - \epsilon_*)\sqrt{2(\delta - \delta_0)}}$$

The second part of the theorem now follows by using the expressions derived in Corollary 7.6 in the last inequality. □□

Proof of Theorem 7.21:

The proof of the theorem follows similar analytical steps as the proof of Theorem 7.13 with the main differences being in the treatment of the L_∞ bounds of the parameter error $\hat{\phi}$ and, of course, the different expressions obtained from the estimator.

Thus, with the existence of solutions established on an interval $[t_0, t_0 + T]$ as in Theorem 7.21 and the same definition of the fictitious signal m_f, let us also define the 'steady-state' parameter error $\hat{\phi}_s$ as the vector with the properties

$$\hat{\phi} = \hat{\phi}_s + \varepsilon \quad ; \quad \|\hat{\phi}_s\|_\infty \leq O\left(K_\phi, \frac{\mu}{\sigma}, \mu\sqrt{\frac{\gamma}{\sigma}}\right)$$

where, as usual, ε is an exponentially decaying term. Such a definition is motivated by the results of Corollary 7.7 and serves in removing the effects of initial conditions from the eventual condition for signal boundedness.

Further, from (7.8) and (7.13) we obtain the following bound for the $L_2(\delta)$-norm of the truncated QU ($\rho = 0$)

$$\|(QU)_t\|_{2,\delta} \leq \|(R)_t\|_{2,\delta} + \|\hat{\phi}_s\|_\infty \gamma_{2,\delta}[QS_r\Lambda_1]\gamma_{2,\delta}\left[\frac{1}{\hat{c}_{0*}} sG_\pi\right] \|(QU)_t\|_{2,\delta}$$

$$+ \|\hat{\phi}_s\|_\infty \gamma_{2,\delta}[QS_r\Lambda_1]\, \gamma_{2,\delta}[G_\pi] \left\|\frac{\mu}{\hat{c}_{0*}^2}\right\|_\infty \|(QU)_t\|_{2,\delta}$$

$$+ \gamma_{2,\delta}[QS_r\Lambda_1]\, g_{2,\delta}\left[\frac{1}{\hat{c}_{0*}}G_\pi\right] \|(|\dot{\hat{\theta}}|\sqrt{m_f})_t\|_{2,\delta}$$

$$+\gamma_{2,\delta}\left[QS_r\Lambda_1\right]\gamma_{2,\delta}\left[G_\pi\right]\left\|\frac{\mu}{\hat{c}_{0*}}\right\|_\infty\|(QU)_t\|_{2,\delta}$$
$$+\gamma_{2,\delta}\left[QS_r\Lambda_1 sG\right]\gamma_{2,\delta}\left[G'_\pi\right]\mu\|(QU)_t\|_{2,\delta}$$
$$+\gamma_{2,\delta}\left[QS_r\Lambda W_m^{-1}\frac{1}{\hat{c}_0}\right]\|(\epsilon_1)_t\|_{2,\delta}$$
$$+\gamma_{2,\delta}\left[QS_r\Lambda W_m^{-1}\frac{1}{\hat{c}_0}W_{m1}\right]g_{2,\delta}\left[W_{m2}G_\pi\right]\|(|\dot{\hat{\theta}}|\sqrt{m_f})_t\|_{2,\delta}$$
$$+C\|(\varepsilon\sqrt{m_f})_t\|_{2,\delta} \qquad (7.17)$$

where, as in the previous theorem, all exponentially decaying contributions are collected under the last term and C is a positive constant. Combining similar terms,

$$\|(QU)_t\|_{2,\delta} \leq \|(R)_t\|_{2,\delta} + H_0\|(QU)_t\|_{2,\delta} + H_1\|(\epsilon_1)_t\|_{2,\delta}$$
$$+H_2\|(|\dot{\hat{\theta}}|\sqrt{m_f})_t\|_{2,\delta} + C\|(\varepsilon\sqrt{m_f})_t\|_{2,\delta} \qquad (7.18)$$

where H_1, H_2 are constants with order of magnitude as given in the proof of Theorem 7.13 and

$$H_0 \leq O\left(\frac{\|\hat{\phi}_s\|_\infty}{a}, \frac{\mu\|\hat{\phi}_s\|_\infty}{a}, \frac{\mu}{a}\right)$$
$$\|\hat{\phi}_s\|_\infty \leq O\left(K_\phi, \frac{\mu}{\sigma}, \mu\sqrt{\frac{\gamma}{\sigma}}\right)$$

Thus, invoking Lemma 7.12, a sufficient condition for m_f to be UB and consequently implying boundedness of all signals in the loop, is

$$\inf_{a,\delta}\left\{H_0 + \left(H_1(RMS\left[\frac{|\epsilon_1|}{\sqrt{m_2}}\right]) + H_2(RMS[\dot{\hat{\theta}}])\right)\frac{1}{\sqrt{2\delta}}\right\} < 1 \qquad (7.19)$$

where, slightly abusing the notation, $RMS[x]$ denotes the root-mean-square value of the signal x. More precisely, if there exist constants C, γ such that,

$$\int_t^{t+T} |x(\tau)|^2\, d\tau \leq C + \gamma^2 T$$

for all $t \geq t_0$ and all $T \geq 0$, then $RMS[x] = \gamma$. Obviously, from Corollary 7.7, we have the following upper bounds

$$RMS\left[\frac{|\epsilon_1|}{\sqrt{m_2}}\right] \leq O\left(\mu, K_\phi\sqrt{\frac{\sigma}{\gamma}}, \frac{\mu}{\sqrt{\gamma\sigma}}\right)$$
$$RMS[\dot{\hat{\theta}}] \leq O\left(\gamma\mu, K_\phi\sqrt{\gamma\sigma}, \mu\sqrt{\frac{\gamma}{\sigma}}, K_\phi\sigma, \mu, \mu\sqrt{\gamma\sigma}\right)$$

APPENDIX VII.

With the rest of the proof being completed as in Theorem 7.13, all that remains is to establish conditions for the validity of (7.19). Expanding the terms of the inequality, we find that (7.19) is composed of terms

$$O\left(\frac{K_\phi}{a}, K_\phi\sqrt{\frac{\sigma}{\gamma}}a^{n^*}, K_\phi\sqrt{\sigma\gamma}a^{n^*}, K_\phi\sigma a^{n^*}, \frac{K_\phi\sqrt{\sigma\gamma}}{a}, \frac{K_\phi\sigma}{a}\right)$$

and various terms, denoted by $f(\mu)$, which are multiplied by μ or μ^2. By inspection, it follows that these terms are continuous and well-defined functions of σ for $\sigma > 0$. Hence, for any finite K_ϕ we may choose 'a' to make the first term smaller than, say, ϵ, and then choose σ_0 such that the rest of the terms are smaller than ϵ, for any $\sigma \in (0, \sigma_0)$. Finally, the term $f(\mu)$ is smaller than ϵ for any $\mu \in (0, \mu_0)$, where μ_0 is a function of σ (as well as a, K_ϕ, γ but these are simply fixed constants at this stage). Choosing $\sigma < \sigma_0$ in the previous step, μ_0 takes a fixed non-zero value and hence the first part of the theorem follows.

It is worthwhile to make some observations at this point, the first one concerning the treatment of terms depending on $\dot{\hat{\theta}}_*$. Owing to the assumption that these terms are uniformly $O(\mu)$-small, one can use a more efficient bounding procedure to include their contribution in the constant H_0. This is exactly the reason why our sufficient condition does not contain the H_2 term which appears in the proof of Theorem 7.13. The same approach is, of course, possible for algorithms using projection although it is not explored in the more general proof of that theorem.

In (7.18), the fourth term includes the part of contribution of $\dot{\hat{\phi}}_s$ that is due to its exponentially decaying component. Since the parameter estimates and their derivatives are UB, it is easy to see that $\dot{\hat{\phi}}_s$ can be expressed in terms of $\dot{\hat{\theta}}, \dot{\hat{\theta}}_*$ and another UB exponentially decaying term. The latter, of course, being integrable disappears from the final stability condition.

It is also quite straightforward to see from the expansion of (7.19) that for any fixed value of σ, boundedness is guaranteed provided that K_ϕ and μ are sufficiently small (see Remark 7.22). However, the similarity of such an argument with what can be obtained from a simple small-gain condition (even for non-adaptive controllers) introduces the need for a more quantitative analysis of the algorithm. And although the results can be improved by using non-zero values of ρ which are subsequently optimized (the details are left to the reader), the usefulness of the approach is questionable, at best, due to the overall conservatism of the extensive bounding procedures. Moreover, this was exactly the reason why we selected to simplify the analysis by taking $\rho = 0$. Using ρ as an additional free parameter can certainly improve the quantitative aspects of our stability condition; but since the analysis is already quite conservative and most of the qualitative aspects remain the same, the merits do not justify the increased complexity.

On the other hand, when the parametric uncertainty becomes sufficiently small, we may work as in the proof of Theorem 7.13 to obtain

$$\|(QU)_t\|_{2,\delta} \leq \|(R)_t\|_{2,\delta} + O\left(K_\phi, \frac{\mu}{\sigma}, \mu\sqrt{\frac{\gamma}{\sigma}}\right)\|(QU)_t\|_{2,\delta}$$
$$+ C\|(\varepsilon\sqrt{m_f})_t\|_{2,\delta}$$

Our previous arguments now show that boundedness is guaranteed for K_ϕ sufficiently small and μ/σ, $\mu/\sqrt{\sigma}$ sufficiently small (for a discussion on the implications of this result, see Remark 7.23).

Finally, the performance characterization in terms of the integral of the square, normalized tracking error follows identically as in the previous theorem except that the expressions for the integrals of the estimation error and parameter derivatives are now obtained from Corollary 7.7 instead. □□

Proof of Theorem 7.24:

The proof of the theorem follows along the lines of Theorem 7.13, after rewriting the I/O description in terms of the nominal closed-loop operators (see also Theorem 4.18 and Corollary 7.10). That is, for $t \in (t_j, t_{j+1})$,

$$QU = QS_r[\tilde{u}] + X_1$$

where

$$\tilde{u} = \frac{\hat{c}_0}{\hat{c}_{0*}}r + \frac{1}{\hat{c}_{0*}}w^T\hat{\phi} + \eta_1$$

$$X_1(t) = c_c(t)\Phi_c(t,t_j)x_c(t_j) - c_o\Phi_o(t,t_j)x_o(t_j)$$
$$+ c_c(t)\int_{t_j}^t \Phi_c(t,\tau)[\tilde{A}_c(\tau)x_c(\tau) + \tilde{b}_c(\tau)\tilde{u}(\tau)]\,d\tau$$
$$+ \tilde{c}_c(t)\int_{t_j}^t \Phi_c(t,\tau)b_c(\tau)\tilde{u}(\tau)\,d\tau + \tilde{d}(t)\tilde{u}(t)$$

and the subscripts 'c', 'o' refer to the actual and nominal state-space representation of the closed loop system and '˜' in the state-space description denotes perturbations due to smooth approximations. Notice that, according to Theorem 4.18, S_r and $\Phi_c(.,.)$ are ES, for sufficiently small ν, μ'. Similarly, y_p can be expressed as

$$y_p = W_m[\tilde{u}] + X_2$$
$$= \frac{1}{\hat{c}_0}\left\{\epsilon_1 - W_{m1}\left[W_{m2}[w^T]\dot{\hat{\theta}}\right] + W_m[\hat{c}_0 r]\right\}$$

where X_2 is a term of the form of X_1. The critical property of both X_1 and X_2 is that when normalized by the fictitious signal m_f, their RMS values are $O\left(\sqrt{\nu}, \sqrt{\mu'}\right)$.[16]

Next, using the decomposition of Theorem 7.13 and taking $L_2(\delta)$ norms of the truncated signals we find

$$\begin{aligned}
\|(QU)_t\|_{2,\delta} \leq\ & \|(R)_t\|_{2,\delta} + H_0 \|(QU)_t\|_{2,\delta} + H_1 \|(\epsilon_1)_t\|_{2,\delta} \\
& + H_2 \|(|\dot{\hat{\theta}}|\sqrt{m_f})_t\|_{2,\delta} + H_3 \|(|\dot{\hat{c}}_{0*}{}^s| + |\dot{\hat{\theta}}_*{}^s|\sqrt{m_f})_t\|_{2,\delta} \\
& + H_4 \|(|E\sqrt{m_f}|)_t\|_{2,\delta} + H_5 \|(|X_2|)_t\|_{2,\delta} + \|(|X_1|)_t\|_{2,\delta} \\
& + C \|(\varepsilon\sqrt{m_f})_t\|_{2,\delta}
\end{aligned}$$

where H_0-H_3 are constants with order of magnitude as in Theorem 7.13, $H_4 \leq O\left(1, 1/a, a^{n^*}\right)$ and $H_5 \leq O\left(a^{n^*}\right)$. The term E results from the differentiation of non-smooth parameters, arising in the expansions of η_1 and $\frac{d}{dt}\left(\frac{1}{\hat{c}_{0*}} G_\pi[QU]\hat{\phi}\right)$, (see Corollaries 2.60 and 7.10) and is upper bounded by a decaying exponential inside the intervals of continuity, i.e.,

$$|E(t)| \leq c e^{-\alpha_c(t-t_j)},\ t \in (t_j, t_{j+1}), \forall j$$

where c, α_c are positive constants independent of j. We should mention that in the case of non-strictly proper MRC structure, one may arrive at a similar expression, but the derivations become even more tedious. The (potential) problem appears in the $sG_\pi[QU]$ term which now involves the differentiation of the plant output. From the expression given above, it is obvious that such an operation cannot be performed directly since the output vector $c_c(t)$ may not be differentiable. Instead, the throughput (y_p) of $G_\pi[QU]$ should be decomposed from the beginning into a nominal part $W_m[\tilde{u}]$ and a perturbation part X_2 and then group the terms accordingly.

It is now straightforward to complete the proof as in the case of smooth parameter variations by applying Lemma 7.12 and using the expressions derived in Corollary 7.10. □□

Proof of Theorem 7.25:

In the light of Theorems 7.21 and 7.24, the proof is only incrementally more difficult in the treatment of the L_∞ bound of $\hat{\phi}$. That is, the parameter

---

[16] See, for example, Corollary 7.10. Notice that this statement is valid for sufficiently small δ, smaller than the ES margin of the nominal closed-loop system, imposing thus a limitation on the selection of the free parameter δ. Moreover, in this case the proportionality constants in $O(\nu), O(\mu')$ may also depend on the size of the parametric uncertainty set, although this is of no further consequence since any dependence on ν, μ' can be made as small as desired by requiring ν, μ' to be sufficiently small.

error is now decomposed into $\hat{\phi} = \hat{\phi}_s + \hat{\phi}_J + \varepsilon$ where $\hat{\phi}_s$ denotes a smooth component satisfying

$$\|\hat{\phi}_s\|_\infty \leq O\left(K_\phi, \frac{\mu}{\sigma}, \mu\sqrt{\frac{\gamma}{\sigma}}, \mu'\sqrt{\frac{\gamma}{\sigma}}\right)$$

and $\hat{\phi}_J$ denotes a piecewise continuous component, upper bounded by a decaying exponential inside the intervals of continuity, i.e.,

$$|\hat{\phi}_J(t)| \leq O\left(K_J, \sqrt{\frac{\gamma}{\sigma}}\right) e^{-\beta(t-t_j)}, \ t \in (t_j, t_{j+1}), \forall j$$

where β is a positive constant depending on σ but independent of j (see Corollary 7.11). The rest of the proof follows exactly the same steps as in the previous theorems, except that the expressions derived in Corollary 7.11 are now used for the RMS values of the normalized estimation error and the derivatives of the estimated parameters.

Note that, to simplify the expressions in the statement of the theorem, we used $O(K_\phi)$ as an upper bound for the size of the jumps K_J. More precisely, using the expressions of Corollary 7.11, the terms describing the effect of the parameter discontinuities on the mean-square, normalized tracking error are

$$O\left(1, \gamma^2, \gamma K_j^2, \frac{K_j^2}{\gamma}, \frac{\sigma^2 K_j^2}{\beta(\sigma)}, \frac{\gamma\sigma}{\beta(\sigma)}\right) \nu T$$

where $\beta < \min[\sigma, 2\alpha_c]$ and $-\alpha_c$ is the rate of exponential stability of the nominal closed-loop system. Since in a typical application σ is selected to be small, we took for simplicity, $\beta = O(\sigma)$ from which the expression given in the theorem follows. It should be noted, however, that the upper bounds of ν, μ and μ' do depend on the selected value of σ. Consequently, care must be exercised when these bounds are used to describe the behavior of the tracking error for variable σ and especially when $\sigma \to 0$. □□

Chapter 8

Adaptive TV PPC

8.1 Introduction

In the previous chapter we presented the design and properties of a class of adaptive controllers, of the MRC type, with the distinguishing characteristic that the controller parameters are updated directly from I/O measurements, without any intermediate identification of the LTV plant or other calculations. This desirable feature, however, was obtained under the somewhat restrictive TV MRC assumptions, i.e., constant and known relative degree and ES zero dynamics. As it was also mentioned in the same chapter, a different class of adaptive control schemes can be designed whereby the LTV plant is first identified and then a controller is designed, on-line, for the identified plant. With this *indirect adaptive control* approach, the designer may choose a control objective and the corresponding controller structure that do not impose conditions on the plant that are too restrictive. Popular examples of such designs from the LTI case are PPC and LQR and their variants, mainly thanks to the relative simplicity of their design equations.

In this chapter, we analyze and discuss the design and properties of an indirect TV adaptive PPC (APPC) as a typical paradigm of the indirect approach. The control objective of the TV APPC is to achieve the TV PPC objective without requiring complete knowledge of the plant parameters. According to the *Certainty Equivalence Principle*, the TV APPC relies on the TV PPC structure which is coupled with one of the LTV plant identification schemes studied in Chapter 6. That is, at every time instant the estimated plant parameters are used to calculate the controller controller parameters by solving the appropriate Diophantine equation. Since the TV PPC assumptions do not require the plant to have fixed relative degree or ES zero dynamics, the

TV APPC has a definite advantage over the TV MRAC of Chapter 7. On the other hand, its main and rather serious drawback is that the controller parameters are calculated for the identified plant and for this to be possible, the identified plant should be strongly controllable and observable. Such an assumption, however, can be restrictive and signal dependent especially when the parametric uncertainty is large. Nevertheless, the TV APPC emerges as an advantageous design in cases where the LTV plant does not satisfy the MRC assumptions and there is sufficient parametric uncertainty to make the design of a fixed TV PPC not feasible but such that the LTV plant is strongly controllable and observable for all possible values of the parameters.[1]

The design and stability properties of a TV APPC are presented in Section 8.2 for smooth and non-smooth parameter variations. In Section 8.3 we employ the Internal Model Principle to design a TV IMP/APPC with command tracking capabilities. We conclude our discussion with examples of TV APPC and TV IMP/APPC designs in Section 8.4

8.2 APPC: Design and Stability Analysis

As mentioned in the Introduction, a TV APPC is obtained by combining a parametric identifier and a TV PPC algorithm, whose fundamental properties have been individually studied in Chapters 6 and 5 respectively. The TV APPC design and analysis is essentially the same for both cases of LTV plants with smooth and non-smooth parameters. Still, for reasons of clarity regarding the involved assumptions, we discuss the case of smooth parameter variations first.

8.2.1 Smooth Parameter Variations

The first step in the design of a TV APPC, based on the Certainty Equivalence Principle, is the design of a parametric identification scheme to identify the plant I/O operator. For this purpose, let us consider a LTV plant with a state-space representation given by (3.1) and satisfying Assumptions 3.1–3.3 and suppose that its order n is known. Invoking Lemma 2.32, the plant admits an I/O representation in the P_L form

$$y_p = G_p^L(s,t)[u_p] = D_p^{-1}(s,t)N_p(s,t)[u_p]$$

which, in turn, is parametrized in terms of a $2n$-dimensional vector θ_* as

$$y_p = G_a(s)[u_p \theta_{1*}] + G_a(s)[y_p \theta_{2*}] \tag{8.1}$$

---
[1] Available or internally generated persistent excitation is another possibility but this is yet unresolved for fast TV plants and is not considered here.

8.2. APPC: DESIGN AND STABILITY ANALYSIS

where $G_a(s) = q_a^T(sI - F_a)^{-1}$ is a filter selected by the designer and $\theta_* = [\theta_{1*}^T, \theta_{2*}^T]^T$. (Note that throughout this chapter the quantities related to the parameter estimator/identifier are assigned the subscript 'a' to distinguish between the identifier and the controller filters.) The parameters θ_* are related to the coefficients of $D_p(s,t)$ and $N_p(s,t)$ by a constant affine transformation, say Q_a, depending on q_a, F_a. Hence, employing the notion of structured parameter variations we assume, without loss of generality, that

$$\theta_*(t) = \Pi(t)\hat{\theta}_*(t)$$

where $\Pi(t) = [\Pi_1(t)|\Pi_2(t)]$ is a known, not necessarily square matrix with smooth UB entries and $\hat{\theta}_*(t)$ is a partially unknown smooth UB vector.

Thus, the LTV plant is described by the parametric model

$$\begin{aligned} y_p &= w_a^T \hat{\theta}_* - \eta \\ w_a &= [G_a(s)\{u_p \Pi_1\} + G_a(s)\{y_p \Pi_2\}]^T \end{aligned}$$

where η is a swapping term (see Section 6.3 for details). Employing the results of Chapter 6, the I/O operator of the LTV plant can be identified by estimating the unknown parameters with a suitable adaptive law. For example, if $\hat{\theta}_*$ satisfies Assumptions 6.3 and 6.10 the following adaptive law can be used

$$\begin{aligned} \dot{\hat{\theta}} &= \mathcal{P}\left(-\gamma \frac{\epsilon_1 w_a}{m^2}\right) \; ; \; \hat{\theta}(t_0) \in \mathcal{M} \\ \epsilon_1 &= w_a^T \hat{\theta} - y_p \\ \dot{m}^2 &= -2\delta_0 m^2 + |QU|^2 + q_e; m^2(t_0) > 0 \\ U &= [u_p, y_p]^T \; ; \; Q = Q^T > 0 \end{aligned} \quad (8.2)$$

with the properties described in Corollary 6.11. On the other hand, if $\hat{\theta}_*$ is UB and $\|\hat{\theta}_*\|_\infty \leq \mu$, an adaptive law with σ-modification is applicable, i.e.,

$$\dot{\hat{\theta}} = -\gamma \frac{\epsilon_1 w_a}{m^2} - \sigma(\hat{\theta} - \hat{\theta}_c) \quad (8.3)$$

with the properties described in Corollary 6.14.

Having obtained the parameter estimates $\hat{\theta}$, the PDO/PIO description of the identified plant is obtained by computing $\theta = \Pi \hat{\theta}$, the estimate of θ_*, and then applying the inverse transformation Q_a^{-1} on θ.[2] This operation results in an I/O description of the LTV plant in terms of the identified plant and

---
[2] The invertibility of the transformation Q_a is guaranteed by the observability of the pair q_a, F_a.

the identification error appearing as a stable-factor perturbation as

$$y_p = D_a^{-1}(s)\hat{N}_p(s,t)[u_p] + D_a^{-1}(s)\{D_a(s) - \hat{D}_p(s,t)\}[y_p] \\ \underbrace{- D_a^{-1}(s)\tilde{N}_p(s,t)[u_p] - D_a^{-1}(s)\tilde{D}_p(s,t)[y_p]}_{e_1 = \epsilon_1 - \hat{\eta}} \quad (8.4)$$

where $D_a(s) = \det(sI - F_a)$ and $(\hat{\cdot})$, $(\tilde{\cdot})$ denote PDO's depending on the parameter estimates θ and the parameter error $\phi = \theta - \theta_*$ respectively.[3]

The next step in the TV APPC design is that of the control law. Since the identified plant I/O operator is obtained in the left form, invoking Lemma 5.1 the controller I/O operator is chosen to meet the TV PPC objective for the identified plant, i.e.,

$$u_p = -N_1(s,t)N_2^{-1}(s,t)[y_p]$$

where $N_1(s,t), N_2(s,t)$ are calculated by solving the Diophantine equation

$$\hat{D}_p(s,t) \star N_2(s,t) + \hat{N}_p(s,t) \star N_1(s,t) = A_*(s,t) \quad (8.5)$$

where $A_*^{-1}(s,t)$ is the desired (ES) PIO and '$\star$' denotes a pointwise operation with respect to $\hat{\theta}$ only. Note that although $\hat{\theta}$ is known, its higher order derivatives are not. As a consequence, (8.5) can only be solved for frozen $\hat{\theta}$. It should be emphasized, however, that '$\star$' does not indicate a complete pointwise operation. The coefficients of $\hat{D}_p, \hat{N}_p$ depend on $\hat{\theta}$ as well as Π. The latter contains known functions with known derivatives and therefore the operator multiplications can be performed by differentiating Π as necessary.

As previously mentioned, the need to solve (8.5), which depends on the estimated plant parameters instead of the actual ones, is responsible for the main APPC drawback. That is, in order to ensure the solvability of (8.5) we assume that

8.1 Assumption: *For any possible frozen $\hat{\theta}$ the estimated PDO's $\hat{D}_p(s,t)$, $\hat{N}_p(s,t)$ are strongly left coprime in $[t_0, \infty)$, uniformly in $\hat{\theta}$.* ∎

Equivalently, the identified plant, with $\hat{\theta}$ frozen, should be strongly controllable and observable, uniformly in $\hat{\theta}$. This assumption is rather standard in the literature of indirect adaptive schemes and, in general, its validity cannot be guaranteed by the properties of the estimator alone. One approach to ensure that Assumption 8.1 is satisfied is to use an estimator with projection and require that the set $\mathcal{M}$ contains no uncontrollable or unobservable

---

[3] As usual, when the initial conditions are not zero, equation (8.4) should be augmented by an exponentially decaying term ϵ. For further details, see Section 6.3.

8.2. APPC: DESIGN AND STABILITY ANALYSIS

points. This procedure is feasible and can be efficient for certain applications though it may be quite restrictive especially when the dimension of the estimated parameter vector is large. In the latter case, the requirement for a convex projection set where the above assumption holds is likely to constrain the parametric uncertainty to be small, in which case the need for an adaptive controller is questionable. Nevertheless, it appears that in this aspect algorithms with projection have an advantage over the σ-modification since for the former sufficient conditions for the validity of Assumption 8.1 can be checked a priori.

Other existing remedies of this problem are injection of persistently exciting signals (slowly TV plants) [G.S.84, GMDD.87] or use of multiple estimators [MGHM.88]. Although these techniques may also be applied to our case, we do not address this problem in the present study.

Finally, the PPC solution $-N_1(s,t)N_2^{-1}(s,t)$ is realized in state-space by following the guidelines of Example 2.38, that is, the control input is obtained by

$$\begin{aligned}
\dot{\omega} &= F\omega + qu_1 \\
u_1 &= p_2^T(t)\omega + (r - y_p) \\
u_p &= p_1^T(t)\omega + p_3(t)u_1
\end{aligned} \quad (8.6)$$

Notice that when solving the Diophantine equation (8.5), the PDO's N_1, N_2 are calculated in the right form while the PPC state-space realization assumes that the PDO's are expressed in the left form. This implies that an intermediate operation should be performed to convert the PDO's N_1, N_2 from the right to the left form. Again, since the derivatives of $\hat{\theta}$ are not available, such an operation is performed with $\hat{\theta}$ being frozen.

The calculations involved in the TV APPC are summarized below:

8.2 TV APPC Algorithm:

1. Obtain $\hat{\theta}$ from the adaptive law and calculate θ as $\theta = \Pi\hat{\theta}$.

2. Determine $\hat{D}_p(s,t)$ and $\hat{N}_p(s,t)$, the estimates of the plant PDO's, using θ in the place of θ_* in the parametric model of the LTV plant.

3. Solve the Diophantine equation (8.5) for the coefficients of $N_1(s,t), N_2(s,t)$ with $\hat{\theta}$ being frozen.

4. Convert the right PDO's $N_1(s,t), N_2(s,t)$ into the left form with $\hat{\theta}$ being frozen and calculate the parameters $p_i(t), i = 1,2,3$ of the control law (8.6). ▽▽

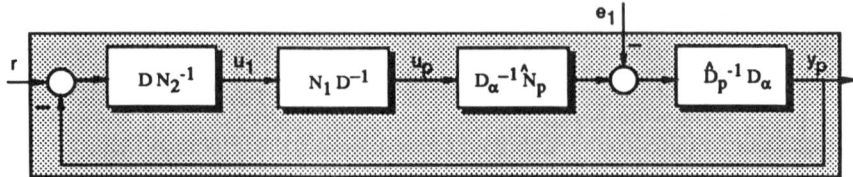

Figure 8.1: Block diagram of the closed-loop plant with the TV APPC.

With the so-defined TV APPC, the closed-loop plant can be described in terms of the estimated parameters θ, instead of θ_*, as shown in Fig. 8.1 (see also Fig. 6.3 in Chapter 6, for more details on the identification part). As pointed out in Chapter 6, such a description involves the identified plant, perturbed by the identification error e_1 (a stable-factor perurbation). Furthermore, the TV APPC guarantees by construction that the shaded part of the closed-loop in Figure 8.1 is ES for $\hat{\theta}$ frozen. Since both the identification error e_1 and the speed of variations of the parameter estimates $\dot{\hat{\theta}}$ are small in the mean-square sense for slow in the mean-square speed of unstructured plant parameter variations, boundedness can be shown by virtue of Lemmas 2.44 and 7.12. This idea is made precise in the following theorems which describe the closed-loop stability properties when the TV APPC 8.2 with an adaptive law using either projection or the σ-modification is used to control the plant (3.1).

8.3 Theorem: *Let K_ϕ be a constant denoting the size of the parametric uncertainty set as in Corollary 6.11 and suppose that the conditions of Corollary 6.11 and Assumption 8.1 are satisfied. Then, for any finite K_ϕ, there exists $\mu_0 > 0$ such that for any $\mu \in [0, \mu_0)$ all signals in the closed-loop of the LTV plant and the TV APPC 8.2 with projection are UB, for any UB command input r, for all $t \geq t_0$.* ▽▽

Proof: In Appendix VIII

8.4 Theorem: *Let K_{θ_*} be a constant denoting the size of the parametric uncertainty set as in Corollary 6.14 and suppose that the conditions of Corollary 6.14 and Assumption 8.1 are satisfied. Then, for any finite K_{θ_*}, there exist $\sigma_0 > 0$ and $\mu_0(\sigma)$[4] such that for any $\sigma \in (0, \sigma_0)$, $\mu_0 > 0$ and for any $\mu \in [0, \mu_0)$ all signals in the closed-loop of the LTV plant and the TV APPC 8.2 with the σ-modification are UB, for any UB command input r, for all $t \geq t_0$.* ▽▽

Proof: In Appendix VIII

---

[4] μ_0, as derived in the proof of the theorem, is a continuous function of σ.

8.2. APPC: DESIGN AND STABILITY ANALYSIS

The above theorems establish the intuitive result we expected. That is, the closed-loop signals are bounded provided that the speed of variation of the unstructured part of the plant parameters is sufficiently small. Their differences are primarily revolving around the measure and description of the uncertainty in the range and speed of variation of the plant parameters. Estimators with projection describe the range of parameters in terms of a projection set and can tolerate small-in-the-mean parameter variations. On the other hand, estimators with the σ-modification impose a 'soft' constraint on the parameter estimates by penalizing their deviation from a nominal point. With this approach the range of the plant parameters is not specified a priori but the admissible variations are restricted to be uniformly slow. Some additional comments regarding these theorems are given next. (For further details, see the proofs in the appendix)

The important quantities for stability/boundedness are the parameter μ, and K_ϕ (or $K_{\theta*}$) representing the effective deviation from the ideal case. While the latter describes the maximum possible range of variations, the former describes how fast a change can occur. As μ approaches zero, K_ϕ may approach ∞ without destroying boundedness, provided of course that Assumption 8.1 holds. Note that, in the structured variations formulation, $\mu = 0$ does not necessarily imply that the plant is LTI. On the other hand, when the parametric uncertainty is sufficiently small, boundedness is preserved independent of the speed of the unstructured parameter variations.

The selection of the adaptation gain (as well as σ for estimators using the σ-modification) involves one of the most obvious and important trade-offs in the design of an adaptive controller. Since both γ and $1/\gamma$ appear in the RMS value of the perturbations $e_1, \dot{\hat{\theta}}$, the adaptation gain should not be chosen either too small or too large. Furthermore, the auxiliary filters in the controller and estimator also affect the range of μ for which the adaptive controller guarantees boundedness but in a more subtle way, through the closed-loop sensitivity operators $e_1 \mapsto u_p, y_p, u_1$. An arbitrary choice of these design parameters may yield an arbitrarily bad performance. With this in mind, the techniques developed in Section 5.2 for the design of overparametrized TV PPC structures as well as the application of the Internal Model Principle (see Section 5.4) can prove quite useful in shaping the properties of the closed-loop sensitivity operators. However, due to the complexity of the analysis, the quantitative aspects of these issues are still elusive and only crude or special-case guidelines are available.

It should be noted that the TV APPC as well as all indirect designs rely on a plant description factorized in the P_L form, since this is the natural form

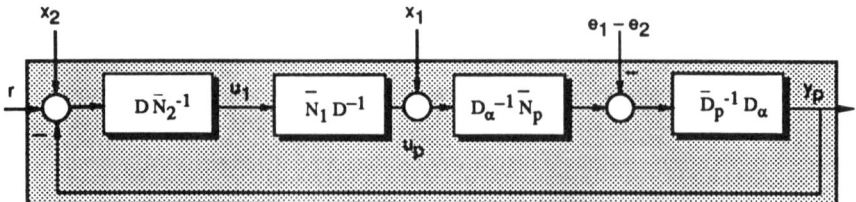

Figure 8.2: Alternative description of the closed-loop plant.

obtained from our identification schemes. It is, in principle, possible to design a TV APPC for the identified plant in the P_R form by first performing the necessary conversion between forms (this is always possible under Assumption 8.1). Such a design may be advantageous from a sensitivity-shaping point of view, but it has the disadvantages of increasing the computational load and complicating the way the identification error perturbs the closed-loop.

Up to this point, our study of the TV APPC was focused only on the boundedness issue. It is also important, however, to discuss its performance with respect to the TV PPC objective. For this, we need to rewrite the closed-loop plant in a slightly different form, shown in Fig. 8.2, with three additional perturbation terms e_2, x_1, x_2. These terms represent errors due to smooth approximations of the controller and plant parameters. Such approximations are needed in order to arrive at a PDO/PIO description of the closed-loop system (since the estimated parameters may not possess as many derivatives as necessary) while they introduce perturbations which are small in the mean-square, normalized sense. The final result states that the closed-loop plant can be viewed as a nominal part with PIO $[A_* + \Delta]^{-1}$ where the coefficients of Δ are small in the mean-square and several perturbations which are small in the mean-square, normalized sense. These perturbations as well as Δ disappear as $\mu \to 0$. (Of course, for the σ-modification, they also depend on σ.)

8.5 Theorem: *Consider the LTV plant (5.1) satisfying the conditions of either Theorem 8.3 or 8.4 and the respective TV APPC. Then the closed-loop plant admits an I/O description of the form*

$$\begin{bmatrix} u_p \\ y_p \\ u_1 \end{bmatrix} = \underbrace{\begin{bmatrix} S_{ru} & S_{eu} & S_{uu} \\ S_{ry} & S_{ey} & S_{uy} \\ S_{r1} & S_{e1} & S_{u1} \end{bmatrix}}_{S_s} \begin{bmatrix} r + x_2 \\ -e_1 + e_2 \\ x_1 \end{bmatrix}$$

where the sensitivity operators in S_s have PIO $D_c^{-1}(s,t)$ satisfying

$$D_c(s,t) = A_*(s) + \Delta(s,t)$$

8.2. APPC: DESIGN AND STABILITY ANALYSIS

and $\Delta(s,t)$ is a PDO of degree at most $2n-2$, with smooth, UB coefficients $O(\|\dot{\hat{\theta}}\|)$. The terms e_2, x_1, x_2 are of the form

$$
\begin{aligned}
e_2 &= G_a(s)[u_p \Pi_1 \tilde{\theta}_1] + G_a(s)[y_p \Pi_2 \tilde{\theta}_2] \\
x_1 &= \tilde{p}_1^\top (sI - F)^{-1} q[u_1] + \tilde{p}_3 u_1 \\
x_2 &= \tilde{p}_2^\top (sI - F)^{-1} q[u_1]
\end{aligned}
$$

where $\tilde{\theta}_i, \tilde{p}_i$ are $O(\|\dot{\hat{\theta}}\|)$. ▽▽

Proof: In Appendix VIII.

Of course, it is possible to use Theorem 8.5 to show the closed-loop boundedness with a purely I/O approach but the derivations are slightly more messy and restricted to a PPC-type scheme, while the approach used in Theorems 8.3 and 8.4 is easily extended to include other controller design techniques.

8.2.2 Non-Smooth Parameter Variations

In contrast to the direct TV MRAC case, the previous results on the TV APPC are easily generalized to admit non-smooth parameter variations. The reason is that the analysis of the indirect schemes relies on the identified plant rather than the actual one. Still, some additional work is necessary in order to encompass practically interesting cases such as discontinuities in the structure matrix Π or infrequent loss of controllability or observability which are more likely to occur when the plant parameters are discontinuous.

Let us consider the plant (3.8), satisfying Assumptions 3.4–3.6. Such a plant admits an I/O description of the form (6.20), parametrized by the parameter vector θ_* (see Chapters 3 and 6 for details). Without loss of generality, we assume that $\theta_* = \Pi \hat{\theta}_*$ where Π is a piecewise smooth, UB matrix and $\hat{\theta}_*$ is a piecewise smooth vector with possible discontinuities at $t \in \{t_j\}_1^\infty$. Employing the results of Chapter 6 regarding the non-smooth parameter case, the I/O operator of the LTV plant can be identified by estimating the unknown parameters with a suitable adaptive law. For example, if $\hat{\theta}_*$ satisfies Assumption 6.3 and its Lipschitz continuous part $\hat{\theta}_*^s$ satisfies Assumption 6.10 the estimator (8.2) can be used with the properties described in Corollary 6.16. Similarly, if $\hat{\theta}_*$ is UB, $\|\dot{\hat{\theta}}_*^s\|_\infty \leq \mu$, the perturbation part of the plant is uniformly small and the minimum time between jumps is large, an adaptive law with σ-modification (8.3) is applicable with the properties described in Corollary 6.18.

As in the previous subsection, the parameter estimates $\hat{\theta}$ are used to compute $\theta = \Pi \hat{\theta}$, from which the PDO/PIO description of the identified plant, parametrized by $\hat{\theta}$ (or equivalently θ) is given by (8.4). Next, a control law could be designed based on the identified plant by solving (8.5) and realized by (8.6). One important difference from the smooth parameter case is that the structure matrix Π may be discontinuous. Moreover, since the continuous $\hat{\theta}$ is an estimate of the discontinuous plant parameters $\hat{\theta}_*$, it is likely that Assumption 8.1 is violated in short time intervals. Therefore, to avoid being overly restrictive we need to modify Assumption 8.1 and the control law, so that the latter is designed in a piecewise sense. That is, we introduce an additional step in design of the TV APPC, given below:

8.6 TV APPC Modified Algorithm:

1. *Obtain $\hat{\theta}$ from the adaptive law and calculate θ as $\theta = \Pi\hat{\theta}$.*

2. *Determine $\hat{D}_p(s,t)$ and $\hat{N}_p(s,t)$, the estimates of the plant PDO's, using θ in the place of θ_* in the parametric model of the LTV plant.*

2.i. *Determine whether the left TV Sylvester matrix of $\hat{D}_p(s,t)$, $\hat{N}_p(s,t)$ is strongly nonsingular, with respect to an a priori selected threshold. If not, set $\theta = \Pi\hat{\theta}_o$ and repeat step 2. Otherwise, set $\theta_o(t) = \theta(t)$ and continue.*

3. *Solve the Diophantine equation (8.5) for the coefficients of $N_1(s,t)$, $N_2(s,t)$ with $\hat{\theta}$ being frozen.*

4. *Convert the right PDO's $N_1(s,t)$, $N_2(s,t)$ into the left form with $\hat{\theta}$ being frozen and calculate the parameters $p_i(t)$, $i = 1, 2, 3$ of the control law (8.6).*   ▽▽

In other words, $\hat{\theta}_o(t)$ is the 'last' estimate for which the the estimated PDO's $\hat{D}_p(s,t)$, $\hat{N}_p(s,t)$ (with $\hat{\theta}_o$ frozen) are strongly left coprime in (t_j, t_{j+1}), uniformly in $\hat{\theta}_o$ and uniformly in j. Hence, using $\hat{\theta}_o$ the controller parameters can be calculated according to the TV PPC procedure, in a piecewise sense. Effectively, this approach reparametrizes the plant in terms of $\hat{\theta}_o$, treating the difference $\hat{\theta} - \hat{\theta}_o$ as an additional perturbation in the identified plant description, which is similar to the identification error e_1. More precisely,

$$y_p = G(s)[u_p\theta_{1o}] + G(s)[y_p\theta_{2o}] - e_1 + e_3 + \varepsilon \qquad (8.7)$$
$$e_3 = G_a(s)[u_p\Pi_1(\hat{\theta}_1 - \hat{\theta}_{1o})] + G_a(s)[y_p\Pi_2(\hat{\theta}_2 - \hat{\theta}_{2o})]$$

We now introduce an assumption to ensure that the perturbation term e_3 is small in a mean-square normalized sense and that the switching between $\hat{\theta}_o$ and $\hat{\theta}$ does not occur too often.

8.2. APPC: DESIGN AND STABILITY ANALYSIS

8.7 Assumption: *There exist constants c, ν' such that for all $t_I \geq t_0$ and $T \geq 0$,*

$$\int_{t_I}^{t_I+T} |(\hat{\theta} - \hat{\theta}_o)(t)|_2^2 \leq c + \nu' T$$

$$n'_I \leq c + \nu' T$$

where n'_I denotes the number of discontinuous switchings in an interval $(t_I, t_I + T)$, between $\hat{\theta}_o$ and $\hat{\theta}$ in the calculations of the controller parameters. ∎

We must emphasize that although Assumption 8.7 is less restrictive than 8.1, allowing the parameters to cross uncontrollability/unobservability surfaces, it still suffers the same major drawback. That is, it is signal dependent and cannot be verified a priori, except for special cases or under some additional excitation conditions.

The stability properties of the modified TV APPC 8.6 are now given by the following theorems.

8.8 Theorem: *Let K_ϕ be a constant denoting the size of the parametric uncertainty set and suppose that the conditions of Corollary 6.16 and Assumption 8.7 are satisfied. Then, for any finite K_ϕ, there exist $\mu_0, \nu_0, \mu'_0, \nu'_0 > 0$ such that for any $\mu \in [0, \mu_0), \nu \in [0, \nu_0), \mu' \in [0, \mu'_0), \nu' \in [0, \nu'_0)$ all signals in the closed-loop of the LTV plant and the TV APPC 8.6 with projection are UB, for any UB command input r, for all $t \geq t_0$.* ▽▽

Proof: The TV APPC 8.6 stabilizes the modified plant (8.7) (without the perturbations e_1, e_3) in a piecewise sense and therefore —under Assumption 8.7— for all t, provided that the average number of discontinuities in Π and $\hat{\theta}_o$ ν, ν' are sufficiently small. Furthermore, since e_3 has similar properties as e_1 in a mean-square, normalized sense, the rest of the proof follows as in Theorem 8.3, but using the expressions of Corollary 6.16 instead. □□

8.9 Theorem: *Let $K_{\theta*}$ be a constant denoting the size of the parametric uncertainty set and suppose that the conditions of Corollary 6.18 and Assumption 8.7 are satisfied. Then, for any finite $K_{\theta*}$, there exist $\sigma_0 > 0$ and $\mu_0(\sigma), \nu_0(\sigma), \mu'_0(\sigma), \nu'_0(\sigma)$ such that for any $\sigma \in (0, \sigma_0), \mu_0 > 0, \nu_0 > 0, \mu'_0 > 0, \nu'_0 > 0$ and for any $\mu \in [0, \mu_0), \nu \in [0, \nu_0), \mu' \in [0, \mu'_0), \nu' \in [0, \nu'_0)$ all signals in the closed-loop of the LTV plant and the TV APPC 8.6 with the σ-modification are UB, for any UB command input r, for all $t \geq t_0$.* ▽▽

Proof: As in Theorems 8.8 and 8.4, but using the expressions of Corollary 6.18 instead. □□

8.2.3 Slowly TV Plants

The approach and tools employed in establishing the results of the previous subsections indicate that a wide variety of controller structures can be used in the design of indirect adaptive control schemes. Indeed, the basic requirements are that the controller has a 'well-behaved' state-space realization and it guarantees that the closed-loop plant is ES for every frozen estimate of the parameter vector in a uniform sense. The idea behind these requirements is to use Lemma 2.42 or its TV extension 2.44 to ensure that the closed-loop plant without the identification error is ES and then use the Bellman-Gronwall Lemma to show boundedness. Note that a 'well-behaved' controller realization is needed in order for Lemma 2.42 to be applicable. For example, this is the case when the controller parameters are Lipschitz functions of the plant parameter estimates which in turn implies that the closed-loop state-space representation is Lipschitz in the estimated parameters (for similar developments see, e.g., [Kre.86, M.G.88, GMDD.87]).

This observation is especially valuable in the special but interesting case of slowly TV plants with completely unstructured parameter variations, where an indirect adaptive controller can rely on a wide variety of controller structures developed for LTI systems (i.e., pointwise designs). In such a case the following result is applicable.[5]

8.10 Corollary: *Suppose that the conditions of Corollary 6.11 are satisfied with $\Pi = I$. Also suppose that, for any possible θ, a control law of constant order and with UB parameters, Lipschitz continuous in θ, is designed so that the frozen closed loop is ES uniformly in θ. Consider the adaptive controller obtained by evaluating the parameters of the control law based on the plant parameter estimates as given by the estimator of Corollary 6.11 at each time instant. Then, there exists $\mu_0 > 0$ such that for any $\mu \in [0, \mu_0)$ all signals in the closed-loop of the LTV plant and are UB, for any UB command input r, for all $t \geq t_0$.* ▽▽

Proof: As in Theorem 8.3. □□

Notice that, the assumptions of the corollary admit a wide class of controllers (e.g., MRC, LQR etc.) which may allow for weaker assumptions than 8.1. Using an LQR-type control, for example, mode cancellation can occur in the identified plant, as long as they are ES. Again, however, this does not solve the intrinsic problem of indirect schemes since, in general, meeting such conditions is not guaranteed by the properties of the estimator alone.

---

[5] An analogous statement can be made for estimators using the σ-modification.

8.3 Command Tracking with the TV APPC

Like the TV PPC studied in Chapter 5, the TV APPC is primarily motivated as a regulation problem and may exhibit very poor performance if used for command tracking purposes (e.g., see Example 5.21). Moreover, the situation becomes even more complicated since the estimated parameters do not necessarily converge to the actual ones, i.e., $\hat{\theta} \to \hat{\theta}_*$ even when the latter is constant. This, in turn, indicates that very little can be said about the PDO's of the closed-loop sensitivity operators, though they all have PIO's 'close' to A_* (see Theorem 8.5). One approach to introduce some tracking capabilities in a TV PPC is via the Internal Model Principle (see Section 5.4). The resulting IMP/PPC is able to track command inputs generated by a prescribed model and although the associated conditions are more restrictive than the PPC ones, they still allow for plants which cannot be controlled by MRC schemes.

In this section we design an adaptive IMP/PPC (IMP/APPC) which, in addition to the TV APPC objective and without the restriction of fully known plant parameters, achieves asymptotic tracking of a class of command inputs whose internal model is specified a priori. Naturally, the design procedure and assumptions are similar to their APPC counterparts, with a few additions outlined below.

As in Section 8.2, we consider a LTV plant with a state-space representation given by (3.1) and satisfying Assumptions 3.1–3.3. For such a plant, suppose that the control objective is

1. The closed-loop PIO is $A_*^{-1}(s, t)$ and

2. The plant output tracks asymptotically UB command inputs r satisfying $\Lambda(s)[r] = 0$,

where $A_*^{-1}(s, t)$ is an ES PIO with UB coefficients and of order $2n + \deg[\Lambda] - 2$.

Under the pertinent assumptions (e.g., see Section 8.2), an identification scheme based on the adaptive law (8.2) or (8.3) is used to identify the plant I/O operator. Further, with the notation of Section 8.2, and applying the TV IMP/PPC design to the identified plant the control input u_p is generated by (5.13)

$$u_p = N_1(s,t) N_2^{-1}(s,t)[r - y_p] - P(s,t) Q^{-1}(s)[y_p]$$

or, in a state-space realization

$$\begin{aligned} \dot{\omega}_1 &= F_1 \omega_1 + q_1 u_1 \quad ; \quad \dot{\omega}_2 = F_2 \omega_2 + q_2 y_p \\ u_1 &= p_2^T \omega_1 + r - y_p \quad ; \quad u_p = p_1^T \omega_1 + p_3 u_1 + p_4^T \omega_2 + p_5 y_p \end{aligned} \quad (8.8)$$

where, $Q^{-1}(s)$ is an ES, TI PIO of order $\deg[\Lambda] - 1$. $N_1(s,t)$, $N_2(s,t)$, $P(s,t)$ are PDO's with UB coefficients of degrees $n + \deg[\Lambda] - 2$, $n + \deg[\Lambda] - 2$, $\deg[\Lambda] - 1$ respectively, obtained as the solution of

$$\hat{D}_p(s,t) \star Q(s) \star \hat{N}_2(s,t) + \hat{N}_p(s,t) \star \hat{N}_1(s,t) = A_*(s,t) \tag{8.9}$$

$$N_2(s,t) = Q(s) \star \hat{N}_2(s,t)$$

$$N_1(s,t) = \hat{N}_1(s,t) - P(s,t) \star \hat{N}_2(s,t)$$

$$X(s,t) \star \Lambda(s) - \hat{N}_p(s,t) \star P(s,t) = \hat{D}_p(s,t) \star Q(s)$$

for some PDO $X(s,t)$ of degree $n - 1$. As usual, '$\star$' denotes a pointwise operation with respect to $\hat{\theta}$ only. Also, in the state-space realization of the controller (8.8), the various parameters are calculated such that:

$$p_1^T(sI - F_1)^{-1} q_1 + p_3 = N_1(s,t) D^{-1}(s)$$

$$p_2^T(sI - F_1)^{-1} q_1 = [D(s) - N_2(s,t)] D^{-1}(s)$$

$$p_4^T(sI - F_2)^{-1} q_2 + p_5 = -P(s,t) Q^{-1}(s)$$

$$D(s) = \det(sI - F_1) \; ; \; Q(s) = \det(sI - F_2)$$

The calculations involved in the TV IMP/APPC are summarized below:

8.11 TV IMP/APPC Algorithm:

1. *Obtain $\hat{\theta}$ from the adaptive law and calculate θ as $\theta = \Pi \hat{\theta}$.*

2. *Determine $\hat{D}_p(s,t)$ and $\hat{N}_p(s,t)$, the estimates of the plant PDO's, using θ in the place of θ_* in the parametric model of the LTV plant.*

3. *Solve the equations (8.9) for the coefficients of $N_1(s,t)$, $N_2(s,t)$, $P(s)$ with $\hat{\theta}$ being frozen.*

4. *Convert the right PDO's $N_1(s,t)$, $N_2(s,t)$, $P(s)$ into the left form with $\hat{\theta}$ being frozen and calculate the parameters $p_i(t)$, $i = 1, \ldots, 5$ of the control law (8.8).* ▽▽

As in the previous section, the solvability of the equations involved in the TV IMP/APPC algorithm cannot be guaranteed in general by the properties of the estimator alone. We, therefore, impose the following conditions on the identified plant.

8.3. COMMAND TRACKING WITH THE TV APPC

8.12 Assumption:

a. *For any possible frozen $\hat{\theta}$ the estimated PDO's $\hat{D}_p(s,t)Q(s)$, $\hat{N}_p(s,t)$ are strongly left coprime in $[t_0, \infty)$, uniformly in $\hat{\theta}$.*[6]

b. *For any possible frozen $\hat{\theta}$ the PDO equation*

$$X(s,t) \star \Lambda(s) - \hat{N}_p(s,t) \star P(s,t) = \hat{D}_p(s,t) \star Q(s)$$

has a solution such that $P(s,t), X(s,t)$ are PDO's of degree $\deg[\Lambda] - 1$, $n - 1$ respectively, with UB coefficients Lipschitz continuous in $\hat{\theta}$, uniformly in $\hat{\theta}$. ∎

Under this assumption, the following theorems describe the properties of the closed-loop plant (3.1) with the TV IMP/APPC.

8.13 Theorem: *Suppose that the conditions of Theorem 8.3 and Assumption 8.12 hold. Then, in addition to the results of Theorem 8.3, there exists $\mu_1 > 0$ such that for any $\mu \in [0, \mu_1)$ and any bounded r such that $\Lambda[r] = 0$, the tracking error $e = y_p - r$ of the LTV plant and the TV IMP/APPC 8.11 with projection satisfies*[7]

$$\int_t^{t+T} e^2(\tau)\, d\tau \leq C_0 + O\left(1, \frac{K_\phi}{\gamma}, \gamma^2, K_\phi \gamma\right) \mu T$$

where C_0 is a constant, for all $t \geq t_0$ and all $T \geq 0$. ▽▽

Proof: In Appendix VIII

8.14 Theorem: *Suppose that the conditions of Theorem 8.4 and Assumption 8.12 hold. Then, in addition to the results of Theorem 8.4, there exist $\sigma_1 > 0$ and $\mu_1(\sigma)$ such that for any $\sigma \in (0, \sigma_1)$, $\mu_1 > 0$ and for any $\mu \in [0, \mu_1)$ and any bounded r such that $\Lambda[r] = 0$, the tracking error $e = y_p - r$ of the LTV plant and the TV IMP/APPC 8.11 with the σ-modification satisfies*

$$\int_t^{t+T} e^2(\tau)\, d\tau \leq C_0 + O\left(\frac{K_{\theta*}^2}{\gamma}, K_{\theta*}^2 \sigma, K_{\theta*}^2 \gamma\right) \sigma T$$
$$+ O\left(1, \gamma^2, \gamma\sigma, \frac{\gamma}{\sigma}, \frac{1}{\gamma\sigma}\right) \mu^2 T$$

where C_0 is a constant, for all $t \geq t_0$ and all $T \geq 0$. ▽▽

---

[6] This is more convenient than requiring $Q(s)$ and $\hat{N}_R(s,t)$ to be strongly left coprime as in Section 5.4 since $Q(s)$ is a fixed LTI PDO.

[7] For simplicity, the statement is made without normalizing the tracking error; consequently, the constants in $O(\cdot)$ depend on the bound of r.

Proof: As in Theorem 8.13, but with the expressions obtained in Corollary 6.14. □□

8.15 Remark: In the case of slowly TV plants where the parameter variations are completely unstructured ($\Pi = I$) and μ is a measure of the total speed of plant parameter variations, the above theorems are also valid (in a qualiatative sense) for PW IMP/APPC designs (e.g., see Remark 5.18). In this case, the PDO swappings in the derivation of an expression for the tracking error require the differentiation of the identified plant and controller parameters. Such an operation results in an $O\left(\|\dot{\theta}\|\right)$-small-in-the-mean-square perturbation which, for slowly TV plants, leaves the rest of the arguments qualitatively unchanged.

Pointwise designs, on the other hand, may be considerably easier to realize and require less stringent assumptions about the identified plant. For example, we may calculate the PDO's $P(s,t), X(s,t)$ using the frozen-$\hat{\theta}$ version of (5.20)

$$\Lambda(s) \star X(s,t) - N_p(s,t) \star P(s,t) = D_p(s,t) \star Q(s)$$

in the design equations (8.9). For the resulting scheme, referred to as PW IMP + TV APPC, the controller parameters are calculated by solving linear algebraic equations only. Furthermore, condition (b.) in Assumption 8.12 is replaced by the easier to check:

b'. *For any possible frozen $\hat{\theta}$ the PDO's $\Lambda(s,t), \hat{N}_p(s,t)$ are strongly left coprime in $[t_0, \infty)$, uniformly in $\hat{\theta}$.*

It is quite straightforward to show that the PW IMP + TV APPC guarantees signal boundedness under the general conditions of Theorems 8.3, 8.4 while, following the arguments used in Theorem 8.13, a small-in-the-mean-square tracking error is obtained, provided that the overall plant is slowly TV. ▽▽

Analogous statements can be made for the case of non-smooth parameters where the same arguments are applicable except, of course, that the expressions for $e_1, \dot{\hat{\theta}}$ are obtained from the appropriate corollaries presented in Chapter 5. The final results contain additional $O(\nu, \mu')$ terms and due to their similarity with the already presented ones are omitted.

Finally, the robustness properties of the various APPC designs with respect to unmodeled dynamics and other disturbances may be established by following similar arguments and procedures including, of course, the derivation of the estimator/identifier properties for the particular problem.

8.4 Examples

We conclude our study of indirect APPC schemes with a few examples illustrating their design procedures and properties. In all cases we consider the LTV plant with I/O description

$$s^2[y_p] + s[a_1 y_p] + a_2 y_p = s[u_p] + b_1 u_p$$

where

$$b_1 = -1 \;;\; a_1 = 20 + 12\sin \mu t \;;\; a_2 = 6\cos \mu t$$

We also suppose that we know a priori that the plant parameters are of the form

$$b_1 = c_{1*} \;;\; a_1 = c_{2*} + c_{4*}\sin \mu t \;;\; a_2 = c_{3*} + c_{5*}\cos \mu t$$

where μ is known (to be specified later), the plant PDO is a priori known to be monic and of degree one and c_{i*} are some unknown constants whose range, say $c_{i,min} \leq c_{i*} \leq c_{i,max}$, is known a priori. For this plant, an adaptive law to estimate the unknown plant parameters is presented in Example 6.15 while various controller designs are discussed in the examples of Chapter 5. In the following, and according to our earlier analysis in this chapter, we combine the estimator and control laws to obtain several adaptive control schemes.

8.16 Example: (*TV APPC Design.*) Before we present our first adaptive control design, let us consider the case where we have only incomplete knowledge of the plant parameters $\hat{\theta}_{i*}$. More precisely, suppose that we design a TV PPC to regulate the output of the plant to zero, but due to the presence of parametric uncertainty the parameters used in the APPC design are

$$b_1 = -0.9 \;;\; a_1 = 23 + 9.6\sin \mu t \;;\; a_2 = 1.2 + 4.8\cos \mu t$$

A simulation of the closed-loop output with $\mu = 5$ and zero initial conditions except $y_p(0) = 10$, is shown in Fig. 8.3. As indicated by this simulation the parametric uncertainty is sufficiently large to cause closed-loop instability.

In order to compensate for such parametric uncertainty, let us employ estimator of Example 6.15 to estimate the partially unknown plant parameters. These are given as functions of the $\hat{\theta}_i$'s by the following equations

$$\begin{aligned} \hat{b}_1 &= \hat{\theta}_1 \\ \hat{a}_1 &= -\hat{\theta}_2 - \hat{\theta}_4 \sin \mu t + 5 \\ \hat{a}_2 &= -\hat{\theta}_3 - \hat{\theta}_5 \cos \mu t + 6 \end{aligned} \quad (8.10)$$

where, in the design of the estimator projection, we assume that $\hat{\theta}_{i*}$ lie in the following intervals:

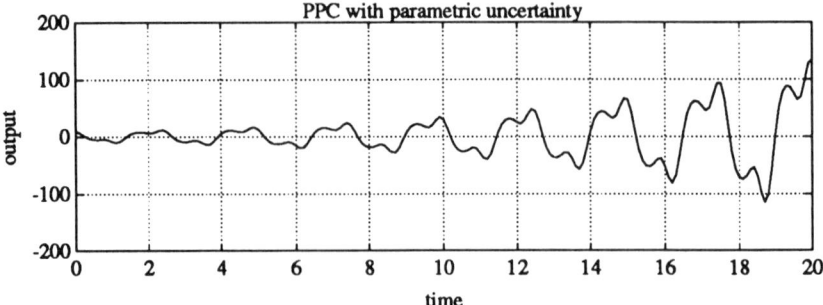

Figure 8.3: Closed-loop response of a TV PPC during regulation, in the presence of parametric uncertainty.

$$\hat{\theta}_* \in \left\{ \begin{array}{cc} [-0.85 & , -1.1] \\ [-25 & , -13] \\ [4.5 & , 8] \\ [-16 & , -8] \\ [-8 & , -4] \end{array} \right\}$$

(see Example 6.15 for additional details).

The next step in the TV APPC design is to calculate the controller parameters from the plant parameter estimates. That is, as in the examples of Chapter 5 and with the same PPC objective, we first solve the Diophantine equation

$$[s^2 + s\hat{a}_1 + \hat{a}_2][s + \psi_1] + [s + \hat{b}_1][s\psi_2 + \psi_3] = s^3 + 6s^2 + 11s + 6 \quad (8.11)$$

for the ψ_i's, considering the $\hat{\theta}_i$'s as frozen (i.e., $\frac{d}{dt}\hat{\theta}_i = 0$). Equation (8.11) can be expressed as a system of linear algebraic equations, as follows

$$\begin{pmatrix} 1 & 1 & 0 \\ \hat{a}_1 & \hat{b}_1 & 1 \\ \hat{a}_2 & 0 & \hat{b}_1 \end{pmatrix} \begin{pmatrix} \psi_1 \\ \psi_2 \\ \psi_3 \end{pmatrix} = \begin{pmatrix} 6 - \hat{a}_1 \\ 11 - \hat{a}_2 + \dot{\hat{a}}_1 \\ 6 + \dot{\hat{a}}_2 \end{pmatrix} \quad (8.12)$$

Under Assumption 8.1[8] $\hat{b}_1^2 + \hat{a}_2 - \hat{b}_1\hat{a}_1 \neq 0$ and therefore we can express the solution for the the ψ_i's as

$$\psi_1 = \frac{A_1\hat{b}_1^2 + A_3 - A_2\hat{b}_1}{\hat{b}_1^2 + \hat{a}_2 - \hat{b}_1\hat{a}_1}$$

---

[8] Its validity for this example can be verified after some straightforward calculations and the properties of estimators with projection.

8.4. EXAMPLES

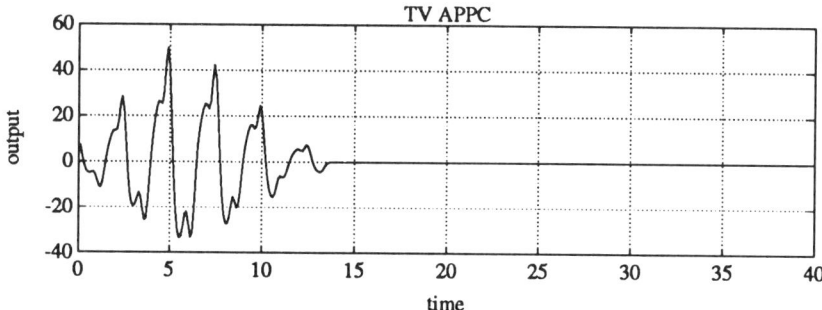

Figure 8.4: Closed-loop response of a TV APPC during regulation. Fast, fully structured parameter variations ($\mu = 5$).

$$\psi_2 = \frac{A_2\hat{b}_1 + A_1\hat{a}_2 - A_3 - A_1\hat{b}_1\hat{a}_1}{\hat{b}_1^2 + \hat{a}_2 - \hat{b}_1\hat{a}_1} \quad (8.13)$$

$$\psi_3 = \frac{A_3\hat{b}_1 + A_2\hat{a}_2 - A_1\hat{a}_2\hat{b}_1 - A_3\hat{a}_1}{\hat{b}_1^2 + \hat{a}_2 - \hat{b}_1\hat{a}_1}$$

where $A_1 = 6 - \hat{a}_1$; $A_2 = 11 - \hat{a}_2 + \dot{\hat{a}}_1$; $A_3 = 6 + \dot{\hat{a}}_2$. Next we calculate the controller parameters, as functions of the $\hat{\theta}_i$'s, from

$$p_1 = \psi_3 - 2\psi_2 + \dot{\psi}_2 \; ; \; p_2 = 2 - \psi_1 \; ; \; p_3 = \psi_2 \quad (8.14)$$

where, again, all calculations are performed with frozen $\hat{\theta}_i$'s. Finally, the parameters p_i are used in the control law

$$u_p = p_1[s+2]^{-1}[u_1] + p_3 u_1 \; ; \; u_1 = p_2[s+2]^{-1}[u_1] + (r - y_p) \quad (8.15)$$

to generate the control input u_p. The regulation response of the above TV APPC is shown in Fig. 8.4 for $\mu = 5$, where in the adaptive law we used

$$\gamma = 100 \; ; \; Q = I \; ; \; \delta_0 = 0.9 \quad (8.16)$$

while the initial conditions for the vector $\hat{\theta}$ correspond to the same plant parameters as in the previous non-adaptive case:

$$\hat{\theta}(0) = [-0.9, -18, 4.8, -9.6, -4.8]^T \quad (8.17)$$

As shown in Fig. 8.4, exact regulation is achieved with the TV APPC even though the plant parameters are varying fast with time. Notice that the initial growth of the output is due to the fact that $\hat{\theta}(0)$ yields a destabilizing compensator.

8.17 Example: (*TV IMP/APPC Design.*) Let us now suppose that in addition to the PPC objective of the previous example, we would like to design a TV IMP/APPC scheme to track constant refernce inputs. The only essential difference of the TV IMP/APPC design from the APPC one is that we first calculate ρ, x by solving

$$s^2 + s\hat{a}_1 + \hat{a}_2 + s\rho + \hat{b}_1\rho = s^2 + xs \qquad (8.18)$$

for ρ, x, where, as usual, the $\hat{\theta}_i$'s are frozen. Equation (8.18) can be written as

$$\begin{aligned} \dot{\rho} &= -\hat{\theta}_1\rho + (\mu\hat{\theta}_4 + \hat{\theta}_5)\cos\mu t + (\hat{\theta}_3 - 6) \\ x &= -\hat{\theta}_2 - \hat{\theta}_4\sin\mu t + 5 + \rho \end{aligned} \qquad (8.19)$$

Under Assumption 8.12 $\hat{\theta}_1 \neq 0$ and we obtain the following closed-form solution for ρ

$$\rho = \frac{\mu^2\hat{\theta}_4 + \mu\hat{\theta}_5}{\mu^2 + \hat{\theta}_1^2}\sin\mu t + \frac{\mu\hat{\theta}_1\hat{\theta}_4 + \hat{\theta}_1\hat{\theta}_5}{\mu^2 + \hat{\theta}_1^2}\cos\mu t + \frac{\hat{\theta}_3 - 6}{\hat{\theta}_1} \qquad (8.20)$$

Note that for this example, if it is known a priori that a_2 has no constant component (i.e., $\hat{\theta}_3 \equiv 6$), the above solution for ρ is well defined and bounded for any bounded value of the estimated parameters, something that simplifies the implementation of the TV IMP/APPC scheme (compare with the TV IMP/PPC where b can be any constant).

Further, the parameters p_i of the controller are calculated in a similar way as in the APPC case and the control input is generated by

$$\begin{aligned} u_p &= \bar{u}_p - \rho y_p \\ \bar{u}_p &= p_1[s+2]^{-1}[u_1] + p_3 u_1 \;\; ; \;\; u_1 = p_2[s+2]^{-1}[u_1] + (r - y_p) \end{aligned} \qquad (8.21)$$

The response of the closed-loop plant with the TV IMP/APPC scheme for a square wave reference input $(10 \leftrightarrow 0)$ is shown in Fig. 8.5. In the simulations we used the same initial conditions for the plant parameter estimates as in the TV APPC example (see eqn. (8.17)) and zero initial conditions for the rest of the state-variables. As in the previous example, notice that the transient effects in the beginning of the adaptation period are highly pronounced. This is due to the large parameter error at $t = 0$ which yields an initially unstable TV PPC closed-loop. However, as $t \to \infty$ the tracking error with a constant reference input converges to zero. Also note that the large transient errors observed after the reference input undergoes a transition $(0 \to 10$ or $10 \to 0)$ are essentially due to the underlying TV IMP/PPC law (compare with Example 5.21). Such errors cannot be reduced by simply introducing an adaptive law but it requires a more careful design and shaping of the closed-loop sensitivity operators. ▽▽

8.4. EXAMPLES

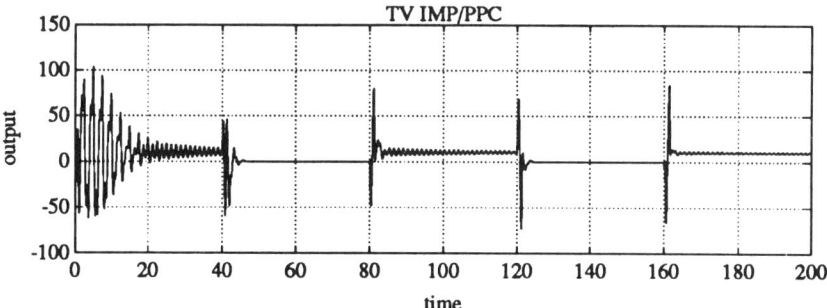

Figure 8.5: Exact asymptotic tracking of step reference inputs with the TV IMP/APPC. Fast, fully structured parameter variations.

8.18 Example: (*PW APPC Design for a Slowly Time-Varying Plant.*) In the previous examples we discussed the case of LTV plants with fully structured parameter variations where, according to the general theoretical results of this chapter, the signal boundedness and the convergence of the tracking error to zero can be established independent of the speed of the parameter variations. On the other hand, when the plant parameters vary in a partially or completely unknown fashion, the boundedness of the closed-loop signals depends heavily on the speed of the unstructured part of the unknown parameters. To illustrate the behavior of the adaptive closed-loop plant under such conditions, let us consider the same plant as in the previous examples, except that we have no knowledge of the form or structure of the time varying parameters. That is, we consider the plant

$$s^2[y_p] + s[a_1 y_p] + a_2 y_p = s[u_p] + b_1 u_p$$

where

$$b_1 = -1 \; ; \; a_1 = 20 + 12\sin \mu t \; ; \; a_2 = 6\cos \mu t$$

and the only a priori available information regarding b_1, a_1, a_2 is their range

$$b_1 \in [-1.2, -0.85] \; , \; a_1 \in [8, 32] \; , \; a_2 \in [-6, 6]$$

Assuming that the variation of the unknown parameters is sufficiently slow, we may employ the estimator of Example 6.15 with $\Pi = I$ to identify the unknown I/O operator of the LTV plant. Thus, the identified plant (excluding the identification error) has the I/O description

$$s^2[y_p] + s[\hat{a}_1 y_p] + \hat{a}_2 y_p = s[u_p] + \hat{b}_1 u_p$$

where,[9] $\hat{b}_1 = \theta_1$, $\hat{a}_1 = -\theta_2 + 5$, $\hat{a}_2 = -\theta_3 + 6$. Using the a priori knowledge of the range of the plant parameters, the estimator projection is selected so that

$$\theta \in \left\{ \begin{array}{ccc} [-0.85 & , & -1.1] \\ [-27 & , & -3] \\ [0 & , & 12] \end{array} \right\}$$

Based on the identified plant, the next step in the (PW) APPC design is to calculate the controller parameters by first solving the Diophantine equation

$$[s^2 + s\hat{a}_1 + \hat{a}_2][s + \psi_1] + [s + \hat{b}_1][s\psi_2 + \psi_3] = s^3 + 6s^2 + 11s + 6 \quad (8.22)$$

with θ_i being frozen, for the ψ_i. As in Example 8.16, equation (8.22) can be expressed as a system of linear algebraic equations, as follows

$$\begin{pmatrix} 1 & 1 & 0 \\ \hat{a}_1 & \hat{b}_1 & 1 \\ \hat{a}_2 & 0 & \hat{b}_1 \end{pmatrix} \begin{pmatrix} \psi_1 \\ \psi_2 \\ \psi_3 \end{pmatrix} = \begin{pmatrix} 6 - \hat{a}_1 \\ 11 - \hat{a}_2 \\ 6 \end{pmatrix}$$

Under Assumption 8.1, $\hat{b}_1^2 + \hat{a}_2 - \hat{b}_1 \hat{a}_1 \neq 0$ and the above system of equations has a well defined the solution for the the ψ_i's. (The solution is the same as in Example 8.16 except that all derivatives are set equal to zero.) The controller parameters are now obtained as functions of the θ_i's, as

$$p_1 = \psi_3 - 2\psi_2; \; p_2 = 2 - \psi_1 \; ; \; p_3 = \psi_2$$

Finally, the parameters p_i are used in the control law

$$u_p = p_1[s+2]^{-1}[u_1] + p_3 u_1 \; ; \; u_1 = p_2[s+2]^{-1}[u_1] + (r - y_p)$$

to generate the control input u_p. Note that, since the plant parameter variations are completely unstructured, the resulting APPC is naturally a PW one.

The above PW APPC, is used to regulate the LTV plant where $\mu = 0.05$. As previously discussed, for a given value of μ, the selection of the adaptation gain γ plays an important role in determining the characteristics of the closed-loop response. Small adaptation gains may be inadequate to track the plant parameter variations while large adaptation gains may cause a deterioration of the closed-loop response, especially in the presence of external disturbances. In our simulations, we use two adaptive gains to account for the different size of parametric uncertainty in the components of θ,[10] that is, θ_1 is updated

---

[9] Note that since $\Pi = I$, $\hat{\theta} = \theta$.
[10] A similar effect can be obtained by introducing constant bias and gain in the estimator filters, as in the MRAC examples.

8.4. EXAMPLES

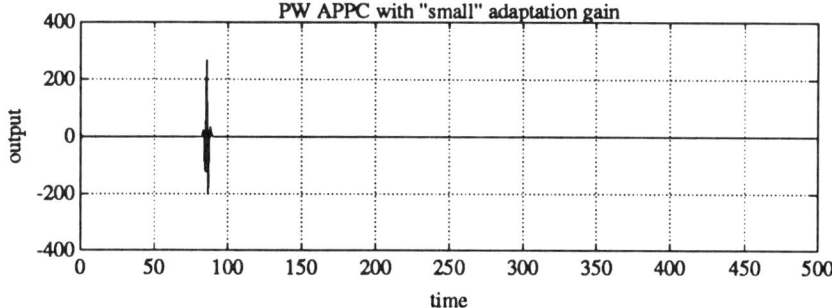

Figure 8.6: Closed-loop response with a PW APPC during regulation.
Fast, fully structured parameter variations, $\mu = 5$.
Large bursting due to a 'small' adaptation gain.

with an adaptation gain γ_1 while θ_2, θ_3 are updated with a gain γ_2. In all cases we use zero initial conditions except for

$$\theta(0) = [-0.9, -18, 4]^T \quad ; \quad y_p(0) = 10$$

In our first simulation, shown in Fig. 8.6 we use

$$\gamma_1 = 1 \quad ; \quad \gamma_2 = 500$$

In this simulation the PW APPC enjoys some initial success in driving the output to zero. Due to the parameter variations, however, the closed loop becomes momentarily unstable (at approximately $t = 82$) creating a large burst.

Increasing the value of γ_2 to 1000, the size of the burst is significantly decreased, as shown in Fig. 8.7. It is interesting to mention that, for this example, only one burst is observed. This is due to the convergence of the estimated parameters to $[-.883, -5.247, 5.86]$, a value that corresponds to a stabilizing compensator.[11] Such a behavior, however, is particular to this example and should by no means be expected in general. That is, the plant may not be stabilizable by single LTI compensator and burst phenomena may not be confined to a finite interval. Finally, it should be mentioned that the terminology 'small' and 'large' adaptation gains is used in a rather loose and qualitative manner. Their distinction essentially depends on the magnitude of the observed derivatives of the parameter estimates, something that is case sensitive and, at this point, quantitatively unclear. ▽▽

---

[11] It can be easily shown that the resulting PPC stabilizes the LTV plant pointwise; consequently the closed-loop is stable for μ sufficiently small.

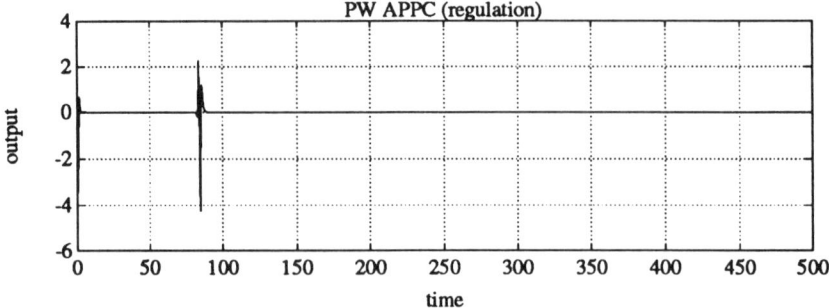

Figure 8.7: Closed-loop response with a PW APPC during regulation.
Unstructured parameter variations, $\mu = 0.05$.
Smaller burst size due to a larger adaptation gain.

8.19 Example: (*PW IMP/APPC Design for a Slowly Time-Varying Plant.*) Let us now suppose that in addition to the PPC objective of the previous example, we would like to design a (PW) IMP/APPC scheme to track constant refernce inputs. One possible design, having the advantage that it does not increase the order of the compensator, is to follow the same steps as in Example 8.17 where, now, θ is considered as frozen. That is we first calculate ρ, x by solving

$$s^2 + s\hat{a}_1 + \hat{a}_2 + s\rho + \hat{b}_1\rho = s^2 + xs$$

for ρ, x. With θ being frozen, the above equation has a simple pointwise solution

$$\rho = (\theta_3 - 6)/\theta_1 \quad ; \quad x = -\theta_2 + 5 + \rho$$

while, by the properties of the projection, Assumption 8.12 is satisfied ($\theta_1 \neq 0$) and both ρ and x are well defined.

Further, the parameters p_i of the controller are calculated in a similar way as in the PW APPC case and the control input is generated by

$$\begin{aligned} u_p &= \bar{u}_p - \rho y_p \\ \bar{u}_p &= p_1[s+2]^{-1}[u_1] + p_3 u_1 \; ; \; u_1 = p_2[s+2]^{-1}[u_1] + (r - y_p) \end{aligned}$$

The response of the closed-loop plant with the PW IMP/APPC scheme for a square wave reference input (10-0) is shown in Figs. 8.8, 8.9 and 8.10 for different adaptation gains[12] and the same initial conditions as in the PW APPC example.

---

[12] $(\gamma_1, \gamma_2) = (1, 500), (1, 1000), (10, 10000)$ respectively.

8.4. EXAMPLES

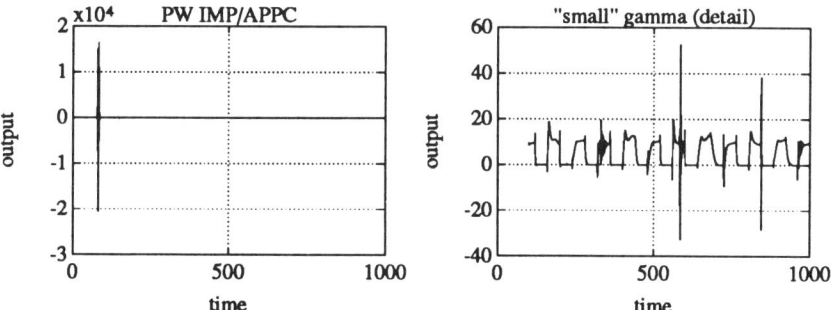

Figure 8.8: Approximate tracking of step reference inputs with the PW IMP/APPC. Unstructured parameter variations, $\mu = 0.05$. Large bursting due to a 'small' adaptation gain.

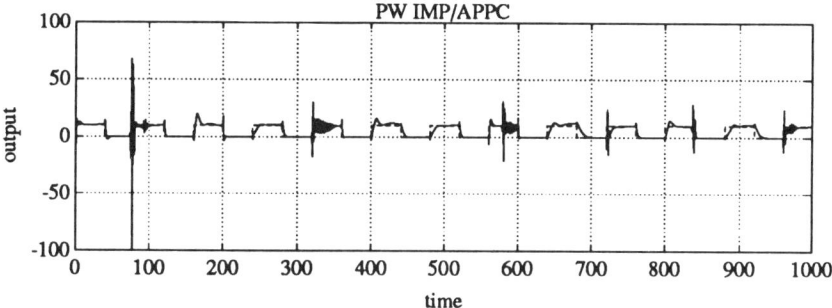

Figure 8.9: Approximate tracking of step reference inputs with the PW IMP/APPC. Unstructured parameter variations, $\mu = 0.05$. Smaller burst size due to a larger adaptation gain.

Since in this case the unknown parameter variations are unstructured, the tracking error no longer converges to zero but to a residual set where it maintains a small in the mean-square sense value. Nevertheless, we observe a decrease of the burst size as the adaptation gain increases. Of course, as mentioned in the previous example, this observation does not imply that an arbitrary increase of the adaptation gain is always beneficial.

Finally, as noted in Example 8.17, the transient errors observed after the reference input undergoes a transition (0-10 or 10-0) depend critically on the underlying control law. Their reduction cannot be achieved by simply 'tuning' the adaptive law but it requires a careful design and shaping of the closed-loop sensitivity operators. ▽▽

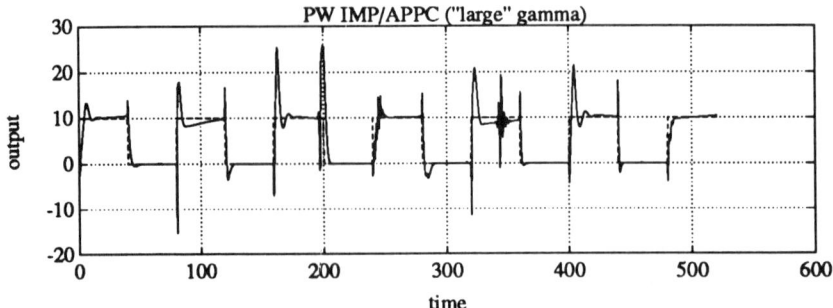

Figure 8.10: Approximate tracking of step reference inputs with the PW IMP/APPC. Unstructured parameter variations, $\mu = 0.05$. Furher reduction of the burst size.

APPENDIX VIII

Proof of Theorem 8.3:

First and as in the proof of Theorem 7.13, observe that under the assumptions of the theorem, the vectorfield of the ODE describing the evolution of the adaptive closed-loop system is locally Lipschitz and therefore there exists $T > 0$ such that the solution exists and is unique on the interval $[t_0, t_0 + T]$. Hence, the assumptions of Corollary 6.11 are satisfied, implying that the parameter estimates are UB, at least on $[t_0, t_0 + T]$.

Further, for $t \in [t_0, t_0 + T]$, define the fictitious signal m_f by

$$\dot{m}_f = -2\delta m_f + |Q_1 U_1|^2 + q_e; \quad m_f(t_0) = m_2(t_0)$$

where $\delta \in (0, \min[\delta_0, \alpha])$, $U_1 = [U^\top, u_1]^\top$ and $Q_1 = diag[Q, q_1]$, $q_1 > 0$. Note that, by construction, $m_f \geq m_2$ on any interval U is well defined.

Next, consider the closed-loop description given by (8.4) and (8.6). Treating r and e_1 as external inputs, the closed-loop is, by design, ES for every frozen $\hat{\theta}$ with rate at most $-a$, the latter depending on the stability margins of A_*^{-1} and D_a^{-1}. Furthermore, the closed-loop system paremeters are Lipschitz continuous functions of $\hat{\theta}$. Hence, by Lemma 2.44, for any $\delta_c \in (0, a)$ there exists $\beta(\delta_c)$ such that if in a subinterval of $[t_0, t_0 + T]$

$$\int_{t_I}^{t_I + T_I} |\dot{\hat{\theta}}(t)|^2 \, dt \leq C_0 + \beta T_I, \quad \forall t_I \geq t_0, \; T_I > 0$$

the closed-loop STM is (uniformly) exponentially decaying with rate at most $-\delta_c$. Invoking Corollary 6.11, $\beta \leq O\left(\gamma^2 \mu, \gamma K_\phi \mu\right)$ and therefore the previous

APPENDIX VIII.

statement is satisfied provided that $\mu < \mu_o(\delta_c, \gamma, K_\phi)$, the latter being positive and well defined for any fixed γ, K_ϕ.

Further, with the external inputs r, e_1, the signal vector U_1 satisfies

$$U_1 = R + S_e[e_1] + E$$

where R is a UB term due to the reference input, E is an exponentially decaying term incorporating the effects of initial conditions and S_e are the (TV) sensitivity operators from e_1 to u_p, y_p, u_1. Note that for $\mu < \mu_o$ these operators have finite $L_2(\delta)$-gains where $\delta < \delta_c$. This is due to the fact that $\hat{\theta}$ belongs to a compact set while the closed-loop is ES by design, uniformly in t_0 and $\hat{\theta}$. Hence, taking norms of the truncated signals in $[t_0, t_0 + T]$,

$$\|(QU_1)_t\|_{2,\delta} \leq \|(R)_t\|_{2,\delta} + H_0\|(e_1)_t\|_{2,\delta} + \|(E)_t\|_{2,\delta} \tag{8.23}$$

where H_0 is the $L_2(\delta)$ gain of QS_e. Although H_0 depends on the trajectory of $\hat{\theta}$, an upper bound can be given in terms of K_ϕ which is finite for finite K_ϕ (this is easily verified using the results of Chapter 2). We may now invoke Lemma 7.12 yielding that a sufficient condition for m_f to be UB in $[t_0, t_0 + T]$ is

$$H_0 \frac{RMS[e_1]}{\sqrt{2\delta}} < 1$$

where, as in Chapter 7, $RMS[x]$ denotes the root-mean-square value of the signal x. Since m_f bounds U_1, the same applies to all the closed loop signals and the solution of the closed-loop system ODE can be extended to eventually cover the whole interval $[t_0, \infty)$ (see also the proof of Theorem 7.13).

Hence, using the results of Corollary 6.11 to evaluate the $RMS[e_1]$ term, boundedness is guaranteed provided that

$$\mu < \mu_o(\delta_c, \gamma, K_\phi)$$
$$H_0^2 O\left(\mu, \frac{K_\phi \mu}{\gamma}, \gamma^2 \mu, \gamma K_\phi \mu\right) < 2\delta \tag{8.24}$$
$$\delta < \delta_c$$

Clearly, for any fixed K_ϕ, γ there exists $\mu_0 > 0$ such that the above inequalities are simultaneously satisfied for $\mu \in [0, \mu_0)$ which completes the proof of the theorem.

Notice that in the stability/boundedness condition (8.24), δ_c is a free parameter and can be chosen so as to yield the maximum possible μ_0. Such an issue is beyond the scope of our study and is not explored here. We should mention, however, that H_0, also depending on δ is the uniform gain of the sensitivity operators with respect to $\hat{\theta}$. To rephrase this statement, consider the

(ES) closed-loop of the LTV plant and the associated TV PPC, parametrized by $\hat{\theta}$. Then, loosely speaking, the closed-loop boundedness depends on the worst-case sensitivity operator with respect to $\hat{\theta}$ ranging in the parametric uncertainty set. This type of result has also been obtained in the LTI case [KSK.89], though here there are some differences because of the time variation of $\hat{\theta}$. Nevertheless, the interesting observation is that the properties of the crucial sensitivity operators depend heavily on the selection of the estimator and controller filters $(D_a(s), D(s))$. These filters, though irrelevant in meeting the TV PPC objective in the ideal case, are important when robustness issues arise, either because of the unstructured variation of the plant parameters or because of modeling errors.

Further, observe that for a given size of the parametric uncertainty set (K_ϕ) the adaptation gain affects the size of μ_0 in two conflicting ways. When γ increases, the term $K_\phi \mu/\gamma$ decreases and μ_0 increases. This term can be interpreted as an effective parametric uncertainty which is normalized by the ratio of the speeds of the actual parameters and their estimates. On the other hand, larger values of γ increase the term $\gamma K_\phi \mu$ and cause μ_0 to decrease. This term is a bound on $\dot{\hat{\theta}}$ which acts as a perturbation since it is not an estimate of $\dot{\hat{\theta}}_*$ and, consequently, may destabilize the closed loop.

Finally, when the parametric uncertainty is sufficiently small, boundedness is preserved independent of the speed of the parameter variations. To show this, consider the expression for e_1 derived in Chapter 6:

$$e_1 = G_a(s)[u_p \phi_1] + G_a(s)[y_p \phi_2]$$

Substituting this expression in (8.23) we obtain a small-gain stability condition of the form $H_0 O(K_\phi) < 1$. □□

Proof of Theorem 8.4:

The proof follows the same steps as that of Theorem 8.3, except that now the expressions of Corollary 6.14 are applicable. Consequently, only the major differences are outlined here.

The first one is that the quantity β is now

$$O(\gamma^2 \mu^2, \gamma \mu^2/\sigma, \gamma \sigma K_{\theta_*}^2, \sigma^2 K_{\theta_*}^2, \mu^2, \gamma \sigma \mu^2)$$

Hence, the closed loop, without the perturbation e_1 is exponentially stable with rate $-\delta_c$ for σ sufficiently small, say $\sigma < \sigma_o$ and $\mu < \mu_o(\sigma)$.[13]

---

[13] The dependence of the various bounds on the rest of the quantities can easily be derived but is suppressed here for simplicity.

APPENDIX VIII. 259

The second difference is that, with the expressions of Corollary 6.14, the final condition for boundedness becomes

$$H_0^2 O\left(\mu^2, \frac{\mu^2}{\gamma\sigma}, \frac{K_{\theta*}^2 \sigma}{\gamma}, \gamma^2\mu^2, \gamma\sigma K_{\theta*}^2, \frac{\gamma\mu^2}{\sigma}, \sigma^2 K_{\theta*}^2, \gamma\sigma\mu^2\right) < 2\delta$$

$$\mu < \mu_o(\sigma)$$
$$\delta < \delta_c$$

As in Theorem 8.3, H_0 depends on the trajectory of $\hat{\theta}$ but an upper bound can be obtained in terms of the bound of the parameter estimates. Observing that for sufficiently small μ the latter approaches $K_{\theta*} + \|\hat{\theta}_*\|_\infty$, a constant independent of σ or γ, the proof of the theorem follows.

Notice that the last argument regarding H_0 is precisely the point where the analysis of slow-in-the-mean parameter variations breaks down. Using the σ-modification, the parameter estimates can grow as large as $O(\gamma/\sigma)$ (see Corollary 6.14) and our stability condition may fail to hold for any choice of μ, γ, σ unless $K_{\theta*}$ is small. It is not clear, however, whether this is a fundamental drawback of the σ-modification or merely a technical problem due to our analysis.

Finally when $K_{\theta*}$ is sufficiently small, $\hat{\phi}$ converges exponentially fast to a set $O(K_{\theta*}/\sigma, K_{\theta*})$ implying that the operator $U \mapsto e_1$ has a $O(K_{\theta*}/\sigma, K_{\theta*})$ gain and therefore, as in Theorem 8.3, boundedness is guaranteed for arbitrarily fast variations of the plant parameters. □□

Proof of Theorem 8.5:

Using Lemma 2.64, consider the vector $\bar{\theta}$ which is a smooth approximation of $\hat{\theta}$ and differentiable as many times as necessary (the number of required derivatives is certainly finite and can be determined from the differentiations involved in the calculation of the controller parameters). Then for $\tilde{\theta} = \hat{\theta} - \bar{\theta}$ we have that $\|\tilde{\theta}\|_\infty \leq O\left(\|\dot{\hat{\theta}}/\alpha\|\right)$ where α is a constant to be chosen.

Next, we use the smooth parameter vector $\bar{\theta}$ to derive smooth approximations of the controller parameters p_i and rewrite the closed-loop equations in terms of PDOs with smooth coefficients. Let $\bar{D}_p(s,t)$, $\bar{N}_p(s,t)$ be the PDOs corresponding to $\bar{\theta}$, as in Lemma 3.10. Thus, the plant I/O description becomes

$$\bar{D}_p(s,t)y_p = \bar{N}_p(s,t)u_p + D_a(s)(e_1 + e_2) \qquad (8.25)$$

where e_1 is the identification error $e_2 = G_a(s)[u_p \Pi_1 \tilde{\phi}_1] + G_a(s)[y_p \Pi_2 \tilde{\phi}_2]$. From the definition of the TV APPC algorithm we have that the controller parameters p_i can be expressed as functions of $\hat{\theta}$ as

$$p_i = Q_a(\hat{\theta})S_\star^{-1}(\hat{\theta})\hat{A}(\hat{\theta}) \tag{8.26}$$

where $Q_a(\cdot)$ is a matrix that relates the coefficients of $N_1(s,t)$, $N_2(s,t)$ with p_i; $S_\star(\cdot)$ is a left Sylvester-like matrix of $\hat{D}_p(s,t)$, $\hat{N}_p(s,t)$; $\hat{A}(\hat{\theta})$ is a vector with the coefficients of $A_\star(s,t) - \hat{D}_p(s,t)s^{n-1}$. Note that no derivatives of $\hat{\theta}$ appear in (8.26) since all operations are performed with $\hat{\theta}$ being frozen. Thus, we define the parameters $\bar{p}_i$ as the smooth approximates of p_i by:

$$\bar{p}_i = Q_a(\bar{\theta})S_\star^{-1}(\bar{\theta})\hat{A}(\bar{\theta}) \tag{8.27}$$

Since $Q_a(\cdot)$, $S_\star(\cdot)$, $\hat{A}(\cdot)$ are uniformly continuous functions in their argument, we can select α such that $|\hat{\theta} - \bar{\theta}|$ is sufficiently small for $|\det[S_\star(\hat{\theta})]| \geq c > 0$ to imply that $|\det[S_\star(\bar{\theta})]| \geq \bar{c}$ for some constant $\bar{c} > 0$. It now follows that the controller parameters $\bar{p}_i$ are well defined by (8.27) and the error $\tilde{p}_i = p_i - \bar{p}_i$ satisfies

$$\tilde{p}_i \leq O(\|\bar{\theta} - \hat{\theta}\|) \tag{8.28}$$

Further, since the controller parameters p_i are calculated from the PDOs $N_1(s,t)$, $N_2(s,t)$ which satisfy

$$\hat{D}_p(s,t) \star N_2(s,t) + \hat{N}_p(s,t) \star N_1(s,t) = A_\star(s,t)$$

with $\hat{\theta}$ being frozen, it follows from the rules of differentiation, the definitions of $\bar{\theta}$ and $\bar{p}_i$ and Lemma 2.64 that

$$\bar{D}_p(s,t)\bar{N}_2(s,t) + \bar{N}_p(s,t)\bar{N}_1(s,t) = A_\star(s,t) + \Delta(s,t)$$

where $\Delta(s,t)$ is a PDO of degree $\leq 2n-2$ and with coefficients of $O(\alpha^i \dot{\hat{\theta}})$, and i ranges between one and the maximum number of differentiations required. Following Lemma 5.2, $[A_\star(s,t)+\Delta(s,t)]^{-1}$ is the PIO involved in all sensitivity operators. Finally, to complete the closed-loop I/O description, we rewrite the controller in terms of the smooth parameters $\bar{p}_i$ and introduce the signals x_1, x_2, depending on the difference $p_i - \bar{p}_i$, as perturbations at the respective nodes, which completes the proof. □□

Proof of Theorem 8.13:

The proof of the theorem follows along the lines of Theorem 8.5 whereby we approximate $\hat{\theta}$ by a smooth vector $\bar{\theta}$ and reparametrize the identified LTV plant in terms of $\bar{\theta}$ and an additional perturbation e_2 due to the difference $\hat{\theta} - \bar{\theta}$.

APPENDIX VIII. 261

Next, the smooth parameters $\bar{\theta}$ are used to derive smooth approximations of the controller parameters, for which our ususal PDO operations can be performed. From the controller design equations (8.9) and under Assumption 8.12, the coefficients of the PDO's $\hat{N}_1(s,t)$, $\hat{N}_2(s,t)$ and $P(s,t)$ are Lipschitz continuous functions of $\hat{\theta}$, say $f_1(\hat{\theta})$, $f_2(\hat{\theta})$ and $f_3(\hat{\theta})$ respectively. Define the smooth-coefficient approximations of these PDO's as the PDO's with coefficients $f_i(\bar{\theta})$, i.e.,

$$\bar{N}_1(s,t) \leftrightarrow f_1(\bar{\theta}) \; ; \; \bar{N}_2(s,t) \leftrightarrow f_2(\bar{\theta}) \; ; \; \bar{P}(s,t) \leftrightarrow f_3(\bar{\theta})$$

Also define the PDO's

$$\bar{\bar{N}}_1(s,t) = \bar{N}_1(s,t) - \bar{P}(s,t)\bar{N}_2(s,t) \; ; \; \bar{\bar{N}}_2(s,t) = Q\bar{N}_2(s,t)$$

and parametrize the state-space realization of the control law

$$u_p = \bar{\bar{N}}_1(s,t)\bar{\bar{N}}_2^{-1}(s,t)[r - y_p] - \bar{P}(s,t)Q^{-1}(s)[y_p]$$

in terms of the parameter vectors $\bar{p}_i$, $i = 1, \ldots 5$, corresponding to p_i. From Lemma 2.64 and the Lipschitz continuity of f_i it follows that $p_i - \bar{p}_i$ is $O(\|\hat{\theta}\|)$ and that the following equations hold:

$$\bar{D}_p(s,t)Q(s)\bar{\bar{N}}_2(s,t) + \bar{N}_p(s,t)\bar{\bar{N}}_1(s,t) = A_*(s,t) + \Delta_1(s,t)$$

$$\bar{X}(s,t)\Lambda(s) - \bar{N}_p(s,t)\bar{P}(s,t) = \bar{D}_p(s,t)Q(s) + \Delta_2(s,t)$$

where $\bar{X}(s,t)$ is the corresponding smooth-coefficient approximation of $X(s,t)$ and $\Delta_i(s,t)$ are PDO's with coefficients $O(\|\dot{\hat{\theta}}\|)$.

After some straightforward calculations the tracking error is expressed as

$$e = \bar{\bar{N}}_2(s,t)[A_*(s,t) + \Delta_1(s,t)]^{-1}\Delta_2(s,t)Q^{-1}(s)[r] + S_e[x_1, x_2, e_1, e_2]^\mathsf{T} + \varepsilon$$

where x_i are the perturbations due to the controller reparametrization and S_e are the corresponding sensitivity operators (see Theorem 8.5). Since the various operators are ES for sufficiently small perturbations (in the mean-square sense), using the expressions of Corollary 6.11, the result follows. □□

Bibliography

[And.77] B.D.O. Anderson, "Exponential Stability of Linear Equations Arising in Adaptive Identification," *IEEE Trans. Autom. Contr.*, AC-22, Feb. 1977.

[A.J.83] B.D.O. Anderson and R. M. Johnstone, "Adaptive Systems and Time Varying Plants," *Int. J. of Control*, 37, No. 2, 367, 1983.

[A++.86] B.D.O. Anderson, R.R. Bitmead, C.R. Johnson, P.V. Kokotovic, R.L. Kosut, I.M.Y. Mareels, L. Praly and B.D. Riedle, *Stability of Adaptive Systems: Passivity and Averaging Analysis*, MIT Press, Cambridge, MA, 1986.

[A.N.89] A.M. Annaswamy and K.S. Narendra, "Adaptive Control of Simple Time-Varying Systems," *Proc. 28th CDC*, 1014-1018, Tampa FL, Dec.1989.

[AMS.58] J.A. Aseltine, A.R. Mancini and C.W. Sarture, "A Survey of Adaptive Control Systems," *IRE Trans. Autom. Contr.*, 3, 102-108, Dec. 1958.

[A.W.88] K.J. Astrom and B. Wittenmark, *Adaptive Control*. Addison-Wesley, Reading, MA, 1988.

[Bel.53] R. Bellman, *Stability Theory of Differential Equations*. McGraw-Hill, New York, 1953.

[Bel.61] R. Bellman, *Adaptive Control Processes — A Guided Tour*. Princeton University Press, Princeton, NJ, 1961.

[Bng.77] G. Bengtsson "Output Regulation and Internal Models– a Frequency Domain Approach," *Automatica*, Vol. 13, pp. 333–345, July 1977.

[BGW.90]   R.R. Bitmead, M. Gevers and V. Wertz, *Adaptive Optimal Control*. Prentice Hall, 1990.

[Bro.70]   R.W. Brockett, *Finite Dimensional Linear Systems*. John Wiley, New York, 1970.

[B.H.69]   A.E. Bryson and Y.C. Ho, *Applied Optimal Control*. Blaisdell Publ. Co., Waltham, MA, 1969.

[C.C.84]   H.F. Chen and P.E. Caines, "On the Adaptive Control of Stochastic Systems with Random Parameters," in *Proc. 23rd CDC*, 33, 1984.

[C.L.55]   E.A. Coddington and N. Levinson, *Theory of Ordinary Differential Equations*. McGraw-Hill, New York, 1955.

[Des.69]   C.A. Desoer, *Notes for a Second Course on Linear Systems*. Van Nostrand Reinhold Co., New York, 1969.

[D.V.75]   C.A. Desoer and M. Vidyasagar, *Feedback Systems: Input-Output Properties*. Academic Press, New York, 1975.

[DLMS.80] C.A. Desoer, R.W. Liu, J. Murray and R. Saeks, "Feedback System Design: The Fractional Representation Approach to Analysis and Synthesis," *IEEE Trans. Automat. Contr.*, AC-25, No. 3, June 1980.

[Ega.79]   B. Egardt, *Stability of Adaptive Controllers*. Springer-Verlag, New York, 1979.

[F.F.84]   A. Feintuch and B.A. Francis, "Uniformly Optimal Control of Linear Time-Varying Systems," *Systems and Control Letters*, 5, 67-71, Oct. 1984.

[Frn.87]   B.A. Francis, *A Course in H_∞ Control Theory*. Springer-Verlag Berlin, Heidelberg, 1987.

[Gaw.87]   P.J. Gawthrop, *Continuous-Time Self-Tuning Control. Volume 1: Design*, John Wiley and Sons, New York, 1987.

[GMDD.87] F. Giri, M. M'Saad, L. Dugard and J.M. Dion, "Pole Placement Direct Adaptive Control for Time-Varying Ill-Modeled Plants," in *Proc. 26th CDC*, Los Angeles, Dec. 1987.

[G.C.86]   O. Gomart and P. Caines, "On the Extension of Robust Global Adaptive Control Results to Unstructured Time-Varying Systems," *IEEE Trans. Automat. Contr.*, AC-31, 370, 1986.

[GHX.84]   G.C. Goodwin, D.J. Hill and X. Xianya, "Stochastic Adaptive Control for Exponentially Convergent Time-Varying Systems," in *Proc. 23rd CDC*, 39, 1984.

[G.S.84]   G.C. Goodwin and K.S. Sin *Adaptive Filtering Prediction and Control*. Prentice Hall, Englewood Cliffs, NJ, 1984.

[G.T.83]   G.C. Goodwin and E.K. Teoh, "Adaptive Control of a Class of Linear Time Varying Systems," in *Proc. IFAC Workshop on Adaptive Systems in Control and Signal Processing*, San Francisco, 1983.

[INS.84]   A. Ilchmann, I. Nurnberger and W. Schmale "Time-Varying Polynomial Matrix Systems," *Int. J. Control*, Vol. 40, No. 2, pp. 329–362, 1984.

[IOP.87]   A. Ilchmann, D.H. Owens and D. Pratzel-Wolters, "Sufficient Conditions for Stability of Linear Time-Varying Systems," *Systems and Control Letters*, 9, 157–163, 1987.

[I.K.83]   P.A. Ioannou and P.V. Kokotovic, *Adaptive Systems with Reduced Models*. Springer-Verlag, New York, 1983.

[I.T.86]   P.A. Ioannou and K.S. Tsakalis, "A Robust Direct Adaptive Controller," *IEEE Trans. Automat. Contr.*, AC-31, No. 11, Nov. 1986.

[I.S.88]   P.A. Ioannou and J. Sun, "Theory and Design of Robust Direct and Indirect Adaptive Control Schemes," *Int. J. Contr.*, V. 47, 775, 1988.

[Jac.61]   O.L.R. Jacobs, "A Review of Self-Adjusting Systems in Automatic Control," *Journal of Electron. Control*, 10, 311–322, 1961.

[Kai.80]   T. Kailath, *Linear Systems*. Prentice Hall, Englewood Cliffs, NJ, 1980.

[Kal.60]   R.E. Kalman, "Mathematical Description of Linear Dynamical Systems," *SIAM J. Control*, 1, pp. 152-192, 1963.

[K.H.79]   E.W. Kamen and K.M. Hafez, "Algebraic Theory of Linear Time-Varying Systems," *SIAM J. Control and Optimization*, vol. 17, no. 4, 500–510, July 1979.

[Kam.79]   E.W. Kamen, "New Results in Realization Theory for Linear Time-Varying Analytic Systems," *IEEE Trans. Autom. Contr.*, vol. AC-24, no. 6, 866–877, Dec. 1979.

[K.K.79]   E.W. Kamen and P.P. Khargonekar, "On the Control of Linear Systems Whose Coefficients are Functions of Parameters," *IEEE Trans. Autom. Contr.*, vol. AC-29, no. 1, 25–33, Jan. 1984.

[K.F.70]   A.N. Kolmogorov and S.V. Fomin, *Introductory Real Analysis*. Dover, Mineola, NY, 1970.

[KSK.89]   J.M. Krause, G. Stein and P. Khargonekar "Toward Practical Adaptive Control," Proc. 1989 Autom. Contr. Conf., 993–998, June 1989.

[K.A.86]   G. Kreisselmeier and B.D.O. Anderson, "Robust Model Reference Adaptive Control," *IEEE Trans. Autom. Contr.*, AC-31, No.2, Feb. 1986.

[Kre.86]   G. Kreisselmeier, "Adaptive Control of a Class of Slowly Time-Varying Plants," *Systems and Control Letters*, Vol. 8, No. 2, Dec. 1986.

[Lan.79]   Y.D. Landau, *Adaptive Control. The Model Reference Approach*. Marcel Dekker, New York, 1979.

[L.S.88]   W. Li and J.-J.E. Slotine, "Estimators for Time-Varying Parameters," *MIT Report*, June 1988.

[Man.87]   V. Manousiouthakis, "On Time-Varying Control," in *Proc. 1987 ACC*, 2050–2053, June 1987.

[M.G.87]   I.M.Y. Mareels and M. Gevers, "Persistency of Excitation Criteria for Linear, Multivariable, Time Varying Systems," *Australian National University Report*, 1987.

[M-S.85]   J.M. Martin-Sanchéz, "Adaptive Control for Time-Variant Processes," in *Proc. 1985, ACC*, 1260, 1985.

[M.M.82]   A. Michel and R. Miller, *Ordinary Differential Equations*. Academic Press, New York, 1982.

[M.G.88]   R.H. Middleton and G.C. Goodwin, "Adaptive Control of Time-Varying Linear Systems," *IEEE Trans. on Autom. Contr.* Vol. 33, No. 2, pp. 150–155, Feb. 1988.

[MGHM.88]   R.H. Middleton, G.C. Goodwin, D. Hill and D. Mayne, "Design Issues in Adaptive Control," *IEEE Trans. Autom. Contr.*, vol. 33, No.1, Jan. 1988.

[M.M.85]   D.R. Mudget and A.S. Morse, "Adaptive Stabilization of Linear Systems with Unknown High-Frequency Gains," *IEEE Trans. Autom. Contr.*, AC-30, 549–554, June 1985.

[N.A.89]   K.S. Narendra and A.M. Annaswamy, *Stable Adaptive Systems*. Prentice Hall, Englewood Cliffs, NJ, 1989.

[N.V.78]   K.S. Narendra and L.S. Valavani, "Stable Adaptive Controller Design – Direct Control," *IEEE Trans. Automat. Contr.*, AC-23, 570, 1978.

[Nus.83]   R.D. Nussbaum, "Some Remarks on a Conjecture in Parameter Adaptive Control," *Syst, Contr. Lett.*, 243–246, Nov. 1983.

[Ohk.85]   F. Ohkawa, "A Model Reference Adaptive Control System for a Class of Discrete Linear Time-Varying Systems with Time Delay," *Int. J. Control*, Vol. 42, No. 5, 1227, 1985.

[P.N.82]   B.B. Peterson and K.S. Narendra, "Bounded Error Adaptive Control," *IEEE Trans. Automat. Control*, Vol. AC-27, No. 6, pp. 1161–1168, 1982.

[P.I.91]   M.M. Polycarpou and P.A. Ioannou, "On the Existence and Uniqueness of Solutions in Adaptive Control Systems," *IEEE Trans. Autom. Contr.*, January 1993.

[Pra.83]   L. Praly, "Robustness of Indirect Adaptive Control Based on Pole Placement Design," in *Proc. IFAC Workshop on Adapt. Syst. Contr. Signal Process.*, San Francisco, 1983.

[Pra.84]   L. Praly, "Robust Model Reference Adaptive Controllers–Part 1: Stability Analysis," in *Proc. 23rd IEEE Conf. Decision Contr.*, 1984.

[Pra.85]   L. Praly, "About Exponential Stability of Operators," *Coordinated Science Laboratory, Report*, University of Illinois, Urbana, 1985.

[RKMN.88]   R. Ravi, P.P. Khargonekar, K.D. Minto and C.N. Nett, "Controller Parametrization Time-Varying Multirate Plants," *Univ. of Minnesota Report*, Oct. 1988 (rev. May 1989).

[RNK.90]   R. Ravi, K.M. Nagpal and P.P. Khargonekar, "The H^∞ Control Problem for Linear Time-Varying Systems," *Univ. of Minnesota Report*, Feb. 1990.

[RVAS.85]   C. Rohrs, L. Valavani, M. Athans and G. Stein, "Robustness of Continuous-Time Adaptive Control Algorithms in the Presence of Unmodeled Dynamics," *IEEE Trans. Autom. Contr.*, 30, 881-889, 1985.

[S.B.89]   S.S. Sastry and M. Bodson, *Adaptive Control: Stability, Convergence and Robustness*. Prentice Hall, Englewood Cliffs, NJ, 1989.

[S.A.91]   J.S. Shamma and M. Athans, "Guaranteed Properties of Gain Scheduled Control for Linear Parameter-Varying Plants," *Automatica*, vol. 27, no. 3, 559–564, May 1991.

[Sil.66]   L.M. Silverman, "Transformation of Time-Variable Systems to Canonical (Phase-Variable) Form," *IEEE Trans. Automat. Contr.*, AC-11, 300, 1966.

[S.M.67]   L.M. Silverman and H.E. Meadows, "Controllability and Observability in Time-Variable Linear Systems," *J. SIAM Control*, Vol. 5, No. 1, 1967.

[Sil.68]   L.M. Silverman, "Synthesis of Impulse Response Matrices by Internally Stable and Passive Realizations," *IEEE Trans. Circuit Theory*, CT-15, 238, Sep. 1968.

[S.A.68]   L.M. Silverman and B.D.O. Anderson, "Controllability, Observability and Stability of Linear Systems," *J. SIAM Control*, Vol. 6, No. 1, 1968.

[Sol.66]   A.V. Solodov, *Linear Automatic Control Systems with Varying Parameters*. American Elsevier, New York, 1966.

[T.I.87] K.S. Tsakalis and P.A. Ioannou, "Adaptive Control of Linear Time-Varying Plants," *Automatica*, Vol. 23, No. 4, pp. 459–468, July 1987.

[T.I.89] K.S. Tsakalis and P.A. Ioannou, "Adaptive Control of Linear Time-Varying Plants: A New Model Reference Controller Structure," *IEEE Trans. Autom. Contr.*, AC-34, 1038–1046, Oct. 1989.

[Tsa.89] K.S. Tsakalis, "Robustness of Model Reference Adaptive Controllers: Input-Output Properties," *Proc. 28th CDC*, 1989.

[Ven.77] Y.V. Venkatesh, *Energy Methods in Time-Varying System Stability and Instability Analyses.* Springer-Verlag, Berlin, 1977.

[Vid.78] M. Vidyasagar, *Nonlinear Systems Analysis.* Prentice Hall, Englewood Cliffs, NJ, 1978.

[Vid.81] M. Vidyasagar, *Input-Output Analysis of Large-Scale Interconnected Systems.* Springer-Verlag, New York, 1981.

[Wil.88] D. Wilson, "Convolution and Hankel Operator Norms for Linear Systems," in *Proc. IEEE 27th CDC*, Austin, TX, 1988.

[X.E.84] X. Xianya and R.J. Evans, "Discrete-Time Adaptive Control of Deterministic Time-Varying Systems," *Automatica*, Vol. 20, No. 3, pp. 309–319, 1984.

[YJB.76] D.C. Youla, H.A. Jabr and J.J. Bongiorno, "Modern Wiener-Hopf Design of Optimal Controllers, Part II: The Multivariable Case," *IEEE Trans. Autom. Contr.*, vol. AC-21, 319–338, June 1976.

[Zam.65] G. Zames, "Nonlinear Time Varying Feedback Systems—Conditions for L_∞ Boundedness Derived Using Conic Operators on Exponentially Weighted Spaces," *Proc. 1965 Allerton Conf.*, 460–471.

[Z.J.89] J. Zhu and C.D. Johnson, "New Results for the Stability Analysis of Time-Varying Linear Systems. Part I: The Case of Reduced Systems," in *Proc. 1989 ACC*, 1494–1499, 1989.

Index

Adaptive Control, 10
    direct, 11, 182
    indirect, 11, 181
Adaptive Pole-Placement Control
    algorithm, 235, 240, 244
    assumptions, 234, 241, 245, 246
    closed-loop description, 236, 238

Bellman-Gronwall Lemma, 58
Bezout Equations, 18, 23

Causality, 42
Certainty Equivalence, 11, 181
Co-eigenvalues, 4
Controllability
    Grammian, 30
    matrix, 30
    pointwise, 102, 136
    strong, 32
    uniform, 30
    uniform complete, 30

Diophantine Equations, 18, 20, 23, 94, 98, 126, 133, 234, 244

Estimation Error, 151, 159, 186, 190
Exponentially Weighted L_p Theory, 48

High-Frequency Gain, 75, 92

Identification Error, 160, 163, 166, 191, 236

Induced Gain, 42, 43, 46, 49, 50
Instability, examples of, 5, 6, 80
Internal Model Principle, 98, 131

Jump Function, 56

L_p Theory, 41
Lyapunov Transformations, 30, 31

Minimum Phase Condition, 90, 92
Model Reference Adaptive Control
    assumptions, 183, 193
    closed-loop description, 186, 197, 199
    control law, 185, 196
    parameter update law, 188, 189, 190, 196
    structured variations in, 184
Model Reference Control, 8
    assumptions, 92
    control law, 94
    IMP, 98
    internal stability of, 99
    mismatch operator, 104, 108, 109
    objective, 91
    piecewise, 106
    pointwise, 101, 104
    realization, 99
    sensitivity operators, 94, 97
    strictly proper, 96

Nominal Part of the plant, 76
Normalization Signals, 53

Observability
  Grammian, 30
  matrix, 31
  pointwise, 102, 136
  strong, 32
  uniform, 31
  uniform complete, 30
Operator Inversion Lemma, 57

Parameter Variations
  fully/partially structured,
    unstructured, 87
Parametric Uncertainty, 148, 150
Peano-Baker Formula, 3
Persistent Excitation, 12, 154
Perturbation Part of the plant, 76
Pole-Placement Control, 9
  control law, 126
  IMP/command tracking, 131
  internal stability, 130
  objective, 124, 125
  piecewise, 135
  pointwise, 136
  realization, 129
  sensitivity operators, 127
  strictly proper, 128
Polynomial Differential Operators, 16
  coprimeness of, 18, 21
  coprimeness of, pointwise, 38, 102, 136
  coprimeness of, strong, 22, 33, 92, 125
  left, 16
  monic, 16
  properties of, 17
  right, 16
Polynomial Integral Operators, 25

  properties of, 28
  stability of, 28, 39
Projection Operator, 152

Relative Degree, 74, 92
Representations
  algebraically equivalent, 30
  topologically equivalent, 30
  uniform, 30

σ-Modification, 155
Smooth Approximation, 59
Stability
  BIBO, 33
  BIBS, 33
  internal, 33
  L_p, 43
  of LTV Systems, 3, 39, 40, 41
  exponential, 3, 39, 40, 41, 78
State Transition Matrix, 2, 25
Structured Parameter Variations, 13, 87, 158
Sylvester Matrix, 19, 21
Swapping Lemma, 55
Switching σ-Modification, 156

Time-Varying Differential Operator, 15
Time-Varying I/O Operators, 25, 26
  factorization (PDO/PIO), 35
  proper, 26
  realization, 27, 32, 35
  parametrization, 81, 82
Truncation Operator, 42

Zero-Input Response, 25
Zero-State Response, 25